AF389338

Multiscale Simulation of Composite Structures: Damage Assessment, Mechanical Analysis and Prediction

Multiscale Simulation of Composite Structures: Damage Assessment, Mechanical Analysis and Prediction

Editor

Stelios K. Georgantzinos

MDPI • Basel • Beijing • Wuhan • Barcelona • Belgrade • Manchester • Tokyo • Cluj • Tianjin

Editor
Stelios K. Georgantzinos
Department of Aerospace
Science and Technology
National and Kapodistrian
University of Athens
Psachna
Greece

Editorial Office
MDPI
St. Alban-Anlage 66
4052 Basel, Switzerland

This is a reprint of articles from the Special Issue published online in the open access journal *Materials* (ISSN 1996-1944) (available at: www.mdpi.com/journal/materials/special_issues/Simulation_Composite_Structure_Analysis).

For citation purposes, cite each article independently as indicated on the article page online and as indicated below:

LastName, A.A.; LastName, B.B.; LastName, C.C. Article Title. *Journal Name* **Year**, *Volume Number*, Page Range.

ISBN 978-3-0365-6543-9 (Hbk)
ISBN 978-3-0365-6542-2 (PDF)

Contents

About the Editor

Stelios K. Georgantzinos

Dr. Stelios K. Georganzinos is an Assistant Professor in the Department of Aerospace Science and Technology (2021) of the School of Sciences of the University of Athens. He holds a Ph.D. in Mechanical and Aerospace Engineering, University of Patras (2010); an MBA, University of Patras (2009); MSc in Mechanical Engineering, University of Thessaly (2007); and a Diploma in Mechanical and Aerospace Engineering, University of Patras (2005). He teaches postgraduate courses and has served and is serving as a supervisor in numerous postgraduate theses and doctoral dissertations. His research interests focus on the design, analysis, and simulation of advanced materials, structures, and systems at multiple scales (nano–micro–macro), as well as additive manufacturing and digitization. His scientific work has received significant international recognition; has published more than 110 papers (most as a principal investigator) in international journals, books, and conferences; and has received numerous citations (2022). He has participated as a researcher and scientific supervisor in numerous international and nationally competitive research projects. He is a regular Editorial Board Member and Guest Editor of international journals with high impact factors. He has served as an organizer of international symposia, as well as a member of scientific committees, chair of thematic areas, and invited speaker at international conferences. He is a reviewer for a significant number of scientific journals, as well as an evaluator of national and international research proposals.

Preface to "Multiscale Simulation of Composite Structures: Damage Assessment, Mechanical Analysis and Prediction"

Composites can be engineered to exhibit high strength, high stiffness, and high toughness. Composite structures have increasingly been used in various engineering applications. In recent decades, most fundamentals of science have expanded in length by many orders of magnitude. Nowadays, one of the primary goals of science and technology seem to be to develop reliable methods for linking the physical phenomena that occur over multiple length scales, particularly from a nano-/microscale to a macroscale. To engineer composites for high performance and to design advanced structures, the relationship between material nano-/microstructures and their macroscopic properties must be established to accurately predict their mechanical performance and failure. Multiscale simulation is a tool that enables studying and comprehending complex systems and phenomena that would otherwise be too expensive or dangerous, or even impossible, to study by direct experimentation and, thus, to achieve this goal.

This reprint assembles high-quality chapters that advance the field of the multiscale simulation of composite structures, through the application of any modern computational and/or analytical methods alone or in conjunction with experimental techniques, for damage assessment or mechanical analysis and prediction.

Stelios K. Georgantzinos
Editor

materials

Editorial

Multiscale Simulation of Composite Structures: Damage Assessment, Mechanical Analysis and Prediction

Stelios K. Georgantzinos 

Laboratory for Advanced Materials, Structures, and Digitalization, Department of Aerospace Science and Technology, National and Kapodistrian University of Athens, 34400 Psachna, Greece; sgeor@uoa.gr

Abstract: Composites can be engineered to exhibit high strength, high stiffness, and high toughness. Composite structures have been used increasingly in various engineering applications. In recent decades, most fundamentals of science have expanded their reach by many orders of magnitude. Currently, one of the primary goals of science and technology seems to be the quest to develop reliable methods for linking the physical phenomena that occur over multiple length scales, particularly from a nano-/micro-scale to a macroscale. The aim of this Special Issue is to assemble high quality papers that advance the field of multiscale simulation of composite structures, through the application of any modern computational and/or analytical methods alone or in conjunction with experimental techniques, for damage assessment or mechanical analysis and prediction.

Keywords: composite structures; multiscale simulation; finite element analysis; damage assessment; mechanical analysis

Citation: Georgantzinos, S.K. Multiscale Simulation of Composite Structures: Damage Assessment, Mechanical Analysis and Prediction. *Materials* **2022**, *15*, 6494. https://doi.org/10.3390/ma15186494

Received: 13 September 2022
Accepted: 16 September 2022
Published: 19 September 2022

Publisher's Note: MDPI stays neutral with regard to jurisdictional claims in published maps and institutional affiliations.

To engineer composites for high performance and to design advanced structures, the relationship between material nano-/micro-structures and their macroscopic properties must be established in order to accurately predict their mechanical performance and failure. Multiscale simulation is a tool that enables the study and comprehension of complex systems and phenomena that would otherwise be too expensive or dangerous, or even impossible, to study by direct experimentation and, thus, to deal with.

The mechanical characterization of textile composites is a challenging task due to their nonuniform deformation and complicated failure phenomena. However, Zhao et al. [1] introduced a three-dimensional mesoscale finite element model to investigate the progressive damage behavior of a notched single-layer triaxially braided composite subjected to axial tension. The damage initiation and propagation in fiber bundles were simulated using three-dimensional failure criteria and the damage evolution law. A traction–separation law was applied to predict the interfacial damage of fiber bundles. The proposed model was correlated and validated by the experimentally measured full field strain distributions and effective strength of the notched specimen. The progressive damage behavior of the fiber bundles was studied by examining the damage and stress contours at different loading stages. Parametric numerical studies were conducted to explore the role of modeling parameters and geometric characteristics on the internal damage behavior and global measured properties of the notched specimen. Moreover, the correlations of damage behavior, global stress–strain response, and the efficiency of the notched specimen were discussed in detail. The results of this paper delivered a throughout understanding of the damage behavior of braided composites and can help in the specimen design of textile composites.

Accelerated construction in the form of steel–concrete composite beams is among the most efficient methods to construct highway bridges. One of the main problems with this type of composite structure, which has not yet been fully clarified in the case of continuous beam, is the crack zone initiation that gradually expands through the beam width. Gautam et al. [2] proposed a semi-empirical model to predict the size of cracks in terms of small box girder deflection and intensity of the load applied on a structure.

A set of steel–concrete composite small box girders were constructed using steel fibrous concrete and experimentally tested under different caseloads. The results were then used to create a dataset of the box girder response in terms of beam deflection and crack width. The dataset obtained was then utilized to develop a simplified formula providing the maximum width of cracks. The results showed that the cracks were initiated in the hogging moment region when the load exceeded 80 kN. Additionally, it was observed that the maximum cracked zone occurred in the center of the beam due to the maximum negative moment. Moreover, the crack width of the box girder at different loading cases was compared with the test results obtained from the literature. A good agreement has been found between the proposed model and experiment results.

With the aim of damage reduction in person subjected to ballistic impact, polymeric, carbon, and glass fibers are commonly used to develop protective systems. However, during recent decades, it has been found that a combination of high strength yarns in different directions generates flexible woven fabrics, which are light and highly resistant at the same time. Therefore, aramid fibers are one of the protection materials most used today, with a growing trend in the industry. Feito et al. [3] investigated the effect of the impact angle of a projectile during low-velocity impact on Kevlar fabrics using a simplified numerical model. The implementation of mesoscale models is complex and usually involves a long computation time, in contrast with practical industry needs to obtain accurate results rapidly. In addition, when the simulation includes more than one layer of composite ply, the computational time increases even in the case of hybrid models. With the goal of providing useful and rapid prediction tools to the industry, a simplified model has been developed in this work. The model offers an advantage in the reduced computational time compared with a full 3D model (around 90% faster). The proposed model has been validated against equivalent experimental and numerical results reported in the literature, with acceptable deviations and accuracies for the design requirements. The proposed numerical model allows for the study of the influence of the geometry on the impact response of the composite. After a parametric study related to the number of layers and the angle of impact, using a response surface methodology, a mechanistic model and a surface diagram were presented to help with the calculation of the ballistic limit.

Concrete-filled steel tubes (CFSTs) show advantageous applications in the field of construction, especially for a high axial load capacity. The challenge in using such structures lies in the selection of many parameters constituting CFST, which necessitates defining complex relationships between the components and the corresponding properties. The axial capacity (Pu) of CFST is among the most important mechanical properties. Nguyen et al. [4] investigated the possibility of using a feedforward neural network (FNN) to predict Pu. Furthermore, an evolutionary optimization algorithm, namely invasive weed optimization (IWO), was used for tuning and optimizing the FNN weights and biases to construct a hybrid FNN–IWO model and to improve its prediction performance. The results showed that the FNN–IWO algorithm is an excellent predictor of Pu, with a value of R^2 of up to 0.979. The advantage of FNN–IWO was also pointed out with gains in accuracy of 47.9%, 49.2%, and 6.5% for root mean square error (RMSE), mean absolute error (MAE), and R^2, respectively, compared with the simulation using the single FNN. Finally, the performance when predicting Pu as a function of structural parameters, such as the depth/width ratio, the thickness of the steel tube, the yield stress of steel, the concrete compressive strength, and the slenderness ratio, was investigated and discussed.

Circular opening steel beams have been increasingly acknowledged in structural engineering because of their many remarkable advantages, including their ability to bridge the span of a large aperture or their lighter weight compared with conventional steel beams. Nguyen et al. [5] investigated and selected the most suitable parameters used in particle swarm optimization (PSO), namely the number of rules (n_{rule}), the population size (n_{pop}), the initial weight (w_{ini}), the personal learning coefficient (c_1), the global learning coefficient (c_2), and the velocity limits (f_v), in order to improve the performance of the adaptive neuro-fuzzy inference system in determining the buckling capacity of circular opening steel beams. This is an important mechanical property in terms of the safety of structures under subjected loads. An available database of 3645 data samples was used for the generation of training (70%) and testing (30%) datasets. Monte Carlo simulations, which are natural variability generators, were used in the training phase of the algorithm. Various statistical measurements, such as root mean square error (RMSE), mean absolute error (MAE), Willmott's index of agreement (IA), and Pearson's coefficient of correlation (R), were used to evaluate the performance of the models. The results of the study show that the performance of ANFIS optimized by PSO (ANFIS-PSO) is suitable for determining the buckling capacity of circular opening steel beams but is very sensitive under different PSO investigation and selection parameters. The findings of this study show that $n_{rule} = 10$, $n_{pop} = 50$, $w_{ini} = 0.1$ to 0.4, $c_1 = [1, 1.4]$, $c_2 = [1.8, 2]$, and $f_v = 0.1$, which are the most suitable selection values for ensuring the best performance for ANFIS-PSO. In short, this study might help in selecting suitable PSO parameters for the optimization of the ANFIS model.

Commonly, nanocomposite material applications are associated with the simultaneous actions with more than one type of loading. Specifically, the investigation of nanocomposites subjected to both thermal as well as mechanical loads is perhaps one of most interesting fields of research, since high-temperature applications are very frequent. Giannopoulos et al. [6] provided a computationally efficient and reliable hybrid numerical formulation capable of characterizing the thermomechanical behavior of nanocomposites, which was based on a combination of molecular dynamics (MD) and the finite element method (FEM). A polymeric material was selected as the matrix—specifically, poly(methyl methacrylate) (PMMA), commonly known as Plexiglas, due to its extensive applications. On the other hand, the fullerene C_{240} was adopted as a reinforcement because of its high symmetry and suitable size. The numerical approach was performed at two scales. First, an analysis was conducted at the nanoscale level by utilizing an appropriate nanocomposite unit cell containing C_{240} at a high mass fraction. A MD-only method was applied to accurately capture all of the internal interfacial effects and, accordingly, its thermoelastic response. Then, a micromechanical, temperature-dependent finite element analysis took place using a representative volume element (RVE), which incorporated the first-stage MD output, to study nanocomposites with small mass fractions, for which a atomistic-only simulation would require a substantial computational effort. To demonstrate the effectiveness of the proposed scheme, numerous numerical results were presented, while the investigation was performed in a temperature range that included the PMMA glass transition temperature, Tg.

Although some work has already been conducted on the vibrations of composite structures reinforced by nanoparticles, there are only a few studies that focused on the vibration behavior of carbon fiber-based laminate composites with pure graphene inclusions. Georgantzinos et al. [7] developed a computational procedure to investigate the vibration behavior of laminated composite structures, including graphene inclusions in a matrix. Concerning the size-dependent behavior of graphene, its mechanical properties were derived using nanoscopic empiric equations. Using the appropriate Halpin–Tsai models, the equivalent elastic constants of the graphene reinforced matrix were obtained. Then, the orthotropic mechanical properties of a composite lamina of carbon fibers and hybrid matrix can be evaluated. Considering a specific stacking sequence and various geometric configurations, carbon fiber-graphene-reinforced hybrid composite plates were modeled using conventional finite element techniques. Applying simply support or clamped boundary conditions, the vibrational behavior of the composite structures was finally extracted. Specifically, the modes of vibration for every configuration were derived, and the effect of graphene inclusions in the natural frequencies was calculated. The higher the volume fraction of graphene in the matrix, the higher the natural frequency for every mode. Comparisons with other methods, where possible, were performed to validate the proposed method.

Giannopoulos and Georgantzinos [8] investigated the thermomechanical effects of adding a newly proposed nanoparticle within a polymer matrix such as polyethylene. A nanoparticle was formed by a typical single-walled carbon nanotube (SWCNT) and two equivalent giant carbon fullerenes that were attached by their nanotube edges through covalent bonds. In this way, a bone-shaped nanofiber that may offer enhanced thermomechanical characteristics when used as a polymer filler, due to each unique shape and chemical nature, was developed. The investigation was based on molecular dynamics simulations of the tensile stress–strain response of the polymer nanocomposites under a variety of temperatures. The thermomechanical behavior of the bone-shaped nanofiber-reinforced polyethylene was compared with that of an equivalent nanocomposite filled with ordinary capped single-walled carbon nanotubes to reach some coherent fundamental conclusions. That study focused on the evaluation of some basic, temperature-dependent properties of the nanocomposite reinforced with these innovative bone-shaped allotropes of carbon.

The flexural strength of Slender steel tube sections is known to achieve significant improvements upon being filled with concrete; however, this section is more likely to fail by buckling under compression stresses. Al Zand et al. [9] investigated the flexural behavior of a Slender steel tube beam that was produced by connecting two pieces of C-sections and filled with recycled-aggregate concrete materials (CFST beam). The C-section's lips behaved as internal stiffeners for the CFST beam's cross section. A static flexural test was conducted on five large-scale specimens, including one specimen that was tested without concrete (hollow specimen). The ABAQUS software was also employed for the simulation and non-linear analysis of 20 additional CFST models in order to further investigate the effects of varied parameters that were not tested experimentally. The numerical model was able to adequately verify the flexural behavior and failure mode of the corresponding tested specimen, with an overestimation of the flexural strength capacity of about 3.1%. Generally, the study confirmed the validity of using the tubular C-sections in the CFST beam concept, and their lips (internal stiffeners) led to significant improvements in the flexural strength, stiffness, and energy absorption index. Moreover, a new analytical method was developed to specifically predict the bending (flexural) strength capacity of the internally stiffened CFST beams with steel stiffeners, which was well-aligned with the results derived from the current investigation and with those obtained by others.

The composite shear wall has various merits over traditional reinforced concrete walls. Thus, several experimental studies have been reported in the literature to study the seismic behavior of composite shear walls. However, few numerical investigations were found in previous literature because of difficulties in the interaction behavior of steel and concrete. Najm et al. [10] presented a numerical analysis of smart composite shear walls that use an infilled steel plate and concrete. The study was carried out using the ANSYS software. The mechanical mechanisms between the web plate and concrete were investigated thoroughly. The results obtained from the finite element (FE) analysis show excellent agreement with the experimental test results in terms of the hysteresis curves, failure behavior, ultimate strength, initial stiffness, and ductility. The results indicate that increasing the gap between the steel plate and the concrete wall from 0 mm to 40 mm improved the stiffness by 18% compared with the reference model, which led to delaying failures in this model. Expanding the infill steel plate thickness to 12 mm enhanced the stiffness and energy absorption at ratios of 95% and 58%, respectively. This resulted in a gradual decrease in the strength capacity of this model. Meanwhile, increasing the concrete wall thickness to 150 mm enhanced the ductility and energy absorption at ratios of 52% and 32%, respectively, which led to restricting the model and reducing the lateral offset. Changing the distance between shear studs from 20% to 25% enhanced the ductility and energy absorption by about 66% and 32%, respectively.

A major part of the computational cost required for determining an optimal material design with extreme properties using a topology optimization formulation is devoted to solving the equilibrium system of equations derived through the implementation of the finite element method (FEM). To reduce this computational cost, among other methodologies, various model order reduction (MOR) approaches can be utilized. Kazakis and Lagaros [11] presented a simple Matlab code for solving the topological optimization for the design of materials combined with three different model order reduction approaches.

The three MOR approaches presented in the code implemented are the proper orthogonal decomposition (POD), the on-the-fly reduced order model construction and the approximate reanalysis (AR) following the combined approximations approach. The complete code, containing all participating functions (including the changes made to the original ones), was also provided.

Funding: This research received no external funding.

Institutional Review Board Statement: Not applicable.

Informed Consent Statement: Not applicable.

Conflicts of Interest: The author declares no conflict of interest.

References

1. Zhao, Z.; Dang, H.; Xing, J.; Li, X.; Zhang, C.; Binienda, W.K.; Li, Y. Progressive Failure Simulation of Notched Tensile Specimen for Triaxially-Braided Composites. *Materials* **2019**, *12*, 833. [CrossRef] [PubMed]
2. Gautam, B.G.; Xiang, Y.Q.; Qiu, Z.; Guo, S.H. A Semi-Empirical Deflection-Based Method for Crack Width Prediction in Accelerated Construction of Steel Fibrous High-Performance Composite Small Box Girder. *Materials* **2019**, *12*, 964. [CrossRef] [PubMed]
3. Feito, N.; Loya, J.A.; Muñoz-Sánchez, A.; Das, R. Numerical Modelling of Ballistic Impact Response at Low Velocity in Aramid Fabrics. *Materials* **2019**, *12*, 2087. [CrossRef] [PubMed]
4. Nguyen, H.Q.; Ly, H.B.; Tran, V.Q.; Nguyen, T.A.; Le, T.T.; Pham, B.T. Optimization of Artificial Intelligence System by Evolutionary Algorithm for Prediction of Axial Capacity of Rectangular Concrete Filled Steel Tubes under Compression. *Materials* **2020**, *13*, 1205. [CrossRef] [PubMed]
5. Nguyen, Q.H.; Ly, H.B.; Le, T.T.; Nguyen, T.A.; Phan, V.H.; Tran, V.Q.; Pham, B.T. Parametric Investigation of Particle Swarm Optimization to Improve the Performance of the Adaptive Neuro-Fuzzy Inference System in Determining the Buckling Capacity of Circular Opening Steel Beams. *Materials* **2020**, *13*, 2210. [CrossRef] [PubMed]
6. Giannopoulos, G.I.; Georgantzinos, S.K.; Anifantis, N.K. Thermomechanical Response of Fullerene-Reinforced Polymers by Coupling MD and FEM. *Materials* **2020**, *13*, 4132. [CrossRef] [PubMed]
7. Georgantzinos, S.K.; Giannopoulos, G.I.; Markolefas, S.I. Vibration Analysis of Carbon Fiber-Graphene-Reinforced Hybrid Polymer Composites Using Finite Element Techniques. *Materials* **2020**, *13*, 4225. [CrossRef] [PubMed]
8. Giannopoulos, G.I.; Georgantzinos, S.K. Thermomechanical Behavior of Bone-Shaped SWCNT/Polyethylene Nanocomposites via Molecular Dynamics. *Materials* **2021**, *14*, 2192. [CrossRef] [PubMed]
9. Al Zand, A.W.; Ali, M.M.; Al-Ameri, R.; Badaruzzaman, W.H.W.; Tawfeeq, W.M.; Hosseinpour, E.; Yaseen, Z.M. Flexural Strength of Internally Stiffened Tubular Steel Beam Filled with Recycled Concrete Materials. *Materials* **2021**, *14*, 6334. [CrossRef] [PubMed]
10. Najm, H.M.; Ibrahim, A.M.; Sabri, M.M.S.; Hassan, A.; Morkhade, S.; Mashaan, N.S.; Eldirderi, M.M.A.; Khedher, K.M. Evaluation and Numerical Investigations of the Cyclic Behavior of Smart Composite Steel–Concrete Shear Wall: Comprehensive Study of Finite Element Model. *Materials* **2022**, *15*, 4496. [CrossRef] [PubMed]
11. Kazakis, G.; Lagaros, N.D. A Simple Matlab Code for Material Design Optimization Using Reduced Order Models. *Materials* **2022**, *15*, 4972. [CrossRef] [PubMed]

Article

A Simple Matlab Code for Material Design Optimization Using Reduced Order Models

George Kazakis [†] and Nikos D. Lagaros *[,†]

Institute of Structural Analysis and Antiseismic Research, School of Civil Engineering, National Technical University of Athens, 9, Heroon Polytechniou Str., Zografou Campus, GR-15780 Athens, Greece; kzkgeorge@gmail.com

* Correspondence: nlagaros@central.ntua.gr; Tel.: +30-210-772-2625

† These authors contributed equally to this work.

Abstract: The main part of the computational cost required for solving the problem of optimal material design with extreme properties using a topology optimization formulation is devoted to solving the equilibrium system of equations derived through the implementation of the finite element method (FEM). To reduce this computational cost, among other methodologies, various model order reduction (MOR) approaches can be utilized. In this work, a simple Matlab code for solving the topology optimization for the design of materials combined with three different model order reduction approaches is presented. The three MOR approaches presented in the code implementation are the proper orthogonal decomposition (POD), the on-the-fly reduced order model construction and the approximate reanalysis (AR) following the combined approximations approach. The complete code, containing all participating functions (including the changes made to the original ones), is provided.

Keywords: topology optimization; microstructure; homogenization; Matlab; reduced order models; reduced basis; on-the-fly construction; POD; approximate reanalysis

Citation: Kazakis, G.; Lagaros, N.D. A Simple Matlab Code for Material Design Optimization Using Reduced Order Models. *Materials* **2022**, *15*, 4972. https://doi.org/10.3390/ma15144972

Academic Editor: Luciano Lamberti

Received: 28 May 2022
Accepted: 14 July 2022
Published: 17 July 2022

Publisher's Note: MDPI stays neutral with regard to jurisdictional claims in published maps and institutional affiliations.

1. Introduction

The basic theory for the implementation of topology optimization in material design was presented first in 1994 by Sigmund [1], followed by Sigmund and Torquato in 1997 [2] and by Gigiansky and Sigmund in 2000 [3]. Since then, many other studies have been published dealing with a variety of different material optimization problem formulations. Neves et al. [4] and Fujii et al. [5] used the density-based approach to design periodic microstructures for optimal elastic properties. Guest and Prévost [6] dealt with the topology optimization of the fluid flows in the design of porous periodic materials. Challis et al. [7], Amstutz et al. [8] and Gao et al. [9] proposed level set-based approaches for the design of microstructures, and Huang et al. in [10,11] presented a Bi-directional Evolutionary Structural Optimization (BESO)-method-based approach for the optimal design of periodic microstructures. A detailed review of the different methodologies in the optimal design of materials together with a description of the variety of the approaches presented so far to deal with the topology optimization of the macro design concurrently with the micro design can be seen in [12].

In the past, due to the increased computational effort required for solving the topology optimization problem, various methodologies along different directions (approximate reanalysis, model order reductions, machine learning, etc.) have been presented. Indicatively, Kirsch and Paralambros [13] first proposed a unified approach to structural reanalysis using the combined approximations in topology optimization. Wang et al. [14] presented a methodology of recycling search spaces in iterative solvers during the optimization procedure. Amir et al. [15] proposed an approximate reanalysis approach in topology optimization based on the combined approximations approach and the use of approximations for dealing with the solution of the analysis problem, generated by a Krylov subspace iterative solver [16]. In addition, in [17], Amir et al. addressed the computational cost of the robust topology optimization formulation. Gogu [18] presented an on-the-fly approach for the construction of the reduced order model. Alaimo in [19] proposed an reduced order model approach where a reduced basis is created based on the functional principal component analysis (FPCA). Ferro et al. [20] proposed a proper orthogonal decomposition (POD) approach where the stages of the SIMP method were used as reduced basis vectors during the optimization procedure. Senne et al. [21] proposed a combination of the approximate reanalysis technique with the sequential piecewise linear programming method, and Xiao et al. [22] proposed a reduced order modeling approach which constructed the reduced basis using the proper orthogonal decomposition (POD) approach. Meanwhile, in the same direction, to reduce the computational effort, various machine learning methodologies have been presented, and the precursor of these was a study by the authors [23].

So far, many Matlab code implementations of the topology optimization formulation have been presented in various publications. For the density-based approach, the first code was the so-called top99 [24] implementation that was followed by the top88 one [25]. Both Matlab codes were dealing with the 2D topology optimization problem formulation. Liu et al. [26] and Ferrari [27] presented an extension of the density-based approach into the 3D space, with Ferrari [27] suggesting code modifications for achieving better performance. Talischi et al. [28] and Chi et al. [29] expanded the 2D and 3D density-based approaches by using the capability to deal with unstructured meshes as well. Amir et al. in [30] presented a code implementation for improving the computational cost of the topology optimization procedure using the multi-grid, preconditioned conjugated gradients solver (MGCG). Huang et al. [31] presented an Evolutionary Structural Optimization (ESO) topology optimization code implementation based on the top99 for the 2D space. Wang et al. [32] and Challis [33] published code implementations that rely on the level set approach for the topology optimization for 2D problem formulations. Otomori et al. [34] and Wei et al. [35] also presented level set-based code implementations for the topology optimization using the reaction diffusion equation and radial basis functions, respectively. In 2019, an integration of a topology-optimization procedure with SAP2000, well-known commercial software for analysis and design of structural systems, was presented by the authors [36]. In addition, Gao et al. in [37] presented IgaTop, a topology optimization formulation using isogeometric analysis.

Subsequently, several code implementations were also presented that were dealing with the homogenization-based topology optimization approaches. Specifically, numerical homogenization implemented for 2D and 3D material design was presented in studies [38,39]. In addition, an energy-based homogenization approach combined with the optimal design of materials was presented in [40]. In this study, a topology-optimization-based Matlab code implementation is presented that deals with the problem of material design at the microstructure level, assisted by model order reduction (MOR) approaches. In particular, the topology optimization procedure is combined with the proper orthogonal decomposition (POD), the on-the-fly reduced order model construction and the approximate reanalysis following the combined approximations approach. Although the code provided covers the case of microscale material design for 2D design domains, it can easily be extended to 3D design domains by modifying the homogenization part to produce the 3D elasticity tensor of the unit cell and extend the problem formulation's description to handle both macro and micro scales. The implementation of the MOR approaches is independent of the dimensionality of the formulation due to being applied in the solution part of the finite element analysis performed at the macro scale.

The layout of the work is composed of four sections accompanied by Sections 1 and 6. In particular, a short description of the optimal design problem of materials is presented in Section 2; subsequently, in Section 3 the theoretical part of the integration of the model order reduction methodologies into the material optimization problem is provided. The detailed description of the most critical parts of the Matlab code's implementation is provided in Section 4, followed by test examples in Section 4 where the ease of use of the code is presented.

2. Optimal Design of Materials

The formulations of the topology optimization (TO) problem used for the design of materials are expressed as the optimal distribution of material volume fraction into the unit cell design domain so that the structural response is optimized. Thus, compared to the original TO problem, the design variables are different, from density values X of the finite elements used to discretize the macro design domain to densities x of the finite elements discretizing the micro-unit cell design domain. In this scope, it can be seen that the optimization procedure performed involves two different scales; the macro scale and the micro one (Figure 1). The design variables as well as the volume constraint are defined at the micro scale, where as the objective function is set on the macro scale; however, it is still expressed as a function of x. The transition between the two scales is achieved through the elasticity tensor by means of the homogenization method [38].

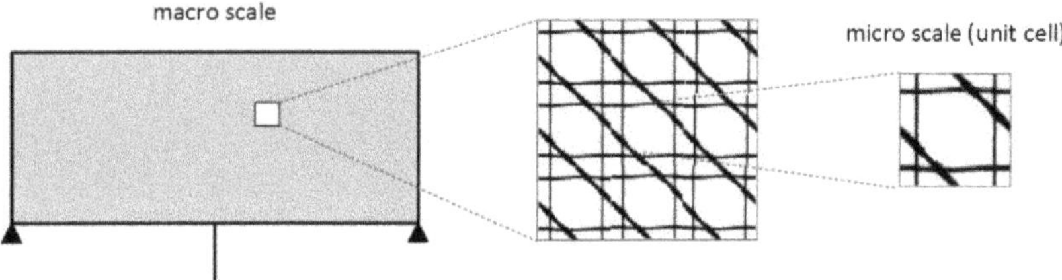

Figure 1. Schematic representation of the periodic unit cell (micro scale) inside the macro structure (macro scale).

The mathematical formulation of the typical topology optimization problem is thus changed according to the following expression of Equation (1).

$$C(x) = F^T \cdot U(x)$$
$$s.t.$$
$$F = K \cdot U(x) \tag{1}$$
$$V(x)/V_0 = f$$
$$0 \leq x_e \leq 1$$

where $C(x)$ is the compliance, F is the load vector, $U(x)$ is the resulting displacements from the structural analysis, $F = K \cdot U$ is the linear system of equations derived from the finite element method, $V(x)$ is the material volume resulting from the densities x, V_0 is the full domain material volume and f is the volume fraction applied as a constraint. The derivative of the objective function is obtained using the adjoint method [41] as described in the following expression of Equation (2):

$$\frac{\partial C}{\partial x_e} = -U^T \cdot \frac{\partial K}{\partial x_e} \cdot U \tag{2}$$

while $\partial K / \partial x_e$ is calculated using the following expression of Equation (3):

$$\frac{\partial K}{\partial x_e} = \frac{\partial C^H}{\partial x_e} \cdot K_0 \tag{3}$$

where C^H is the homogenized elasticity tensor. According to the homogenization theory, the elasticity tensor is obtained by applying unit strains globally to the unit cell domain as well as locally to the finite elements used to discretize the unit cell domain and then using the following expression of Equation (4).

$$C_{i,j}^H = \frac{1}{V} \sum_{e=1}^N \int_{x_e} (u_e^{0(i)} - u_e^{(i)}) \cdot k_e^0 \cdot (u_e^{0(j)} - u_e^{(J)}) dV_e \tag{4}$$

where superscript 0 denotes the globally applied strains, thus u^0 are the displacement fields resulting from the globally applied unit strains, u are the displacement fields resulting from the locally applied unit strains and C^H is the elasticity tensor. If Equation (4) is then differentiated with respect to y_e, the derivative of the elasticity tensor can be obtained using the following expression of Equation (5):

$$\frac{\partial C^H}{\partial x_e} = \frac{\partial E}{\partial x_e} \cdot \frac{1}{V} \int_{x_e} (u_e^{0(i)} - u_e^{(i)}) \cdot k_e^0 \cdot (u_e^{0(j)} - u_e^{(J)}) dV_e \tag{5}$$

Furthermore, the derivative of the Young modulus with respect to each unit cell element density is obtained using the modified SIMP approach [42]. Thus, the Young modulus as a function of density value x is defined using the following expression of Equation (6):

$$E(x) = E_{min} + x^p \cdot (E_0 - E_{min}) \tag{6}$$

and the derivative from the expression of Equation (7):

$$\frac{\partial E}{\partial x_e} = p \cdot x^{p-1} \cdot (E_0 - E_{min}) \tag{7}$$

3. Model Order Reduction in Material Optimization

The main focus of the model order reduction (MOR) approaches is to reduce the computational cost required for solving the linear system of equations formulated from the finite element method (FEM). This is achieved by creating a reduced basis model and finding an approximate solution instead of solving the full-order system of equations. The creation of the reduced basis system of equations is accomplished by substituting first the displacement vector U in the finite element equilibrium system of equations with an approximation vector $\Phi \cdot y$. Matrix $\Phi = \{\Phi_i, \ldots, \Phi_m\}$ consists of the reduced basis vectors, and its first dimension corresponds to the dimension of vector U. Its second dimension refers to a small number (e.g., 5 to 10) chosen as the number of the reduced basis vectors. The left and right hand sides of the resulting equation is then multiplied by Φ^T, as shown in the following expressions of Equation (8).

$$\begin{aligned} F = K \cdot U &\rightleftharpoons \\ F \approx K \cdot \Phi \cdot y &\rightleftharpoons \\ \Phi^T \cdot F \approx \Phi^T \cdot K \cdot \Phi \cdot y &\rightleftharpoons \\ F_{eff} \approx K_{eff} \cdot y & \end{aligned} \tag{8}$$

where F is the load vector, U is the displacement vector, K is the stiffness matrix, F_{eff} is the reduced approximation of the load vector, K_{eff} is the reduced approximation of the stiffness matrix and y is the reduced basis displacement vector. In general, to assess the accuracy of the projected solution, the residual load can be obtained and divided by the norm of the original load vector, as shown in the following Equation (9).

$$e^2 = \frac{\|K \cdot \Phi \cdot a - F\|^2}{\|F\|^2} \tag{9}$$

Thus, the most important part of the procedure is the creation of the reduced basis vectors that represent the main variation in most of the reduced model approaches. In the following subsections, the methodology as well as the creation of the reduced basis vectors of three MOR approaches will be presented. The three approaches implemented are the proper orthogonal decomposition (POD), the on-the-fly reduced order model construction and the approximate reanalysis following the combined approximations approach. In most MOR approaches, the approximation of the displacement field is taken into account in the calculation of the sensitivities through the expression of Equation (10).

$$\frac{\partial C}{\partial x_e} = -y^T \cdot \Phi^T \cdot \frac{\partial K}{\partial x_e} \cdot \Phi \cdot y - \sum_{i=1}^{N_b} \lambda_i^T \cdot \frac{\partial K_i}{\partial x_e} \cdot U_i \tag{10}$$

where the first term of the expression corresponds to the sensitivity calculated from the approximate solution and the second part denotes the adjustment term which corrects the sensitivity calculation, taking into account that the solution is a approximation. The term λ_i is the solution vector of the following expression of Equation (11) for each reduced basis vector.

$$K_i \cdot \lambda_i = 2 \cdot y_i \cdot (F - K \cdot \Phi \cdot y) \tag{11}$$

U_i used in Equation (10) and K_i of Equation (11) denote the displacement vector and stiffness matrix of each reduced basis vector, respectively. Aiming to simplify the code implementation of the adopted MOR approaches into the optimal material design procedure that is described below, the sensitivity adjustment term in not taken into account.

3.1. Proper Orthogonal Decomposition

According to the proper orthogonal decomposition (POD) approach, the construction of the reduced basis vectors is achieved by means of the singular value decomposition (SVD) factorization methodology. In particular, a small number (e.g., five to ten) of optimization iterations is performed first that is equal to the number of the reduced basis vectors, in which the full-scale system equations are solved and the resulting displacement vectors are stored in the matrix A. Thus, matrix A is composed of different snapshots of the displacement field in the early phases of the optimization procedure. Before applying the SVD factorization methodology to the matrix A, the mean of the displacement snapshot of the final reduced basis vector is subtracted from A, as described in [22]. Then, SVD methodology is applied to matrix A and three different matrices are generated, as shown in the following Equation (12):

$$A = \overline{U} \cdot \Sigma \cdot V^{'} \tag{12}$$

Matrix $\overline{U}$ in general contains information about the spacial correlation of the snapshots of matrix A. Matrix Σ is a diagonal matrix containing the weight coefficients denoting the importance of each column of matrix $\overline{U}$, and finally V contains the corresponding time dynamics of each of the columns of matrix $\overline{U}$. The columns of matrix $\overline{U}$ are also called the POD modes and are used as the reduced basis vectors Φ_i consisting of matrix Φ. Thus, in the POD approach, the reduced basis matrix Φ coincides with the first matrix (i.e., $\overline{U}$) of the SVD of matrix A.

$$\Phi = \overline{U} \tag{13}$$

Subsequently, given the creation of the reduced basis matrix Φ in each optimization iteration, the displacement vector is obtained using the constructed reduced model and then projected to the full scale. The accuracy of each new solution in validated using Equation (9), and if the deviation is too large, a full-scale finite element analysis (FEA) is performed and the matrix A is updated with the new snapshot of displacements (mth column of matrix A), removing the earliest generated one (i.e., the first column of matrix A). A new SVD is performed on the updated variant of A and a new reduced basis matrix Φ is used for the next iterations.

3.2. On-the-Fly Reduced Order Model Construction

Similarly to the POD approach, according to the on-the-fly approach a number of optimization iterations are performed first in order to generate displacements snapshots of the early optimization stages. Then, the reduced basis vectors are created based on the Gram–Schmidt orthogonalization methodology which is applied onto the displacement snapshots of the early optimization stages, as follows: for the first reduced basis vector, only the first displacement snapshot is utilized, in which a normalization is performed following the expression of Equation (14):

$$\Phi_1 = \frac{U_1}{\|U_1\|} \tag{14}$$

For the next reduced basis vectors, the following Gram–Schmidt orthogonalization procedure is applied, taking into account all previous reduced basis vectors, as shown in the following expression of Equation (15).

$$\widehat{\Phi}_{i+1} = U_{i+1} - \sum_{j=1}^{i} \langle U_{i+1}, \Phi_j \rangle \Phi_j \tag{15}$$

Subsequently, the new $\widehat{\Phi}_{i+1}$ is normalized (as denoted in Equation (16) and the resulting vector is added to the reduced basis matrix.

$$\Phi_{i+1} = \frac{\widehat{\Phi}_{i+1}}{\|\widehat{\Phi}_{i+1}\|} \tag{16}$$

Following the same steps as described for the POD approach, when the reduced basis matrix Φ is constructed, the subsequent optimization iterations rely on approximate displacement fields obtained using the reduced basis matrix and the accuracy of every new reduced basis based FEA in assessed using the expression of Equation (9). If the accuracy is not acceptable, a new reduced basis vector is created by means of a full-scale FEA and using the previously described procedure. Then, matrix Φ is updated by removing the earliest generated reduced basis vector (i.e., first column of matrix Φ) and adding the new one as the mth column of matrix Φ.

3.3. Approximate Reanalysis

In contrast to the POD and on-the-fly approaches, in the approximate reanalysis, one the reduced basis vectors is not created based on displacement snapshots obtained from the initial optimization iterations. Instead, new reduced basis vectors are created in each optimization iteration. These reduced basis vectors are based only on a single snapshot of the displacement field obtained by solving the full-scale system of equations; the displacement field snapshot is updated during the optimization procedure. The fist reduced basis vector is equal to the displacement snapshot used as the basis of the reduced order model, and thus is defined using the following expression of Equation (17):

$$\Phi_1 = U_1 = K_0^{-1} \cdot F \tag{17}$$

Using Φ_1 and K_0 as the basis of each reduced basis matrix Φ, at each iteration, a new set of reduced basis vectors is constructed. Each vector is obtained using the following expression of Equation (18):

$$K_0 \cdot U_i = F - \Delta K \cdot U_{i-1} \quad i = 2 \ldots i_{max} \tag{18}$$

where $\Phi_i = U_i$, ΔK is the difference between the original stiffness matrix K_0 and the one corresponding to the current iteration. The size of the reduced basis is not the same for every iteration; after the creation of each reduced basis vector the accuracy of the solution is validated using Equation (9), and if it is below a certain threshold, the accuracy of the solution is accepted. A maximum size of reduced basis vectors is also provided. The update of the first reduce basis vector is usually performed after either a fixed number of iterations or after the change in the design variables or the compliance is significant. For a more detail review of the approximate reanalysis approach, the reader is referred to [15].

4. The Matlab Code Implementation

Part of the implementation of the methodologies described previously into a Matlab code is based on two existing codes. For the homogenization part, the basis was the Matlab code presented by Andreassen in [38], whereas for the topology optimization part, the basis was the Matlab code presented also by Andreassen in [25]. For efficiency, in the following sections only the parts of the code modified and the logic behind these modifications will be presented, starting from the part of the homogenization method and then to the topology optimization part. The code implementation presented here is composed of nine Matlab files. These are: homogenize function (i.e., *homogenize.m* Matlab file) that implements the homogenization procedure, elementMatVec function (i.e., *elementMatVec.m* Matlab file) that is used to compute the element load vectors and stiffness matrix, Q4elementStiffnessMatrix function (i.e., *Q4elementStiffnessMatrix.m* Matlab file) that is used for performing a similar role to the elementMatVec function, interpolate function (i.e., *interpolate.m* Matlab file) that performs the SIMP interpolation scheme, UCOpt function (i.e., *UCOpt.m* Matlab file) that performs the material topology optimization procedure, and three additional Matlab files containing the procedures of the corresponding MOR approaches, i.e., *pod.m* Matlab file containing the pod function, *onthefly.m* Matlab file containing the onthefly function and *ar.m* Matlab file containing the ar function.

4.1. Homogenization Code Implementation (Matlab File "homogenize.m")

In this section, the modifications made to the homogenization Matlab files will be presented. For a more in-depth description of the functionalities of the original Matlab homogenization code, the reader is referred to [38]. There were two main modifications of the current implementation compared to the original function, denoted as homogenize, that is used for implementing the homogenization method, originally presented by Andreassen [38]. The first modification refers to the transition from the lame parameters to the Poisson ratio and Young modulus parameters, and the second one to the addition of the derivative of the homogenized tensor dC^H/dy_e with respect to the densities at the unit cell level.

Input parameters: The input arguments of the new implementation of the homogenize function are the following:

```
function [CH,DCH] = homogenize(lx, ly, E, nu, dE, phi)
```

where lame parameters as well as the mapping parameters are replaced by matrix E containing the values of the Young modulus for every finite element used to discretize the unit cell domain, matrix dE contains the derivative of the Young modulus based on the modified SIMP approach, following the expression of Equation (7) and the Poisson ratio nu that is the same for all finite elements. In addition, an extra output argument was added to the method called DCH which is the derivative of the elasticity tensor from the expression of Equation (5), calculated using the parameter matrix dE.

Initialization: Due to the elimination of the mapping variable, the number of elements along the directions of the abscissa and ordinate of the unit cell are taken from the size of matrix E, and thus *Line* 4 (of the original code in [38] function) was slightly modified to take the number of elements from matrix E, as follows:

```
[nely, nelx] = size(E);
```

By using the Young modulus and Poisson ratio, the need for decomposing into two parts the loading vectors and stiffness matrix as described in [38] is not required. Thus, function elementMatVec was modified to return three loading vectors corresponding to the three different unit strains, as shown in the following expression of Equation (19) and the element stiffness matrix computed using the Poisson ratio parameters without the Young modulus.

$$f_e^i = \int_{V_e} B_e^T \cdot C_e \cdot e^i dV_e \tag{19}$$

Thus, *Line* 9 (of the original code in [38]) was modified to have the following form:

```
9  [ke, fe] = elementMatVec(dx/2, dy/2, phi, nu);
```

Assembly of the stiffness matrix and loading vectors: elementMatVec function (see Matlab file "*elementMatVec.m*") was modified to compute the element loading vectors and stiffness matrix using the Poisson ratio by changing the first *Line* of the function to compute the elasticity tensor from Poisson ratio and Young modulus of one instead of the Lame parameters as presented below:

```
2  A = [1 nu 0; nu 1 0; 0 0 (1−nu)/2];
3  C = 1/(1−nu^2)*A;
```

Thus, in the last *lines* of the function where the loading vectors and stiffness matrix are computed, the *lines* are

```
34  % Element matrices
35  ke = ke + weight*(B' * C * B);
36  % Element Loads
37  fe = fe + weight*(B' * C * diag([1 1 1]));
```

In the assembly of the global stiffness matrix part of the homogenize function, *Lines* 34 and 35 (of the original code in [38]) are removed due to Young modulus already being a matrix, and *Line* 37 (of the original code in [38]) is modified to multiply the element stiffness matrix with a vector of the Young modulus, as shown below:

```
37  sK = ke(:)*E(:).';
```

Moving now to the creation of the global loading vector, *Line* 41 (of the original code in [38]) is replaced by a simple multiplication of the element loading vector with the element Young modulus.

```
41  sF = fe(:)*E(:).';
```

Due to the element loading vectors as well as the element stiffness matrix no longer being separated into two parts, *Lines* 53 and 54 (of the original code in [38]) are removed from the code. In addition, an extra line is added bellow the initialization of the elasticity tensor, initializing the derivative of the elasticity tensor. During the iterative procedure performed from *Lines* 64 to 75 (of the original code in [38]), the parameters *sumLambda* and *sumMu* are replaced with the parameter *sumYoung* which is obtained in the same way using *ke* instead of *keLambda* and *keMu*. An extra procedure is added to compute the derivative of the elasticity tensor in which the variable *sumYoung* is multiplied with the derivative of the Young modulus and then added to the cell variable *DCH*. *DCH* is a cell variable of a size of the number of elements, and contains the 3×3 derivative of the elasticity tensor of each element. The new iterative procedure is presented below:

```
64  for i =1:3
65    for j = 1:3
66      sumYoung = ((chi0(:,:,i) − chi(edofMat+(i−1)*ndof))*ke).*...
67      (chi0(:,:,j) − chi(edofMat+(j−1)*ndof));
68      sumYoung = reshape(sum(sumYoung,2), nely, nelx);
```

```
69     % Homogenized  elasticity  tensor
70     CH(i,j) = 1/cellVolume*sum(sum(E.*sumYoung));
71     finalSum = dE.*sumYoung;
72     for k=1:nely
73         for l=1:nelx
74             DCH{k,l}(i,j) = 1/cellVolume*finalSum(k,l);
75         end
76     end
77   end
78 end
```

4.2. Topology Optimization Code Implementation

In this section, all modification applied to the top88 code published in [25] will be presented. The aim of all modifications was to transfer the code implementation from the conventional topology optimization formulation into the optimal design of materials. The new function used to perform the optimization procedure is called UCOpt. In this function, in addition to the input parameters already present in the original top88 code, four extra parameters were added. Due to the two different scales (micro and macro), two parameters (i.e., lx, ly) representing the dimensions along the directions of the abscissa and ordinate of the macro domain, respectively, were added for the case of macro scale, and two parameters (i.e., nlx, nly) representing the number of elements along the directions of the abscissa and ordinate of the unit cell were added for the micro scale. Thus, the resulting function is presented below:

```
2 function UCOpt(lx,ly,nelx,nely,nlx,nly,volfrac,penal,rmin,ft)
```

As for the creation of the stiffness matrix variable KE, a new function is utilized. This function takes into consideration the length of the finite element along the directions of the abscissa and ordinate in the form of dx and dy, as well as the elasticity tensor instead of the Young modulus and the Poisson ratio used in the original code. This change is applied to enable the creation of the stiffness matrix from the homogenized elasticity tensor created by the homogenization function. In addition, the creation of the element stiffness matrix is performed inside the optimization procedure before the finite element analysis. Moving to the initialization of the design variables, in *Lines* 40 to 47 (of the UCOpt function), the initialization of the design variable is performed, in which instead of mapping the volume fraction to all densities and circle of zero densities is created in the centre of the unit cell, and all other densities are set to one, as shown in Figure 2.

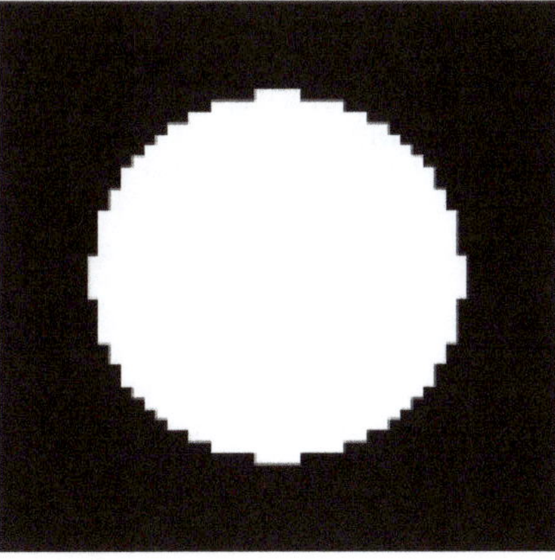

Figure 2. Initial unit cell.

This is achieved using the following iterative procedure:

```matlab
40  x = ones(nly,nlx);
41  for i = 1:nlx
42     for j = 1:nly
43        if sqrt((i-nlx/2-0.5)^2+(j-nly/2-0.5)^2) < min(nlx,nly)/3
44           x(j,i) = 0;
45        end
46     end
47  end
```

Moving to the optimization loop, two additional steps are added before performing the finite element analysis part. In the first step, a function called interpolate is utilized to compute the Young modulus and its derivative with respect to the design variables using the expressions of Equation (6) for the element Young modulus and Equation (7) for the corresponding derivative. During the second step, the resulting Young modulus and its derivative are provided to the homogenize function, which in turn produces the elasticity tensor and its derivative for each element consisting the unit cell. Moving to the finite element analysis part, the function Q4elementstiffnessMatrix is utilized to obtain the element stiffness matrix from the homogenized elasticity tensor, which in turn is used to perform the finite element analysis and compute the macro-domain displacements. For the computation of the sensitivities, an iterative procedure is utilized, looping for each element of the micro domain to create a different stiffness matrix for each unit cell element based on each element's derivative of the homogenized elasticity tensor. Then, the variable *ce* is computed in the same manner as in the original code, resulting in the computation of the derivative *dc*.

```matlab
67  for i = 1:nly
68     for j = 1:nlx
69        dKE = Q4elementStiffnessMatrix(lx/nelx/2,ly/nely/2,90,DCH{i,
              j});
70        ce = reshape(sum((U(edofMat)*dKE).*U(edofMat),2),nely,nelx);
71        c = c + sum(sum(ce));
72        dc(i,j) = - sum(sum(ce));
73     end
74  end
```

4.3. Model Order Reduction: Code Implementation

In this section, the implementation of the three model order reduction approaches will be presented. Aiming to create an easy integration of the three approaches into the UCOpt function presented in the previous section, the usage of the class structure is opted for the three MOR approaches. Thus, each approach is created as a single Matlab class object containing three common functions denoted as solve, fea and counts, respectively. The solve function is implemented differently for each class, while it is used by the UCOpt function in order to compute every set of displacements during the optimization procedure. The two other functions are the same for all three classes and they are used to perform the full-scale finite element analyses (function fea) and to return the number of full- and reduced-scale iterations performed (function counts). Since the class properties are modified during the optimization procedure, all MOR classes inherit from the handle a Matlab class. In order to use the MOR classes in the UCOpt function, an extra parameter is used called *p*, representing the MOR class, while *Line* 62 (of the pod class) is modified to call the solve function of the *p* class, as presented below:

```matlab
62  %    U(freedofs) = K(freedofs,freedofs)\F(freedofs);
63  U(freedofs) = p.solve(K(freedofs,freedofs),F(freedofs));
```

4.3.1. POD: Code Implementation (Matlab File "pod.m")

The pod class was developed for the code implementation of POD approach, which except for the constructor function, requires seven properties and three functions. Out of these properties, three refer to iteration trackers, i.e., parameters labeled as *loop, fll* and *rdc* tracking the total number of TO iterations performed, the total number of full finite element analyses and the total number of reduced basis finite element analyses, respectively. The forth parameter refers to the tolerance *tol* that represents the residual force tolerance used as a criterion for updating the reduced basis vectors after the creation of the reduced basis. The remaining three parameters correspond to the number of the reduced basis vectors Nb, the reduced basis matrix fi and a matrix containing the displacement snapshots A that is used to create the reduced basis matrix.

The implementation of the POD approach is performed inside the solve function. The solve function is separated into two main sections. The first section is executed during the first iterations for creating the first variant of the reduced basis matrix fi, as shown below:

```
24  U = obj.fea(K,F);
25  obj.A(:,obj.loop) = U;
26  if obj.loop == obj.Nb
27      obj.A = obj.A - mean(U);
28      [obj.fi,~,~] = svd(obj.A,'econ');
29  end
```

The second section is executed after the first creation of the reduced basis. In particular, in this section the reduced displacement field y is calculated, projecting it to the full scale of the displacement field U. Then, it is determined if the forces residual is acceptable. If the forces' residual is deemed not acceptable, then the reduced basis is updated using a new set of displacement snapshots. The implementation is presented below:

```
31  y = obj.fi'*K*obj.fi \ obj.fi'*F;
32  U = obj.fi*y;
33  dF = K*U-F;
34  res = norm(dF);
35  if res > obj.tol
36      U = obj.fea(K,F);
37      obj.A(:,1) = [];
38      obj.A(:,obj.Nb) = U;
39      obj.A = obj.A - mean(U);
40      [obj.fi,~,~] = svd(obj.A,'econ');
41  else
42      obj.rdc = obj.rdc + 1;
43  end
```

For the creation of the reduced basis matrix fi, the svd function of Matlab is utilized selecting the $'econ'$ option for generating an economy-size decomposition of matrix A.

4.3.2. On-the-Fly: Code Implementation (Matlab File "onthefly.m")

In the implementation of the on-the-fly reduced order model approach, the number of properties required by the corresponding class is reduced from seven to six, basically removing only the displacement snapshot matrix A. All other properties remain the same as those used in the POD implementation of the corresponding class. In the same manner as in the POD class, the implementation of the on-the-fly approach requires the use of the solve function. The on-the-fly implementation is also separated into two main parts. In the first one where the reduced basis matrix fi is computed, the norm function of Matlab is utilized to perform the normalization of the displacement field vector, whereas the procedure is exactly as described in Section 3.2 where the theoretical description of the on-the-fly reduced order model approach is provided.

```
22  if  obj.loop  ==  1
23    U  =  obj.fea(K,F);
24    obj.fi(:,obj.loop)  =  U/norm(U);
25  elseif  obj.loop  <=  obj.Nb
26    U  =  obj.fea(K,F);
27    Uorth  =  U  −  obj.fi*(obj.fi'*U);
28    obj.fi(:,obj.loop)  =  Uorth/norm(Uorth);
29  else
```

In the second part of the on-the-fly implementation, the procedure mirrors that of the POD implementation where instead of the svd function of Matlab, the update of the reduced basis matrix is performed as described in the expressions of Equations (15) and (16) in *Lines* 42 to 45 (of the onthefly class).

```
30  y  =  obj.fi'*K*obj.fi  \  obj.fi'*F;
31  U  =  obj.fi*y;
32  dF  =  K*U−F;
33  res  =  norm(dF);
34  if  res  >  obj.tol
35    U  =  obj.fea(K,F);
36    obj.fi(:,1)  =  [];
37    Uorth  =  U  −  obj.fi*(obj.fi'*U);
38    obj.fi(:,obj.Nb)  =  Uorth/norm(Uorth);
39  else
40    obj.rdc  =  obj.rdc  +  1;
41  end
```

4.3.3. Approximate Reanalysis: Code Implementation (Matlab File "ar.m")

To keep the same structure of the code implementation for the approximate reanalysis approach as that of the previously presented two MOR approaches, the displacement snapshots are updated in a fixed number of iterations without taking into account the change in the objective function or the design variables. For the implementation of the approximate reanalysis approach, two new properties were added compared with the implementation of the on-the-fly approach. These properties correspond to the stiffness matrix $K0$ of the full-scale FEA and to a counter rf that keeps record of how often a new full-scale FEA will be performed.

As far as the solve function goes, its first part, i.e., *Lines* 32 to 35 (of the ar class), deals with the initialization of the reduced basis matrix fi. As discussed earlier, at the beginning of this section the criterion for the update of this procedure is simplified. More specifically, the update of the reduced basis matrix takes place in the first iteration and then after a fix number of iterations specified by the class variable rf, as shown bellow:

```
25  if  (obj.loop  ==  1)  ||  (mod(obj.loop,obj.rf)  ==  0)
26    U  =  obj.fea(K,F);
27    obj.fi(:,1)  =  U;
28    obj.K0  =  K;
29  else
```

In the second part of the solve function, the reduced displacement vector is obtained. In more detail, in *Line 37*, the difference between the stiffness matrices (dK) is computed. Then, a while loop is implemented (see *Lines* 41 to 49 of the ar class) which builds the reduced basis matrix fi until either the maximum number of reduced basis vectors is reached or the residual is smaller than the tolerance value tol predefined.

```
30  else
31    dK  =  K−obj.K0;
32    U  =  obj.fi(:,1);
```

```
33    res = 100;
34    s = 1;
35    while (res > obj.tol) && (s <= obj.Nb)
36        s = s + 1;
37        U = obj.K0\(F - dK*U);
38        obj.fi(:,s) = U;
39        y = obj.fi'*K*obj.fi \ obj.fi'*F;
40        U = obj.fi*y;
41        dF = K*U-F;
42        res = norm(dF);
43    end
44    obj.rdc = obj.rdc + 1;
45 end
```

5. Test Examples

In this section, three simple test examples will be presented in order to demonstrate the ease of use of the proposed topology optimization Matlab code and how the three MOR approaches are integrated in order to assist the search procedure. The first test example refers to a simple bridge problem, the second one refers to the cantilever beam problem and the third one corresponds also to a cantilever beam problem with the load applied at the central right side of the domain. The macro domains for all test examples are schematically presented in Figure 3. In all test examples, the number of the iterations required, the number of full-scale FEAs and the final objective function value achieved are presented when the three MOR approaches are implemented, as well as the case without the application of any MOR approach.

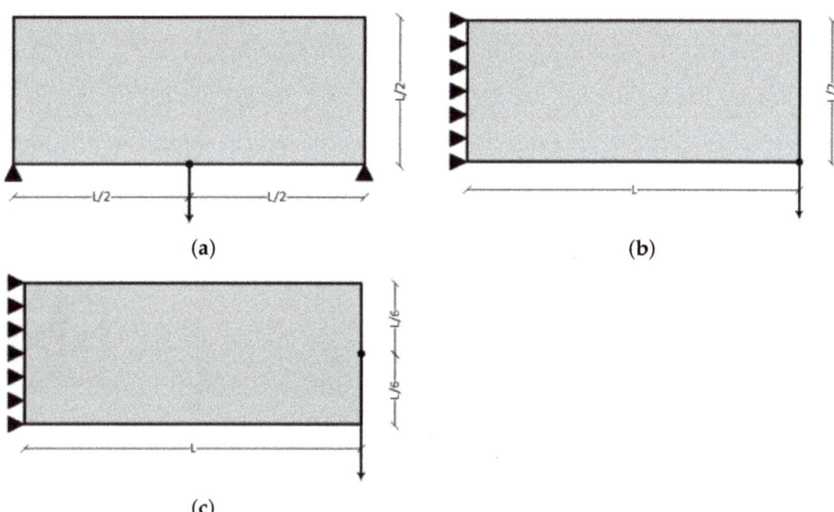

Figure 3. Test examples considered. (**a**) Bridge test example. (**b**) Cantilever beam 1 test example. (**c**) Cantilever beam 2 test example.

5.1. Bridge Test Example

The implementation of the MBB beam test example with respect to the load vector and fixed degrees of freedom is the default implementation of the UCOpt function. The optimization parameters were a grid of 300×150 finite elements in the directions of the abscissa and ordinate for the discretization of the macro domain and a grid of 50×50 finite elements in the directions of the abscissa and ordinate for the discretization of the micro domain. A target volume fraction of 40%, penalization factor for the SIMP approach of 3, filter radius of 1.5 and application only of a sensitivity filter (option $ft = 1$) were chosen. As far as the three MOR approaches go, the size of the reduced basis was chosen to be 8 for the POD and on-the-fly approaches and 10 for the approximate reanalysis, the tolerance for the update was set equal to 0.01 for all approaches and the update frequency for the approximate reanalysis was set to every six iterations. The script implementation for the POD-assisted optimization implementation is presented below:

```
p = pod(8, 0.01); % for the proper orthogonal decomposition
    approach
UCOpt(10,10,200,100,50,50,0.5,3,1.5,1,p);
```

For the implementation of the other two MOR approaches, changes are only required in the first *line* where the MOR is created, as follows:

```
p=onthefly(8, 0.01); % for the on-the-fly approach
```

and

```
p=ar(10,0.01,6);      % for the approximate reanalysis approach
```

The results obtained of every MOR approach as well as the implementation without MOR assistance are presented in Table 1.

Table 1. Bridge test example: Results of each MOR approach as well as the classic implementation.

Approach	Total itrns	Full FEAs	Compliance
FEA	26	26	105.14
POD	26	8	105.14
on-the-fly	26	8	105.14
AR	26	5	105.14

On the results of topology optimization achieved, in terms of unit cell structure, for the various implementations (with and without MOR), minor differences are observed, while the compliance value resulting from the four implementations is the same. Thus, only one of the results achieved, and in particular, the one obtained by means of the POD approach, is presented in Figure 4.

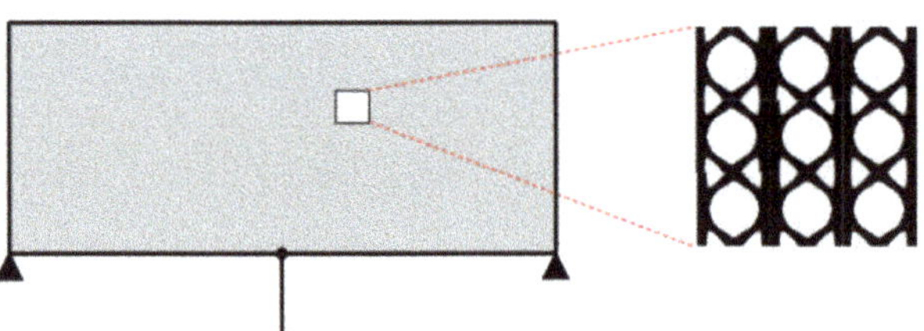

Figure 4. Optimized periodic unit cell for the bridge test example macro structure.

5.2. Cantilever Beam 1 Test Example

For the implementation of the first cantilever beam test example, changes to the load vector and fixed degrees of freedom should be made. In more detail, *Lines* 13 and 14 of the UCOpt function should be changed, as presented below:

```
F = sparse(2*(nelx+1)*(nely+1),1,-1,2*(nely+1)*(nelx+1),1);
fixeddofs = 1:2*(nely+1);
```

The optimization parameters were a grid of 200×100 finite elements in the directions of the abscissa and ordinate for the discretization of the macro domain and a grid of 50×50

finite elements in the the directions of the abscissa and ordinate for the discretization of the micro domain. Similarly to the first test example, a target volume fraction of 50%, penalization factor for the SIMP approach of 3, filter radius of 1.5 and application only of the sensitivity filter (i.e., option $ft = 1$) were chosen. For the MOR approaches, the size of the reduced basis was chosen to be 4 for the POD and on-the-fly approaches and 10 for the approximate reanalysis, the tolerance for the update was set equal to 0.01 for all approaches and the update frequency for the approximate reanalysis was set to every five iterations. The results obtained for the first cantilever beam test example are presented below in Table 2.

Table 2. Cantilever Beam 1: Results of each MOR approach as well as the classic implementation.

Approach	Total TOP itrns	Full FEAs	Compliance
FEA	107	107	257.3
POD	104	20	257.3
on-the-fly	101	21	257.3
AR	107	22	257.3

Similarly to the first test example, on the results of topology optimization achieved, in terms of unit cell structure, for the various implementations (with and without MOR) minor differences are observed, while the compliance value resulting from the four implementations is the same. Thus, only one of the results achieved, and in particular, the one obtained by means of the on-the-fly approach, is presented in Figure 5.

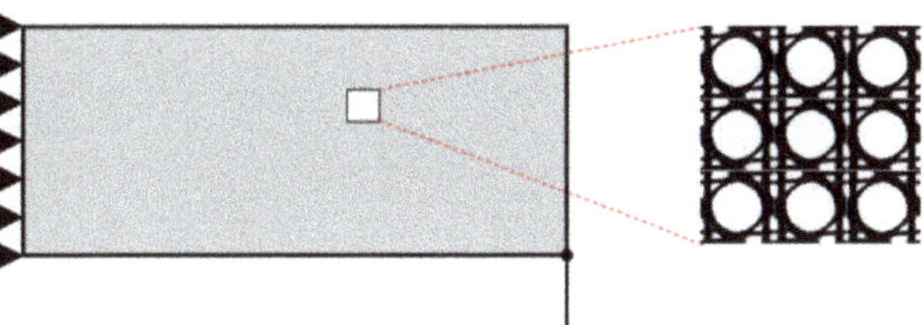

Figure 5. Optimized periodic unit cell for the cantilever beam 1 test example macro structure.

5.3. Cantilever Beam 2 Test Example

For the implementation of the second cantilever beam test example (labeled as cantilever beam 2), changes to the load vector and fixed degrees of freedom should be applied. In more detail, *Lines* 13 and 14 (of the UCOpt) function should be changed as presented below:

```
13 F = sparse(2*((nely+1)*nelx + (nely+1) - ceil(1/2*nely))
       ,1,-1,2*(nely+1)*(nelx+1),1);
14 fixeddofs = 1:1:2*(nely+1);
```

The optimization parameters were a grid of 300×100 finite elements in the directions of the abscissa and ordinate for the discretization of the macro domain and a grid of 50×50 finite elements in the the directions of the abscissa and ordinate for the discretization of the micro domain. Similarly to the first test example, a target volume fraction of 50%, penalization factor for the SIMP approach of 3, filter radius of 1.5 and application only of the sensitivity filter (i.e., option $ft = 1$) were chosen. For the MOR approaches, the size of the reduced basis was chosen to be 4 for the POD and on-the-fly approaches and 10 for the approximate reanalysis, the tolerance for the update was set equal to 0.01 for all approaches and the update frequency for the approximate reanalysis was set to every five iterations. The results obtained for the second cantilever beam test example are presented below in Table 3.

Table 3. Cantilever Beam 2: Results of each MOR approach as well as the classic implementation.

Approach	Total TOP itrns	Full FEAs	Compliance
FEA	84	84	1369.0
POD	83	16	1369.1
on-the-fly	84	16	1368.9
AR	84	17	1368.9

Similarly to the remarks reported for the first two examples presented before, on the results of topology optimization achieved, in terms of unit cell structure, for the various implementations (with and without MOR) observed minor differences are observed, while the compliance value resulting from the four implementations is the same. Thus, only one of the results achieved, and in particular, the one obtained by means of the approximate reanalysis approach, is presented in Figure 6.

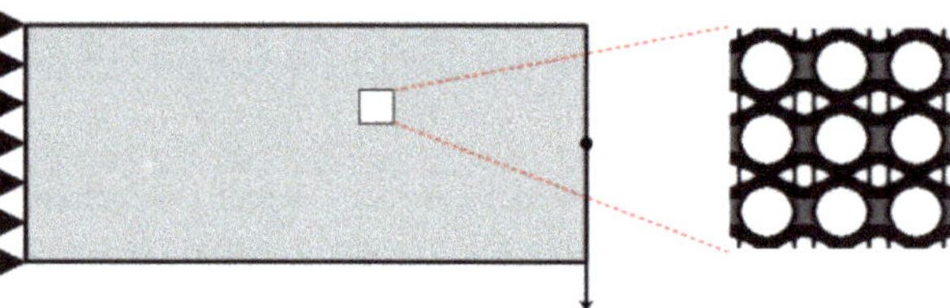

Figure 6. Optimized periodic unit cell for the cantilever beam 2 test example macro structure.

6. Conclusions

The scope of this work is to present an open source numerical implementation of a methodology dealing with the optimal design of material structure using the theories of topology optimization and homogenization as well as the application of reduced order models. The code presented is written in Matlab, and two of the functions are partially based on existing well-known codes, published on the topology optimization and homogenization formulations. The implementation of the three model order reduction (MOR) approaches is simple, and the aim is to provide the means to integrate such models in any type of topology optimization problem formulation. Although the code implementation of the topology optimization part is based on a 2D space variant, it can easily be extended to the 3D space as well without the need to modify any of the three MOR classes presented in this study. The authors would be happy to receive suggested improvements that can be implemented in the public domain of the *UCOpt* codes.

Author Contributions: Conceptualization, G.K. and N.D.L.; methodology, G.K. and N.D.L.; software, G.K.; validation, G.K.; formal analysis, G.K.; investigation, G.K.; writing—original draft preparation, G.K. and N.D.L.; writing—review and editing, G.K. and N.D.L.; visualization, G.K.; supervision, N.D.L.; project administration, N.D.L.; funding acquisition, N.D.L. All authors have read and agreed to the published version of the manuscript.

Funding: This research has been financed by the ADDOPTML project: "ADDitively Manufactured OPTimized Structures by means of Machine Learning" (No: 101007595).

Institutional Review Board Statement: Not applicable.

Informed Consent Statement: Not applicable.

Data Availability Statement: Please address the corresponding author by his e-mail address nlagaros@central.ntua.gr. The complete code, containing all participating functions (including the changes made to the original ones), is listed in https://github.com/nikoslagaros/TOPcodes, (accessed on 14 June 2022) .

Acknowledgments: This research has been supported by the ADDOPTML project: "ADDitively Manufactured OPTimized Structures by means of Machine Learning" (No: 101007595) belonging to the Marie Skłodowska-Curie Actions (MSCA) Research and Innovation Staff Exchange (RISE) H2020-MSCA-RISE-2020.

Conflicts of Interest: The authors declare no conflicts of interest.

Abbreviations

The following abbreviations are used in this manuscript:

BESO	Bi-directional Evolutionary Structural Optimization
FE	Finite Element
FEA	Finite Element Analysis
FEM	Finite Element Method
MOR	Model Order Reduction
POD	Proper Orthogonal Decomposition
SIMP	Solid Isotropic Material with Penalization
STO	Structural Topology Optimization
SVD	Singular Value Decomposition
TO	Topology Optimization
TOP	Topology Optimization Problem

References

1. Sigmund, O. Materials with prescribed constitutive parameters: An inverse homogenization problem. *Int. J. Solid Struct.* **1994**, *31*, 2313–2329. [CrossRef]
2. Sigmund, O.; Torquato, S. Design of materials with extreme thermal expansion using a three-phase topology optimization. *J. Mech. Phys. Solids* **1997**, *45*, 1037–1067. [CrossRef]
3. Gibiansky, L.V.; Sigmund, O. Multiphase composites with extremal bulk modulus. *J. Mech. Phys. Solids* **2000**, *48*, 461–498. [CrossRef]
4. Neves, M.M.; Rodrigues, H.; Guedes, J. Optimal design of periodic linear elastic microstructures. *Comput. Struct.* **2000**, *76*, 421–429. [CrossRef]
5. Fujii, D.; Chen, B.; Kikuchi, N. Composite material design of two-dimensional structures using the homogenization design method. *Int. J. Numer. Methods Eng.* **2001**, *50*, 2031–2051. [CrossRef]
6. Guest, J.K.; Prévost, J.H. Design of maximum permeability material structures. *Comput. Method Appl. Mech. Eng.* **2007**, *196*, 1006–1017. [CrossRef]
7. Challis, V.J.; Roberts, A.P.; Wilkins, A.H. Design of three dimensional isotropic microstructures for maximized stiffness and conductivity. *Int. J. Solids Struct.* **2008**, *45*, 4130–4146. [CrossRef]
8. Amstutz, S.; Giusti, S.M.; Novotny, A.A.; Souza Neto, E.A. Topological derivative for multi-scale linear elasticity models applied to the synthesis of microstructures. *Int. J. Numer. Methods Eng.* **2010**, *84*, 733–756. [CrossRef]
9. Gao, J.; Li, H.; Gao, L.; Xiao, M. Topological shape optimization of 3D micro-structured materials using energy-based homogenization method. *Adv. Eng. Softw.* **2018**, *116*, 89–102. [CrossRef]
10. Huang, X.; Radman, A.; Xie, Y. Topological design of microstructures of cellular materials for maximum bulk or shear modulus. *Comput. Mater. Sci.* **2011**, *50*, 1861–1870. [CrossRef]
11. Huang, X.; Zhou, S.; Xie, Y.; Li, Q. Topology oprimization of microstructures of cellular materials and composites for macrostructures. *Comput. Mater. Sci.* **2013**, *67*, 397–407. [CrossRef]
12. Wu, J.; Sigmund, O.; Groen, J. Topology optimization of multi-scale structures: A review. *Struct. Multidiscip. Optim.* **2021**, *63*, 1455–1480. [CrossRef]
13. Kirsch, U.; Papalambros, P.Y. Structural reanalysis for topological modifications—A unified approach. *Struct. Multidiscip. Optim.* **2001**, *21*, 333–344. [CrossRef]
14. Wang, S.; Sturler, E. Paulino, G.H. Large-scale topology optimization using preconditioned Krylov subspace methods with recycling. *Int. J. Numer. Methods Eng.* **2006**, *69*, 2441–2468. [CrossRef]
15. Amir, O.; Bendsøe, M.P.; Sigmund, O. Approximate reanalysis in topology optimization. *Int. J. Numer. Methods Eng.* **2008**, *78*, 1474–1491. [CrossRef]
16. Amir, O.; Stolpe, M.; Sigmund, O. Efficient use of iterative solvers in nested topology optimization. *Struct. Multidiscip. Optim.* **2009**, *42*, 55–72. [CrossRef]
17. Amir, O.; Sigmund, O.; Lazarov, B.S.; Schevenels, M. Efficient reanalysis techniques for robust topology optimization. *Comput. Methods Appl. Mech. Eng.* **2012**, *245-246*, 217–231. [CrossRef]
18. Gogu, C. Improving the efficiency of large scale topology optimization through on-the-fly reduced order model construction. *Int. J. Numer. Methods Eng.* **2014**, *101*, 281–304. [CrossRef]
19. Alaimo, G.; Auricchio, F.; Bianchini, I.; Lanzarone, E. Applying functional principal components to structural topology optimization. *Int. J. Numer. Methods Eng.* **2018**, *115*, 189–208. [CrossRef]

20. Ferro, N.; Micheletti, S.; Perotto, S. POD-assisted strategies for structural topology optimization. *Comput. Math. Appl.* **2019**, *77*, 2804–2820. [CrossRef]
21. Senne, T.A.; Gomes, F.A.M.; Santos, S.A. On the approximate reanalysis technique in topology optimization. *Optim. Eng.* **2019**, *20*, 251–275. [CrossRef]
22. Xiao, M.; Lu, D. Breitkopf, P.; Raghavan, B.; Dutta, S.; Zhang, W. On-the-fly model reduction for large-scale structural topology optimization using principal components analysis. *Struct. Multidiscip. Optim.* **2020**, *62*, 209–230. [CrossRef]
23. Kallioras, N.A.; Kazakis, G.; Lagaros, N.D. Accelerated topology optimization by means of deep learning. *Struct. Multidiscip. Optim.* **2020**, *63*, 1185–1212. [CrossRef]
24. Sigmund, O. A 99 line topology optimization code written in Matlab. *Struct. Multidiscip. Optim.* **2001**, *21*, 120–127. [CrossRef]
25. Andreassen, E.; Clausen, A.; Schevenels, M. Lazarov, B.; Sigmund, O. Efficient topology optimization in MATLAB using 88 lines of code. *Struct. Multidiscip. Optim.* **2010**, *43*, 1–16. [CrossRef]
26. Liu, K.; Tovar, A. An efficient 3D topology optimization code written in Matlab. *Struct. Multidiscip. Optim.* **2014**, *50*, 1175–1196. [CrossRef]
27. Ferrari, F.; Sigmund, O.; Guest, J.K. Topology optimization with linearized buckling criteria in 250 lines of Matlab. *Struct. Multidiscip. Optim.* **2021**, *63*, 3045–3066. [CrossRef]
28. Talischi, C.; Paulino, G.H.; Pereira, A.; Menezes, I.F.M. PolyTop: A Matlab implementation of a general topology optimization framework using unstructured polygonal finite element meshes. *Struct. Multidiscip. Optim.* **2012**, *45*, 329–357. [CrossRef]
29. Chi, H.; Pereira, A.; Meneze, I.F.M.; Paulino, G.H. Virtual element method (VEM)-based topology optimization: An intergrated framework. *Struct. Multidiscip. Optim.* **2020**, *62*, 1089–1114. [CrossRef]
30. Amir, O.; Aage, N.; Lazarov, B.S. On multigrid-CG for efficient topology optimization. *Struct. Multidiscip. Optim.* **2014**, *419*, 815–829. [CrossRef]
31. Huang, X.; Xie, Y. A futher review of ESO type methods for topology optimization. *Struct. Multidiscip. Optim.* **2010**, *41*, 671–683. [CrossRef]
32. Wang, M.; Wang, X.; Guo, D. A level set method for structural topology optimization. *Comput. Methods Appl. Mech. Eng.* **2003**, *192*, 227–246. [CrossRef]
33. Challis, V.J. A discrete level-set topology optimization code written in Matlab. *Struct. Multidiscip. Optim.* **2010**, *41*, 453–464. [CrossRef]
34. Otomori, M.; Yamada, T.; Izui, K.; Nishiwaki, S. Matlab code for a level-set based topology optimization method using a reaction diffusion equation. *Struct. Multidiscip. Optim.* **2014**, *51*, 1159–1172. [CrossRef]
35. Wei, P.; Li, Z.; Li, X.; Wang, M.Y. An 88-line MATLAB code for the parameterized level set method based topology optimization using radial basis functions. *Struct. Multidiscip. Optim.* **2018**, *58*, 831–849. [CrossRef]
36. Lagaros, N.D.; Vasileiou, N.; Kazakis, G. A C# code for solving 3D topology optimization problems using SAP2000. *Optim. Eng.* **2019**, *20*, 1–35. [CrossRef]
37. Gao, J.; Wang, L.; Luo, Z.; Gao, L. IgaTop: An implementation of topology optimization for structures using IGA in MATLAB. *Struct. Multidiscip. Optim.* **2021**, *64*, 1669–1700. [CrossRef]
38. Andreassen, E.; Andreasen, C.S. How to determine composite material properties using numerical homogenization. *Comput. Mater. Sci.* **2014**, *83*, 488–495. [CrossRef]
39. Dong, G.; Tang, Y.; Zhao, Y. A 149 Line Homogenization Code for Three-Dimensional Cellular Materials Written in Matlab. *J. Eng. Mater. Technol.* **2019**, *141*, 555. [CrossRef]
40. Xia, L.; Breitkopf, P. Design of materials using topology optimization and energy-based homogenization approach in Matlab. *Struct. Multidiscip. Optim.* **2015**, *52*, 1229–1241. [CrossRef]
41. Bendsøe, M.P.; Sigmund, O. *Topology Optimization: Theory, Methods and Applications*, 2nd ed.; Springer: Berlin/Heidelberg, Germany, 2004. [CrossRef]
42. Sigmund, O. Morphology-based black and white filters for topology optimization. *Struct. Multidiscip. Optim.* **2007**, *33*, 401–424. [CrossRef]

Article

Flexural Strength of Internally Stiffened Tubular Steel Beam Filled with Recycled Concrete Materials

Ahmed W. Al Zand [1,*], Mustafa M. Ali [1,2], Riyadh Al-Ameri [3], Wan Hamidon Wan Badaruzzaman [1], Wadhah M. Tawfeeq [4], Emad Hosseinpour [1] and Zaher Mundher Yaseen [5]

[1] Department of Civil Engineering, Universiti Kebangsaan Malaysia (UKM), Bangi 43600, Malaysia; p86577@siswa.ukm.edu.my (M.M.A.); wanhamidon@ukm.edu.my (W.H.W.B.); emadhoseinpour@gmail.com (E.H.)
[2] Department of Civil Engineering, University of Technology, Baghdad 10066, Iraq
[3] School of Engineering, Faculty of Science Engineering and Built Environment, Deakin University, Geelong, VIC 3216, Australia; r.alameri@deakin.edu.au
[4] Faculty of Engineering, Sohar University, Sohar 311, Oman; WTawfeeq@su.edu.om
[5] New Era and Development in Civil Engineering Research Group, Scientific Research Center, Al-Ayen University, Thi-Qar 64001, Iraq; yaseen@alayen.edu.iq
*   Correspondence: ahmedzand@ukm.edu.my

Citation: Al Zand, A.W.; Ali, M.M.; Al-Ameri, R.; Badaruzzaman, W.H.W.; Tawfeeq, W.M.; Hosseinpour, E.; Yaseen, Z.M. Flexural Strength of Internally Stiffened Tubular Steel Beam Filled with Recycled Concrete Materials. *Materials* **2021**, *14*, 6334. https://doi.org/10.3390/ma14216334

Academic Editor: Stelios K. Georgantzinos

Received: 26 September 2021
Accepted: 18 October 2021
Published: 23 October 2021

Publisher's Note: MDPI stays neutral with regard to jurisdictional claims in published maps and institutional affiliations.

Abstract: The flexural strength of Slender steel tube sections is known to achieve significant improvements upon being filled with concrete material; however, this section is more likely to fail due to buckling under compression stresses. This study investigates the flexural behavior of a Slender steel tube beam that was produced by connecting two pieces of C-sections and was filled with recycled-aggregate concrete materials (CFST beam). The C-section's lips behaved as internal stiffeners for the CFST beam's cross-section. A static flexural test was conducted on five large scale specimens, including one specimen that was tested without concrete material (hollow specimen). The ABAQUS software was also employed for the simulation and non-linear analysis of an additional 20 CFST models in order to further investigate the effects of varied parameters that were not tested experimentally. The numerical model was able to adequately verify the flexural behavior and failure mode of the corresponding tested specimen, with an overestimation of the flexural strength capacity of about 3.1%. Generally, the study confirmed the validity of using the tubular C-sections in the CFST beam concept, and their lips (internal stiffeners) led to significant improvements in the flexural strength, stiffness, and energy absorption index. Moreover, a new analytical method was developed to specifically predict the bending (flexural) strength capacity of the internally stiffened CFST beams with steel stiffeners, which was well-aligned with the results derived from the current investigation and with those obtained by others.

Keywords: stiffened CFST beam; recycled concrete; finite element; flexural strength; cold-formed tube

1. Introduction

The usefulness of various types of concrete-filled steel tube (CFST) has been proven in modern composite structural projects since they have achieved better performances (in terms of strength, ductility and stiffness) than the corresponding conventional concrete/steel members under axial and flexural loading [1–4]. In general, the structural performance of concrete-filled steel composite members, under different loading scenarios, has been experimentally and numerically examined in several studies; for example, those presented in [5–9].

Particularly, the adequacy of using the Slender steel tube section (Class 4; as per Eurocode classification) in the concept of CFST beams has been previously investigated [10–14]. This section (Slender) has achieved more strength improvement percentages than those of the Noncompact and Compact sections, after being filled with concrete materials, as compared to their corresponding hollow tube sections [10,15]. In addition, the steel tubes of CFST members with Slender cross-sections usually have lighter self-weight than the Noncompact and Compact sections; thus, they are more favourable in terms of reducing the impact of cost in construction projects. However, the Slender tube section is more likely to buckle outward at the compression zone stress because of Slender structural sections [16,17]. Therefore, internal steel stiffeners were provided to delay and restrict the outward tube's buckling of the Slender CFST beams' cross-section [18–21]. For example, Al Zand et al. [19] experimentally and numerically examined the effect of using varied shapes of internal steel stiffeners that were welded along all sides of high-slenderness cold-formed square CFST beams. Their investigation showed that these CFST beams' flexural stiffnesses and strength capacities were improved by about 22–29% when a single stiffener was welded along each tube's internal sides, and these values were increased by a further 48–59% when double steel stiffeners were provided. Furthermore, in some cases, the CFST beam's cross-sections were externally stiffened, either by using additional steel plates that were fixed mechanically along the beam's flanges [5], or by preparing a cold-formed tube's cross-section with V-grooves that were provided along the sides of the CFST beam [22]. However, the welding of additional steel stiffeners for the stiffening of the CFST beam's cross-section could require an additional fabrication process and extra labour costs. Thus, using pre-fabricated cold-formed tubular steel sections (C-sections/C-Purlins) without extra welding and/or mechanical fixation could be considered as a new concept for use in the stiffened Slender CFST beam system, particularly for lightweight flooring structures. For example, in Malaysia, the cost of 1 ton of fabricated steel sections is about 1000 USD, and the equivalent cost of hot-rolled sections is about 1200 USD, while the cost of 1 ton of concrete is about 40–50 USD. This material's cost comparison shows that it is very useful if the engineers can utilise CFST beams that have been prepared from cold-formed sections filled with concrete in the lightweight composite structures, even though the self-weight of these composite beams can increase by 5–8 times as compared to the equivalent steel beams due to the effect of concrete infill material.

The research topic of CFST has received considerable attention over the past few years from a wide range of structural engineering scholars. A dependent stress–strain model was developed for the CFST column [23]. The flexural behaviour of steel-fibre-reinforced self-stressing recycled-aggregate CFST was studied by [24]. The behaviour of heated CFST stub columns, containing steel fibre and tire rubber, was tested by [25]. Reinforced CFST with steel bars was studied experimentally in terms of its seismic behaviour [26] Several other related research studies on the advanced application of CFST can be reviewed through [27,28]. Nevertheless, this review of the literature evidenced the importance of the enhancement of concrete properties for the achievement of better reliability and robustness in construction projects [29–31].

In recent years, engineers have been increasingly utilizing the recycled aggregate generated from the demolition of existing materials and waste materials in the field of construction [24,32–34]. There have been several studies on the behaviour of Recycled-Aggregate Concrete (RAC) material, which is prepared by crushing the old structural elements of concrete. For example, variations in the RAC replacement percentages (0%, 25% and 50% of the raw aggregate) were adopted in the concrete infill material for CFST members under flexural and compression loads [11,24,35–40]. The results achieved were quite similar to those obtained for CFST members filled with normal concrete, albeit with slightly lower strength capacities [38]. Recently, Liu et al. [24] used RAC and sulphoaluminate cement in varied concrete mixtures, which were reinforced with steel fibre and used as infilling material for 54 CFST beams. Their experimental study reveals that the reinforcement of the recycled-aggregate concrete mixture with 1.2% steel fibre can achieve similar flexural behaviour to that of the CFST beams that are filled with normal concrete Furthermore, using the lightweight concrete infill material can significantly reduce the overall self-weight of the CFST members [17,41–43], which is significant given that the self-weight represents one of the most important challenges in terms of adopting composite members of this type in modern structures. Generally, using expanded polystyrene (EPS) beads as a replacement for the raw aggregate is the most effective method that is usually used to achieve lightweight concrete mixtures [19,44–46]. For instance, EPS can also be sourced from waste polystyrene materials (recycled materials). However, the performance of cold-formed steel tubes (the Slender tube section) filled with lightweight and recycled concrete material under pure flexural loading has not yet been extensively investigated. In such cases, as explained earlier, the overall self-weight and the cost impact constitute important factors in the Engineer's decision to adopt this system in modern composite structures. Therefore, at present, finding an alternative CFST composite structural design concept is a matter of urgency.

The main objectives of this research can be highlighted as follows: The first aim was to examine the suitability of using the new concept of cold-formed tubular cross-sections fabricated from two pieces of galvanized C-sections (face-to-face connection) in the Slender CFST beam's system, in which the lips of these C-sections are expected to behave as internal steel stiffeners along the top and bottom flanges. Second, this research project aimed to explore the behaviour of the proposed Slender CFST beams, specifically when recycled-aggregate concrete mixture is used instead of the normal concrete mixture, in order to reduce the self-weight and cost of the beam's section. Lastly, to date, all of the theoretical methods for predicting the flexural strength of CFST beams were developed according to different standards, and they were mostly intended to be applied for beams with conventional tube sections (unstiffened sections). As such, this study aimed to develop a new analytical method that can theoretically predict the flexural strength capacity (ultimate bending moment; M_u) of the internally stiffened CFST beam's cross-section with varied numbers/sizes of steel stiffeners; this constitutes one of the main novelties of the current research work. Therefore, five CFST specimens were prepared for this purpose and were experimentally tested under pure flexural loading, including one specimen without concrete filling material (hollow tube section). The finite element (FE) ABAQUS software was used to develop and analyse an additional 20 models that were designed to examine the effect of further parameters that were not explored experimentally.

2. Experimental Approach

2.1. Preparation of Samples

In order to suggest a new concept for the design of Slender steel tube cross-sections that are internally stiffened, in this research, five specimens were produced from two steel C-sections connected by means of tack welding to obtain the suggested steel tubular section, as shown in Figure 1. These prefabricated tubular sections had 20 mm lips that functioned as internal stiffeners along the top and bottom flanges of the tubular beams. A single specimen was tested as a hollow tube beam (HB), while the other four tubes' specimens were filled with different concrete mixtures (filled beam; FB). The raw coarse aggregate of the concrete mixtures was replaced with different recycled-aggregate materials (EPS and RAC), since one of the main objectives of this research was to reduce the self-weight and cost of the mixture. The replacement percentages of raw coarse aggregate (by volume) were equal to 0% (filled beam designation; FB-RC0), 30% (FB-RC30), 50% (FB-RC50), and 70% (FB-RC70), as presented in Table 1. All steel tubes specimens were placed vertically, after which the concrete material was poured from the top ends in multiple stages, while the bottom ends were temporary sealed to prevent water leakage.

2.2. Material Properties

Steel tube: Three coupons were cut from the cold-formed C-sections and prepared in accordance with the ASTM-E8/E8M-2009 standard. The average results of the yield tensile strength (f_y), maximum elongation (%), ultimate tensile strength (f_u), and elastic modulus (E_s) were 489 N/mm^2, 27.4%, 558 N/mm^2, and 201 $\times$ 10^3 N/mm^2, respectively, and these results were obtained from the direct tensile of coupons test.

Concrete mixtures: the recycled aggregates used in the suggested concrete mixtures were EPS beads (4–6.3 mm) with a density of 9.5 (kg/m^3), and RAC (4.75–16 mm) with a density of 1278 kg/m^3; these densities were much lower than that of the raw coarse aggregate (1490 kg/m^3). As explained earlier, the first concrete mixture was prepared without the use of any recycled aggregate (normal concrete; 0% replacement aggregate: RC0). The second concrete mixture (RC30) was prepared by replacing 30% (by volume) of the raw coarse aggregate by EPS beads only. Meanwhile, the concrete mixtures RC50 and RC70 were also made by replacing 30% of the raw aggregate with EPS beads and adding another 20 and 40% (in terms of total volume) of RAC, respectively, bringing the total replacement to 50 and 70%, respectively. In addition, silica fume (SF) material was used to enhance the bonding performance between the cement and EPS beads, as previously advised in [45]. Lastly, for all concrete mixtures, a water/cement (w/c) ratio of 0.46 was used together with the superplasticizer liquid (Real Flow 611). The proportions of the concrete mixtures are presented in Table 2. For each concrete mixture, three samples of concrete cubes (150 mm) were prepared, cured for up to 28 days, and tested in accordance with BS 1881: 1983.

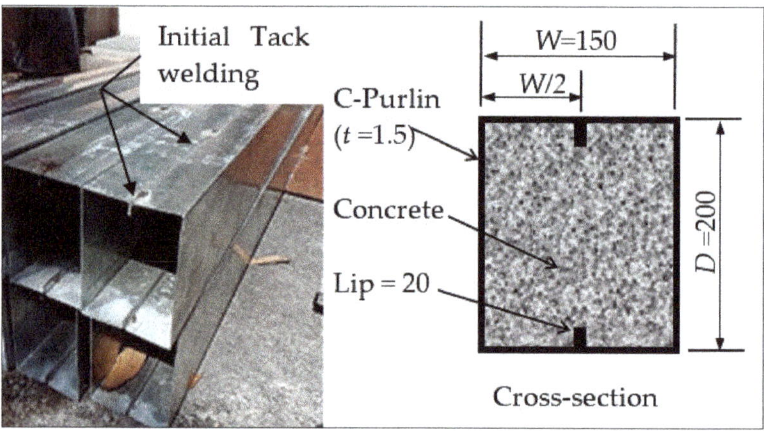

Figure 1. Cold-formed steel C-sections (all dimensions in mm).

Table 1. Designations and results of the tested specimens.

Specimen Designations	$D \times B \times t$ (mm)	L_e (m)	f_y (MPa)	f_{cu} (MPa)	E_c (GPa)	M_u (kN·m)	K_i (kN·m^2)	K_s (kN·m^2)	EAI (kN·mm)
HB	200 × 150 × 1.5	2.8	489	-	-	14.1	1207	1121	201.8
FB-RC0	200 × 150 × 1.5	2.8	489	26.2	21.1	57.7	2216	1924	5996
FB-RC30	200 × 150 × 1.5	2.8	489	14.6	14.5	53.7	2157	1751	5539
FB-RC50	200 × 150 × 1.5	2.8	489	14.1	13.8	52.6	2194	1726	5493
FB-RC70	200 × 150 × 1.5	2.8	489	13.7	13.2	51.5	2066	1690	5439

Table 2. Proportions of the concrete mixtures (kg/m^3).

Mixture Designations	Cement	Fine Agg.	Coarse Agg.	Silica Fume	EPS (%)	EPS	RAC (%)	RAC	Water	Slump (mm)	Density
RC0	390	700	1115	-	-	-	-	-	180	128	2295
RC30	350	700	781	40	30	2.1	-	-	180	147	1881
RC50	350	700	558	40	30	2.1	20	190	180	151	1813
RC70	350	700	335	40	30	2.1	40	380	180	154	1772

2.3. Test Setup

All specimens were tested under four-point static loads, as shown in the schematic of the test setup in Figure 2. A hydraulic jack with a maximum capacity of 500 kN was used to apply the static load on the suggested CFST specimens. Utilizing linear variable differential transducers (LVDTs) type KYOWA, Osaka, Japan, the vertical deflections of the specimens were recorded at different locations. The strain gauges SG1 to SG5 were fixed vertically at the mid-span of each CFST specimen. A data logger was used for collection of the data from the load cell, strain gauges, and LVDTs during the tests, and these data were recorded in a computerized system.

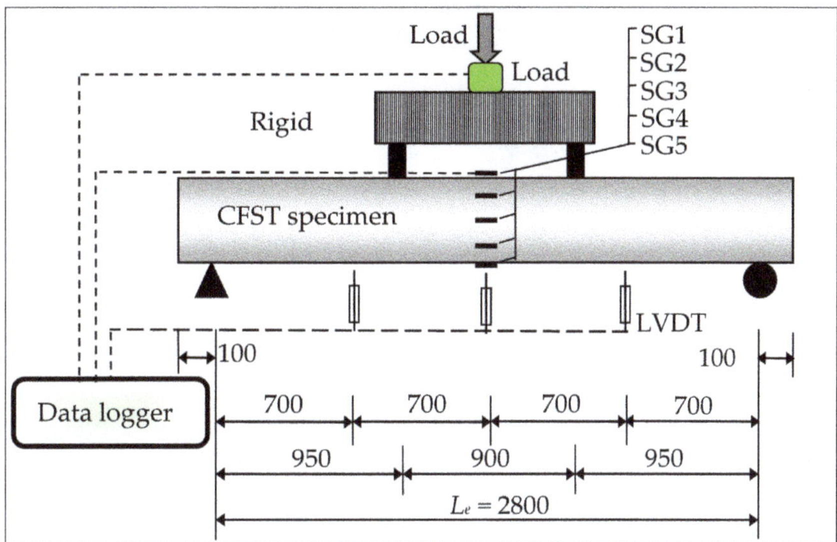

Figure 2. Schematic of specimens' test setup (all dimensions in mm).

3. Discussion of Experimental Results

3.1. Failure Modes

All specimens were tested beyond their strength capacities in order to study their extreme failure performance. The hollow steel tube specimen (HB) showed the inward tube's buckling failure at the support points only, wherein the failure of tube's webs was increased gradually by increasing the applied load, as shown in Figure 3a. Meanwhile, all filled specimens (FB-RC0, FB-RC30, FB-RC50, and FB-RC70) showed almost typical failure modes regardless of the types of their concrete mixtures. It was found that, for all filled specimens, outward buckling failure started occurring at the top tube's flange, between the two-point loads of these filed specimens, particularly when the applied load values reached about 70–80% of their ultimate loading capacity, as shown in Figure 3b. This performance was due to the effects of the concrete infill materials, which mainly prevented the occurrence of inward buckling failure of the steel tube. Figure 3c presents all filled specimens after testing at the extreme failure limits.

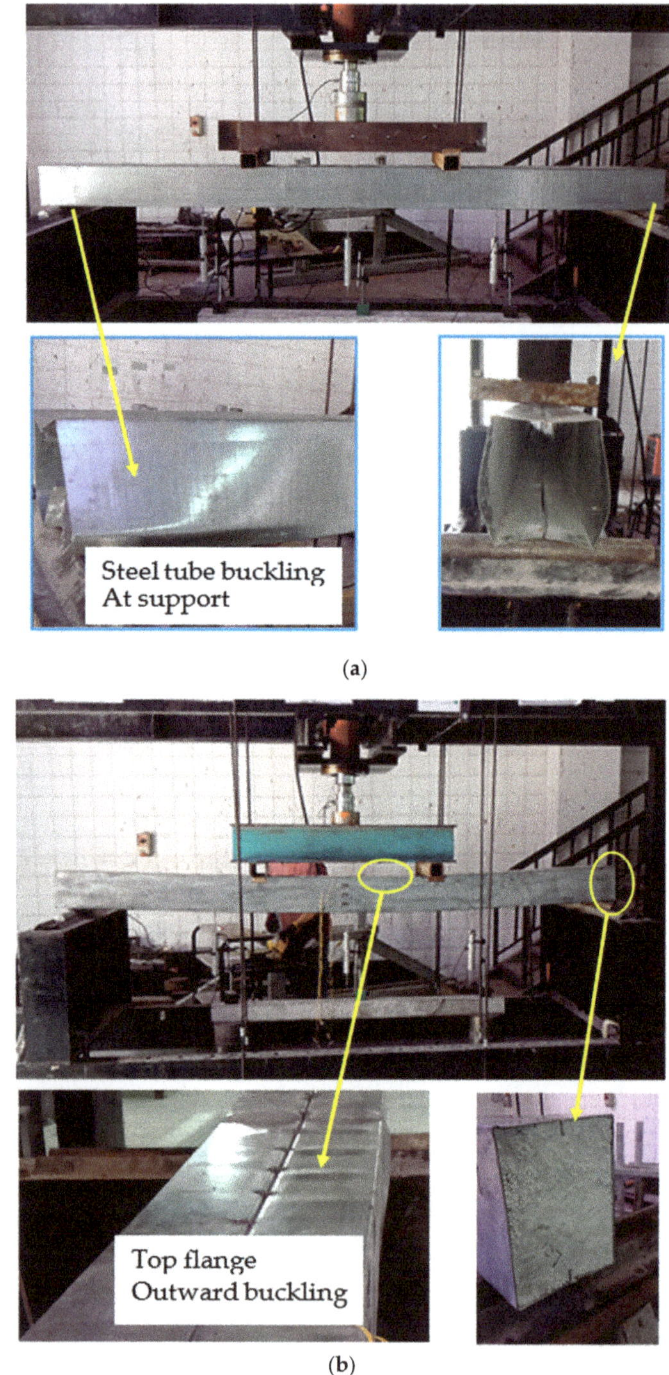

Figure 3. *Cont.*

(c)

Figure 3. Typical failure modes: (**a**) unfilled specimen (HB); (**b**) filled specimen (FB-RC30); (**c**) all filled specimens after testing.

Furthermore, the 20 mm lips located at the top and bottom edges of the prefabricated cold-formed C-sections were well bonded with the concrete core, thus implying that these lips functioned sufficiently as internal stiffeners for the top and bottom flanges of the steel tubes of these filled specimens. Therefore, the outward buckling failure occurred only at the half-width of their top flanges (see Figure 3b). These lips achieved similar contributions to those of the additionally welded steel stiffeners that were included in previous research to stiffen the Slender CFST members [19,47]. Moreover, there was no slippage failure between the steel tube and the concrete core observed at both opened ends of the tested filled specimens. Unlike the unfilled specimen (HB), the filled specimens (FB-RC0, FB-RC30, FB-RC50 and FB-RC70) displayed a smooth deflection behaviour during the loading stages, which was quite similar to the half-sine curve, as illustrated in Figure 4. This performance allowed the CFST beams to distribute the point loads with almost uniform stress along the beam's span, which was very similar to the beams that were loaded in a uniform loading scenario during the practical application of the members. Based on this finding, the above hypothesis was confirmed insofar as the currently suggested prefabricated cold-formed tubular steel beam filled with recycled concrete materials performed very similarly to the conversional CFST beams, which were tested earlier, for example, in [11,17,19,48].

3.2. Flexural Behaviour and Strength Capacity

This section discusses the flexural behaviour, ultimate moment capacity (M_u), and the moment vs. strain relationships of the tested specimens. The relationships between the bending moment and the deflection at mid-span for the tested specimens are compared in Figure 5, including the hollow specimen (HB). In general, for the concrete filled specimens, the moment–deflection curves continued to show linear behaviour until an archive of about 50–60% of their M_u values, after which these curves behaved as elasto-plastics up to the loading limits of about 70–80% of their M_u values (at this point, buckling failure of the top flange started to occur). Thus, due to further outward buckling failure, the same moment–deflection curves showed fully plastic behaviour with continual slow decreases. The M_u value of the filled specimens was recorded at the deflection limit of $L_e/50 - L_e/60$ [49], which was about 46 mm to 56 mm. However, the moment–deflection curve of the unfilled specimen (HB) showed an almost linear behaviour until the peak loading point (M_u) was achieved, at which the bottom flanges started to buckle inwardly at the supporting points; furthermore, as a result of extreme tube buckling failure, the loading curve began to descend.

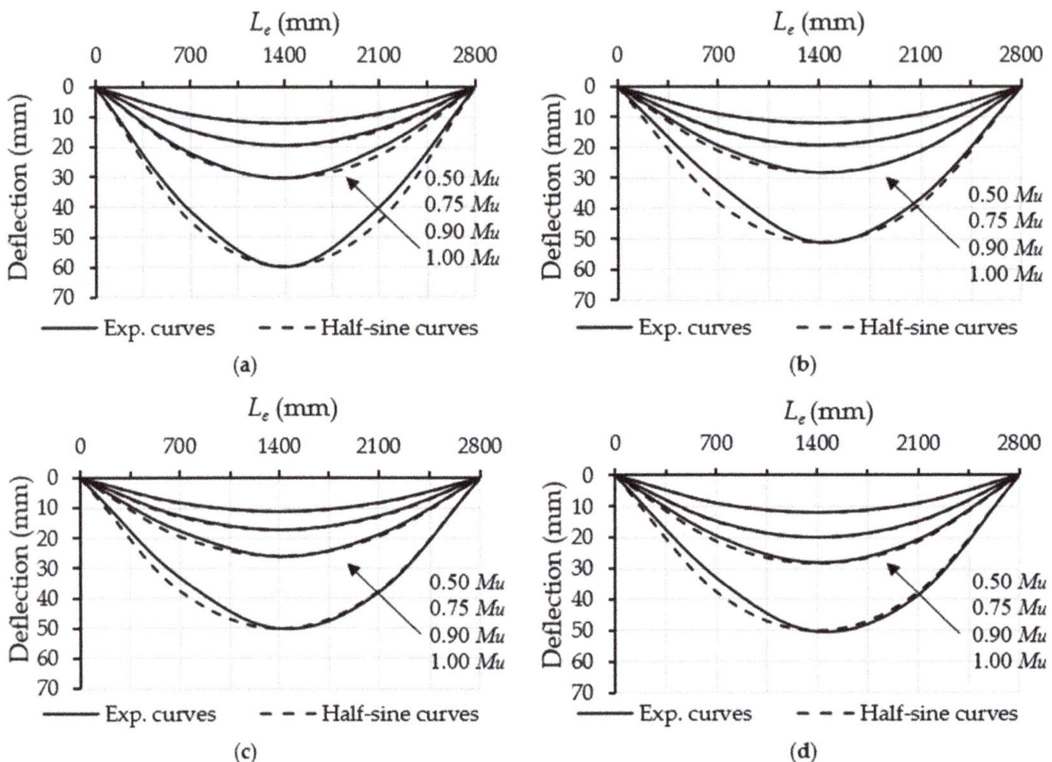

Figure 4. Deflection shapes of the filled specimens: (**a**) FB-RC0; (**b**) FB-RC30; (**c**) FB-RC50; (**d**) FB-RC70.

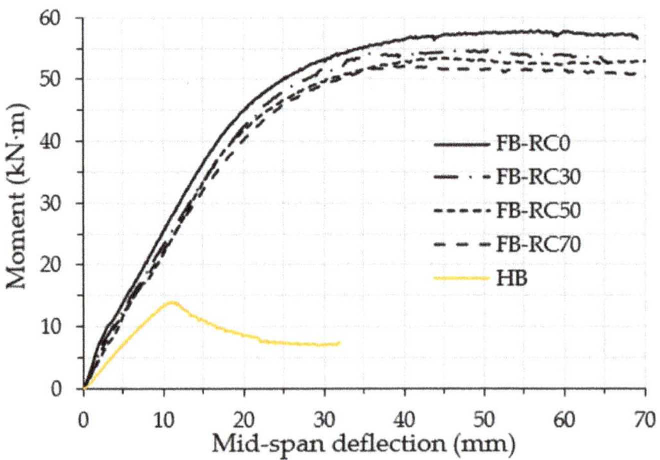

Figure 5. Moment vs. mid-span deflection relationships.

The M_u values of the tested specimens are presented in Table 1. The HB specimen achieved the ultimate flexural capacity of 14.1 kN·m; this capacity increased to 57.7 kN·m when the same fabricated steel tubular specimen was filled with normal concrete (FB-RC0), achieving an improvement of 409%. The tubular specimen filled with varied recycled concrete achieved M_u values that were slightly lower than those of the FB-RC0 specimen (57.7 kN·m); these values were 53.7 kN·m, 52.6 kN·m and 51.5 kN·m, for specimens FB-RC30, FB-RC50 and FB-RC70, respectively. In addition, it was noticed that, regardless of the aggregate replacement percentages, the concrete filled tubular specimens achieved much higher flexural strength capacity values as compared to the capacity of the hollow specimen (HB), with a strength improvement of about 363 to 381%. It is worth mentioning that the self-weight of the tube beam was substantially increased due to the addition of the filled concrete; however, the flexural strength was remarkably enhanced, as stated above (363–409%). From the structural engineering point of view, this represents a very important finding regardless of the increment of the weight, and, in this case, the role of the structural engineer, who can prospectively determine scenarios in which this would be reliable for the targeted purpose of construction projects. On the other hand, the cost was also found to be a vital element in the construction design when we compared the concrete cost with the steel cost; this was the case because, in global terms, the cost of 1 ton of steel is about 15–20 times the cost of 1 ton of concrete material.

Furthermore, the moment vs. longitudinal tensile strain relationships obtained at the mid-span of the steel tube of the filled specimens are presented in Figure 6. In this figure, it can be seen that the moment–strain relationships showed a fairly typical behaviour for all of the filled specimens regardless of the percentages of replacement aggregate in the concrete material [11,24]. The strain gauges SG1 and SG2 (see Figure 2) showed gradually increasing negative values (compression stress) with the increasing of the bending loads, while the strain gauges located at the bottom-half of beam's cross-section (SG4 and SG5) showed gradual increases in their tension-related stress values. In contrast, the strain gauge SG3 (located in the middle of the specimen's mid-span) showed a slight increase in tension-related stress, confirming that a positive correlation was discovered with the upward movement of the neutral axis in the tested filled specimens as the bending loads increased. Figure 7, as an example, shows the maximum strains distributed at the tube's mid-span depth along its cross-section for specimens FB-RC0 and FB-RC30 during a variety of loading stages. In the current study, the aforementioned moment–strain relationships of the fabricated filled-steel tubular beams behaved similarly to those obtained for the conventional cold-formed CFST beams [11,48].

3.3. Flexural Stiffness

The initial stiffness (K_i) and serviceability stiffness (K_s) levels of tested specimens are usually measured from the moment vs. mid-span curvature relationships [50–53], which are based on moment values of $0.2M_u$ and $0.6M_u$, respectively. The K_i and K_s values of the tested specimens are presented in Table 1. The HB specimen achieved the lowest flexural stiffness values as compared to those of the filled specimens (FB), which were 1207 kN·m^2 and 1121 kN·m^2 for K_i and K_s, respectively. These values increased to 2216 kN·m^2 and 1924 kN·m^2 for the FB-RC0 specimen due to the influence of the normal concrete infill material. In general, similar flexural stiffness improvement behaviours were recorded for all of the filled recycled concrete materials (FB-RC30, FB-RC50 and FB-RC7), but with slightly lower values than that of FB-RC0, since they had lower concrete modulus (E_c) values. For example, compared to the FB-RC0 specimen (0% aggregate replacement), the FB-RC70 specimen, which was filled with recycled concrete (30% EPS plus 40% RAC) achieved lower K_i and K_s values (−6.7% and −12.1%, respectively). From the above discussion, it can be concluded that even when using the recycled aggregate in the concrete infill materials of the cold-formed CFST beams, the flexural stiffness can nevertheless be significantly improved, to a far greater degree than that of the corresponding hollow tube beam.

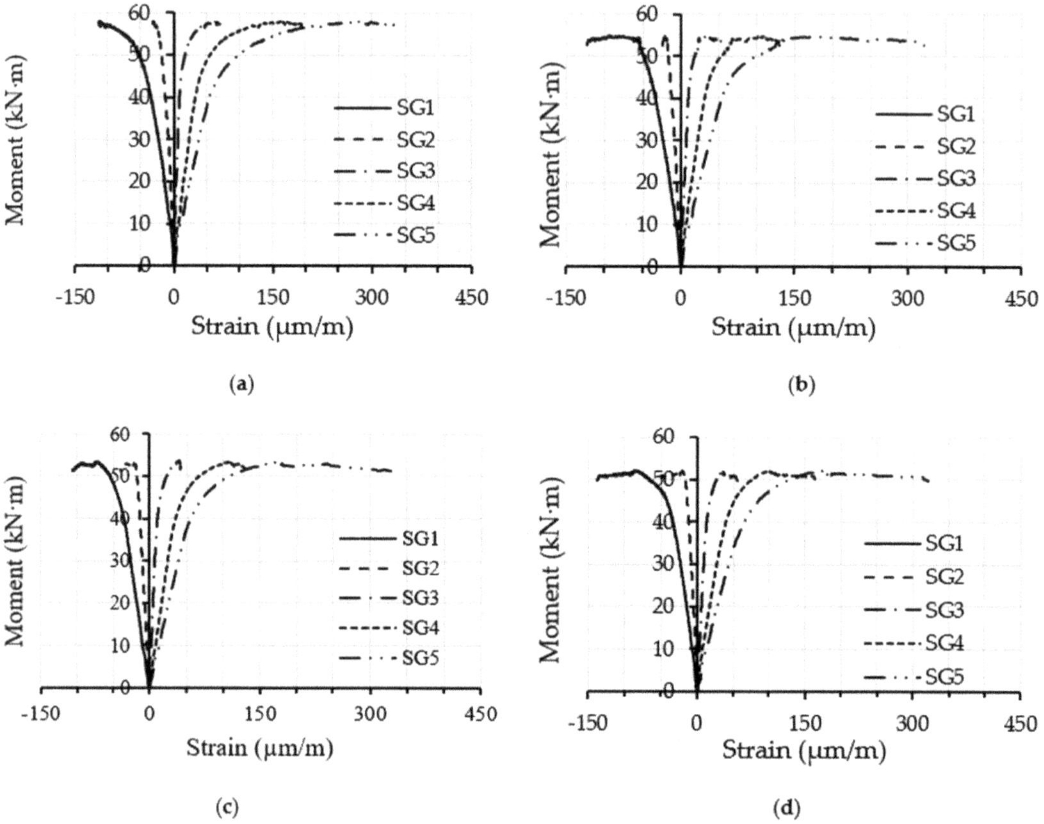

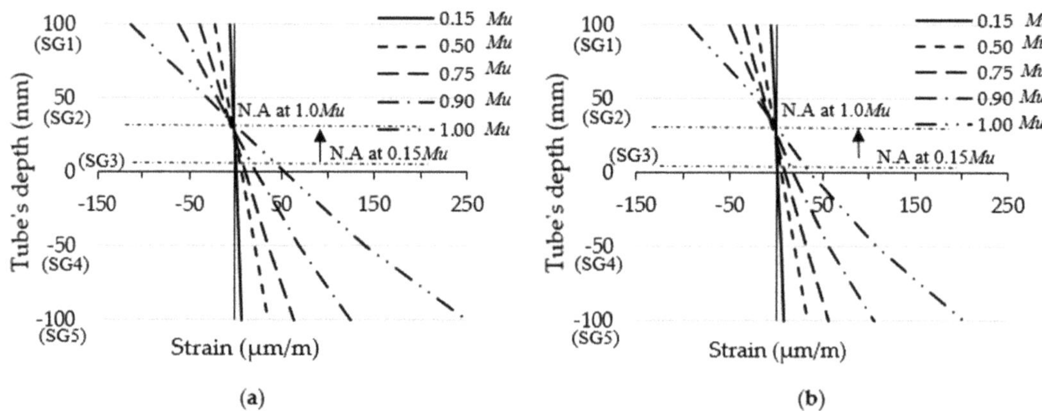

Figure 6. Moment vs. longitudinal strain relationships of the filled specimens: (**a**) FB-RC0; (**b**) FB-RC30; (**c**) FB-RC50; (**d**) FB-RC70.

Figure 7. Depth of steel tube cross-section vs. longitudinal strain relationships: (**a**) FB-RC0; (**b**) FB-RC50.

3.4. Energy Absorption Index

The ability of CFST beams to dissipate energy compared to that of the hollow steel tube beams was confirmed in several previous research studies [19,54]. Most commonly, the energy absorption index (EAI) of CFST beams was estimated from the cumulative area under the load vs. the deflection curves, up to the point at which the maximum beam strength limit had been reached [22,54,55]. The EAI values of the tested specimen in the current research are presented in Table 1; these values clearly confirm that the energy dissipation ability of the newly fabricated hollow specimen was extremely enhanced when filled with normal/recycled concrete materials. For example, the EAI value of the HB specimen increased from 201.8 kN·mm to 5996 kN·mm (29.7 times) when it was filled with normal concrete (FB-RC0). Moreover, compared to the tubular specimens that were filled with normal concrete (0% replacement aggregate), the specimens that were filled with recycled aggregate (EPS and RAC) had a slightly lower ability to dissipate energy, which can be considered a logical outcome since they achieved slightly lower loading capacities (see Figure 6). It is interesting to note that the EAI values of specimens FB-RC30, FB-RC50, and FB-RC70 were approximately 5539 kN·mm (−7.6%), 5493 kN·mm (−8.3%), and 439 kN·mm (−9.2%), respectively, as compared to the value of the FB-RC0 specimen (5996 kN·mm). These values represent significant achievements when compared to the value obtained for the corresponding unfilled HB specimen (201.8 kN·mm), being approximately 26.9 to 27.5 times higher

4. Numerical Approach

4.1. Development and Verification of the Numerical Model

The suggested CFST beam was further investigated in terms of its flexural behaviour by using the ABAQUS 6.14 software to conduct non-linear finite element (FE) analysis. In order to reduce the time conception of the analyses of the models, a typical 3D quarter model was built to simulate the actual tested CFST specimen, as shown in Figure 8, which had the advantages of the symmetric cross-section and the beam's loading scenario already having been prepared [15,19,56–58]. The actual test loading scenario was implemented in the FE modelling scenario by allowing the nodes that were positioned at the upper tube's flange (loading point) to gradually move downwards during the FE analysis, using the incremental downward displacement option (displacement control approach), which is available in the software. The nodes placed at the support's location were restricted from horizontal and vertical movement, but they were allowed to freely rotate around the X-axis to simulate the roller support. Then, the loading value of the FE model was obtained from the reaction forces that were observed at the support's nodes [19,56].

The main component materials of the currently developed FE CFST models are the concrete infill and the steel tube (double C-sections). For the concrete component, the C3D8R element type was used, which has eight nodes integrated with six degrees of freedom, while, for the steel tube components, the type S4R shell element was used. The penalty friction coefficient of 0.75 was adopted in the current FE analysis to realise the mechanical interaction between the surfaces of the steel and concrete parts. In general, several parameters that mainly affected the proper friction coefficient values, such as the loading type, the size/shape of the beam's cross-section, and the properties of the materials were selected [10,19,56–58]. Thus, in the current study, a convergence study was adopted in order to establish a suitable friction coefficient value (0.75), where several preliminary FE CFST models, which had varied friction coefficient values ranging from 0.4 to 0.9, were utilised. A finer mesh size was used for the distance between the two applied loads to sufficiently represent the failure modes of the analysed CFST models.

Consequently, the material properties of the steel tube and concrete components of the FE model were the same as those of the corresponding tested specimen. Generally, the concrete material is considered to be a brittle material since it cracks under tension-related stress and is crushed under compression stress [15]. Thus, the "Concrete Damage Plasticity" approach was adopted for the current FE models in order to ensure the compressive and tensile performance of the concrete infill component [15,57,59,60]. The elastic modulus and Poisson's ratio of the steel component were identified in the elastic-isotropic section, while the plastic-isotropic option was used to identify the steel-yielding strength and the relevant strain values. For both the steel and concrete materials of the developed FE CFST models, the constitutive stress vs. strain relationships were estimated by adopting the same expressions that were used earlier in [15,56].

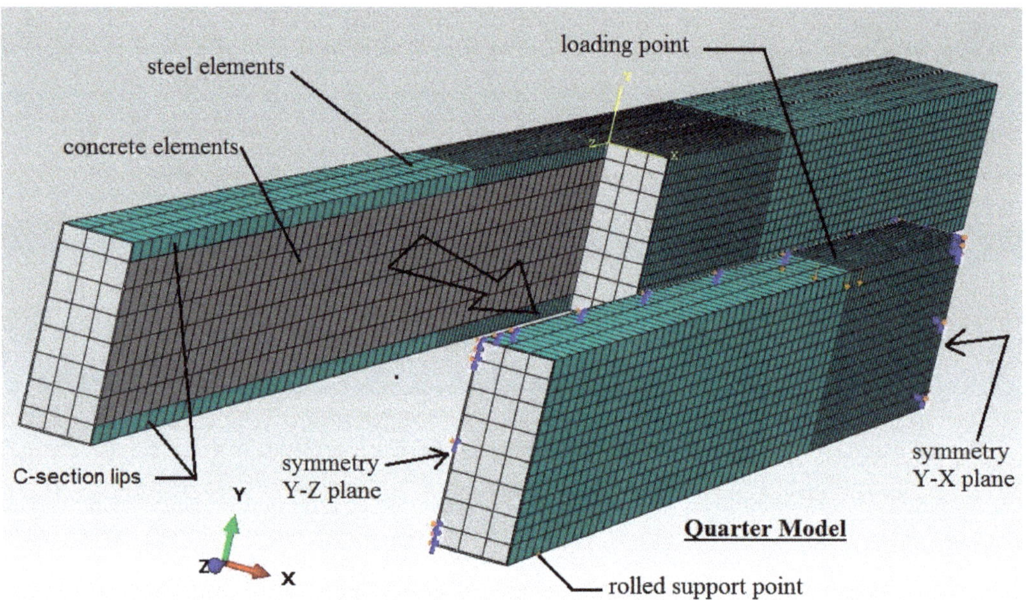

Figure 8. Typical finite element CFST model.

The validity of the current FE models was confirmed via the corresponding experimental results; almost all of the filled specimens with varied concrete mixtures achieved similar behaviours and close flexural strength capacities. Thus, the FB-RC30 specimen (a steel tubular beam filled with a recycled concrete mixture of 30% EPS beads) was selected to verify the relevant FE model. The numerical analysis showed that the moment vs. mid-span deflection relationship of the FB-RC30 (FE) model sufficiently agreed with that of the corresponding tested specimen (FB-RC30), as shown in Figure 9a. The flexural strength capacity value of the numerical model of FB-RC30 was overestimated by about 3.1% (55.4 kN·m) as compared to that obtained from the corresponding tested specimen, which was an acceptable degree of deviation. In addition, the outward buckling failure occurring at the top-half cross-section of the tested FB-RC30 specimen, for the distance between the two-point loads, was numerically simulated by the relevant FE model, as shown in Figure 9b.

4.2. Parametric Studies

After confirming the validity of the developed CFST model, additional models were built and analysed to investigate the influence of further parameters. In particular, the effects of varied tube thickness (1.0 mm to 3.0 mm), concrete infill strength (14.6 MPa to 55.0 MPa), steel yield strength (275 MPa to 550 MPa), and steel tube depth (150 mm to 250 mm) were studied, and these were categorized into groups A, B, C and D, respectively. Table 3 presents the model's designation along with the physical properties adopted for the 20 CFST models, including the M_u, K_i and K_s values derived from their FE analyses.

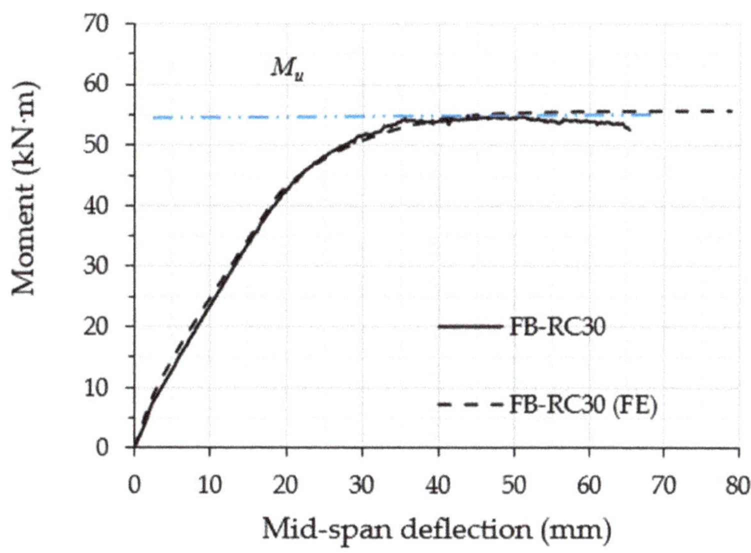

(**a**)

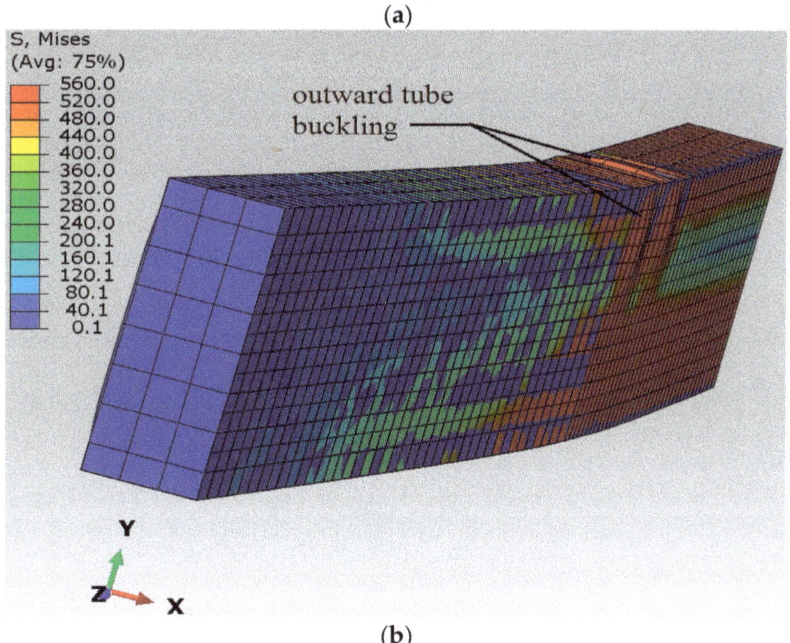

(**b**)

Figure 9. Verification of the FB-RC30 FE model with the corresponding tested specimen: (**a**) flexural behavior; (**b**) numerical failure mode.

Table 3. FE model designations and analysis results.

Models Designation	$D \times B \times t$ (mm)	f_y (MPa)	f_{cu} (MPa)	E_c (GPa)	As (mm²) $\times 10^2$	Is (mm⁴) $\times 10^6$	Ac (mm²) $\times 10^4$	Ic (mm⁴) $\times 10^7$	M_u (kN·m)	K_i (kN·m²)	K_s (kN·m²)
FB1-A	$200 \times 150 \times 1.0$	489.0	14.6	16.2	7.72	4.87	2.92	9.52	38.1	2093	1929
# FB2-A	$200 \times 150 \times 1.5$	489.0	14.6	16.2	11.5	7.23	2.88	9.28	55.4	2375	2125
FB3-A	$200 \times 150 \times 2.0$	489.0	14.6	16.2	15.3	9.54	2.85	9.05	72.4	2511	2345
FB4-A	$200 \times 150 \times 2.5$	489.0	14.6	16.2	19.0	11.8	2.81	8.82	88.8	3024	2815
FB5-A	$200 \times 150 \times 3.0$	489.0	14.6	16.2	22.7	14.0	2.77	8.60	105.0	3575	3285
# FB1-B	$200 \times 150 \times 1.5$	489.0	14.6	16.2	11.5	7.23	2.88	9.28	55.4	2375	2125
FB2-B	$200 \times 150 \times 1.5$	489.0	25.0	21.2	11.5	7.23	2.88	9.28	58.7	2491	2219
FB3-B	$200 \times 150 \times 1.5$	489.0	35.0	25.0	11.5	7.23	2.88	9.28	60.3	2699	2339
FB4-B	$200 \times 150 \times 1.5$	489.0	45.0	28.4	11.5	7.23	2.88	9.28	61.0	2819	2529
FB5-B	$200 \times 150 \times 1.5$	489.0	55.0	31.4	11.5	7.23	2.88	9.28	62.2	3033	2798
FB1-C	$200 \times 150 \times 1.5$	275.0	14.6	16.2	11.5	7.23	2.88	9.28	36.0	2293	1978
FB2-C	$200 \times 150 \times 1.5$	350.0	14.6	16.2	11.5	7.23	2.88	9.28	39.3	2274	1998
FB3-C	$200 \times 150 \times 1.5$	420.0	14.6	16.2	11.5	7.23	2.88	9.28	48.3	2302	2011
# FB4-C	$200 \times 150 \times 1.5$	489.0	14.6	16.2	11.5	7.23	2.88	9.28	55.4	2375	2125
FB5-C	$200 \times 150 \times 1.5$	550.0	14.6	16.2	11.5	7.23	2.88	9.28	59.9	2432	2173
FB1-D	$150 \times 150 \times 1.5$	489.0	14.6	16.2	10.0	3.74	2.15	3.85	37.0	1287	978
FB2-D	$175 \times 150 \times 1.5$	489.0	14.6	16.2	10.8	5.32	2.52	6.17	46.1	1814	1436
# FB3-D	$200 \times 150 \times 1.5$	489.0	14.6	16.2	11.5	7.23	2.88	9.28	55.4	2375	2125
FB4-D	$225 \times 150 \times 1.5$	489.0	14.6	16.2	12.3	9.51	3.25	13.3	66.0	3286	2686
FB5-D	$250 \times 150 \times 1.5$	489.0	14.6	16.2	13.0	12.2	3.62	18.3	77.1	4407	3530

Verified FE model with the corresponding tested FB-RC30 specimen.

4.2.1. Performance of Bending Behaviour

In Figure 10, the moment vs. mid-span deflection curves of the analysed FE models are presented independently for each group. These curves showed elastic behaviour at the initial loading stage up to a certain limit, followed by plastic behaviour until the ultimate strength capacity of the CFST model was achieved. The models with varied tube thicknesses and depths showed a major influence on their moment–deflection curves at both the elastic and plastic loading stages, as shown in Figure 10a,d for the FE models in groups A and D, respectively. This is a logical flexural behaviour since the cross-section parameters (the steel area and moment of inertia) of the suggested CFST models were increased as a result of enhancements in their tubes' thicknesses and/or depths. However, the use of varied tube yield strengths did not bring about a major impact on the models' moment–deflection behaviours at the elastic stage, but a major effect was found at their plastic stage only, as shown in Figure 10c for the models in group C. Moreover, very limited improvement was recorded for the moment–deflection curves' behaviour when only the compressive strength of the infill material increased, as shown in Figure 10b.

4.2.2. Performance of Stiffness

The stiffness values of the analysed FE models are given in Table 3. The K_i and K_s values improved significantly with the increases in the steel tube's thickness and/or the depth of the studied models (group A and D), while very limited improvements were recorded when only the concrete compressive strength was increased (models in Group C). For example, the FB2-A model achieved K_i and K_s values of 2375 kN·m² and 2125 kN·m², respectively. These values were improved by about 50–55% (3575 kN·m² and 3285 kN·m²) when only the tube's thickness increased from 1.5 mm to 3.0 mm (FB5-A). Moreover, the stiffness K_i and K_s values of the FB1-B model were improved by about 18–19% (2819 kN·m² and 2529 kN·m²) when only the f_{cu} value increased by about three times (from 14.6 MPa to 45 MPa).

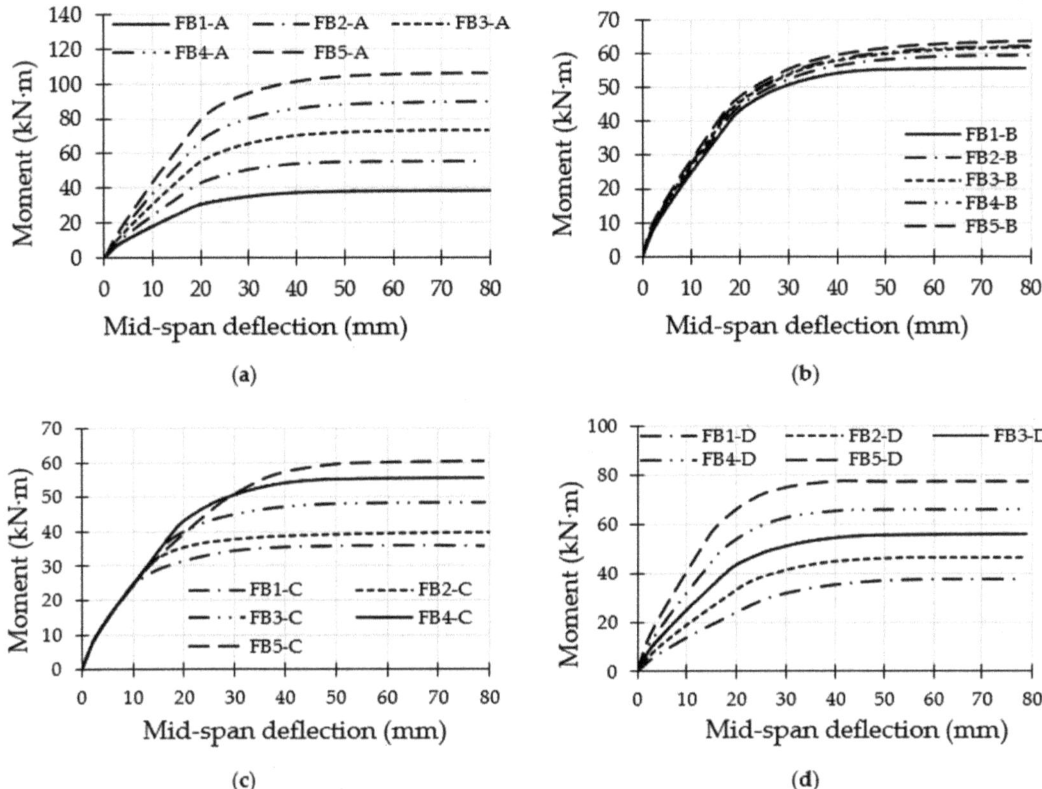

Figure 10. Moment vs. mid-span deflection of CFST models with varied parameters: (**a**) group A; (**b**) group B; (**c**) group C; (**d**) group D.

4.2.3. Performance of Bending Strength

The ultimate bending strength capacities (M_u) of the analysed FE models are given in Table 3. In addition, they are independently compared for each group in Figure 11. Generally, compared to all of the studied parameters, increasing the tube's thickness (group A) led to major improvements in the suggested CFST models' M_u values; these improvements were even more significant than the effects of tube's depth (Group D). Meanwhile, increasing the strength of the concrete infill led to limited improvements in their M_u values. This can be considered a reasonable outcome since the tube's thickness directly increased the overall area of the steel cross-section, including the internal stiffeners (the lips of the C-sections) at the top and bottom beam's flanges. Similar outcomes have been recorded in other studies for the conventional CFST beams that were investigated here [10,14,61]. For example, the FB2-A control model with 1.5 mm thickness achieved an M_u value of 55.4 kN·m; this value was increased to about 30.6% (72.4 kN·m) and 60.3% (88.8 kN·m) when only the tube's thickness was increased to 2.0 mm and 2.5 mm, respectively. Meanwhile, the same M_u value (55.4 kN·m) for the FB1-B model, in which concrete infill of 14.6 MPa strength was used, was increased by about 12.2% (62.2 kN·m) when a three-times-higher compressive strength was used (55 MPa; model FB5-B).

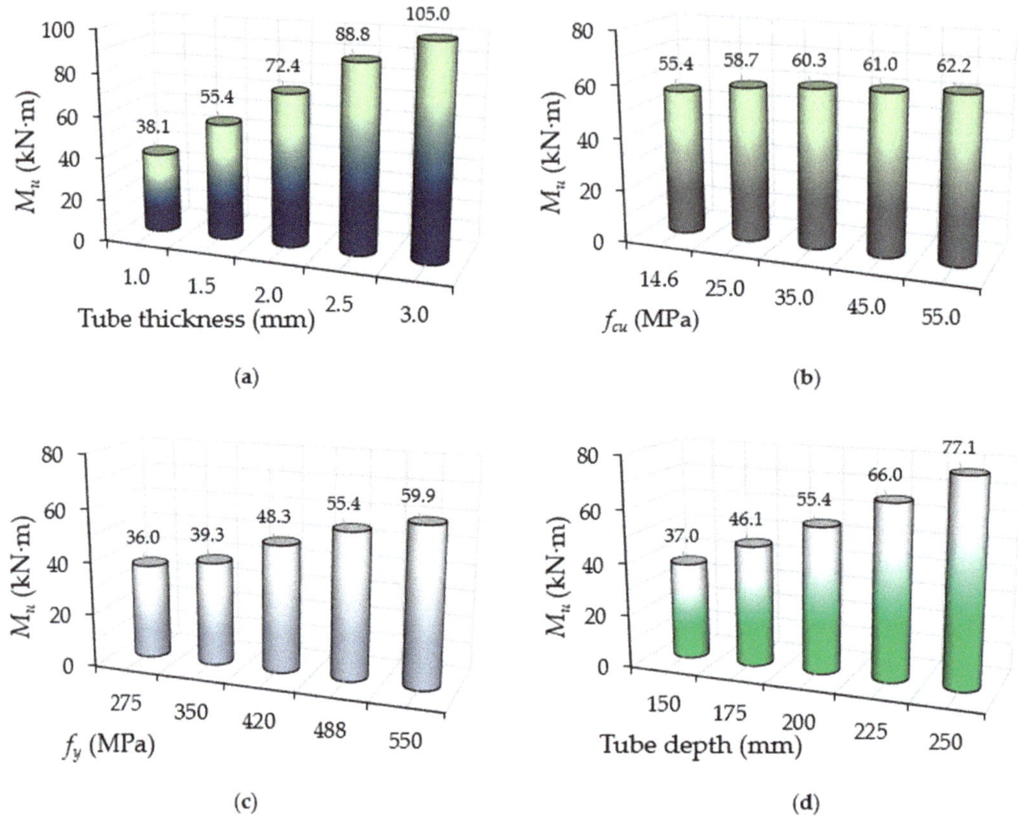

Figure 11. Effects of the M_u values of CFST models with varied parameters: (**a**) group A; (**b**) group B; (**c**) group C; (**d**) group D.

5. Design Guidelines

5.1. Evaluation of the Obtained Flexural Stiffness

This section evaluates the currently known flexural stiffness values (K_i and K_s) obtained from existing experimental and numerical approaches. The theoretical expressions that are presented in the AIJ-1997 [62], EC4-2004 [63], and ANSI/AISC 360-10 [64] standards are used:

$$K = E_S I_S + C1 E_C I_C \tag{1}$$

where E_s and I_s are the modulus of elasticity and moment of inertia, respectively, for the steel part. E_c and I_c are the modulus of elasticity and moment of inertia, respectively, for the concrete part. The $C1$ is a reduction factor for the concrete stiffness part, which is taken to be 0.6 in EC4-2004, and 0.2 in AIJ-1997. However, in the ANSI/AISC 360-10 standard, the $C1$ value is estimated to be $0.6 + 2A_s/(A_s + A_c)$, but should not exceed 0.9. The E_c value in Equation (1) is 9500 $(f_{ck} + 8)^{1/3}$, 4700 $(f_c)^{0.5}$ and 21,000 $(f_c/19.6)^{0.5}$ for the EC4-2004, AISC-2010 and AIJ-1997 standards, respectively. Furthermore, the theoretical methods developed by Han et al. [51] and Al Zand et al. [56] for the independent prediction of the values of flexural stiffness at the initial and serviceability levels (K_i and K_s) were adopted.

The predicted values of flexural stiffness ($K_{predicted}$), and the experimentally and numerically obtained values ($K_{obtained}$), are compared in Figure 12. Generally, the obtained flexural stiffness values at the initial loading stage (K_i) are slightly overestimated by the theoretical prediction, which is acceptable since it is within ±20% [50,52]. However, the AIJ-1997 standard achieves the lowest predicted stiffness values (*K-AIJ*) as compared to the other standards and methods, since this standard uses the lowest concrete stiffness redaction factor (*C1* = 0.2). Additionally, the EC4-2004 and ANSI/AISC 360-10 standards showed a more conservative prediction for the stiffness values (*K-EC4* and *K-AISC*) at the serviceability level (see Figure 12b), given that they use a single-expression formula (Equation (1)) to estimate the flexural stiffness value of the CFST members, unlike the expression methods of Han-2006 and Al Zand-2020, in which the flexural stiffness values of CFST beams are estimated at two different loading stages (two independent levels: K_i and K_s).

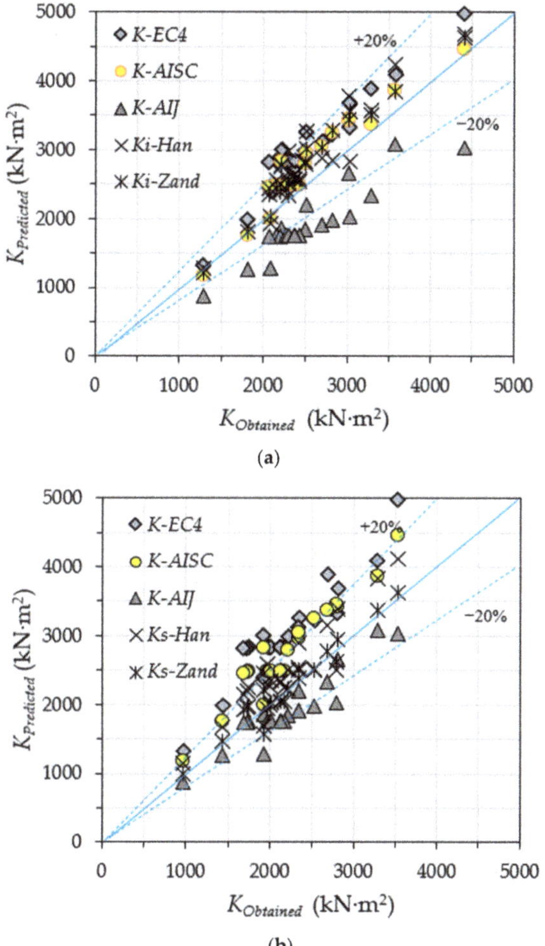

Figure 12. Verification of the obtained flexural stiffness: (**a**) *Ki*; (**b**) K_s.

5.2. Evaluation of the Obtained Flexural Strength

The ultimate flexural strength (M_u) values obtained for the tested and analysed CFST models were evaluated, using the existing theoretical methods given by EC4-2004 [63], Han-2004 [61], and Al Zand-2020 [56], to verify the findings of the current study. In Table 4, the predicted M_u values of the currently investigated beams and models are compared, using the above theoretical methods (M_{u-EC4}, M_{u-Han} and M_{u-Zand}), with those obtained from the current experimental and numerical investigations, including six models analysed by others [19]. Generally, the existing methods showed a conservative prediction of the M_u values of the investigated CFST beams and models, with reasonable coefficient of variation (COV) values, since these methods were mainly developed for conventional CFST beams (section of beams without internal steel stiffeners). In Table 4, it can be seen that $M_{u\text{-}EC4}$ achieved a mean value (MV) of 0.643. However, $M_{u\text{-}Han}$ and $M_{u\text{-}Zand}$ achieved higher MVs, which were 0.740 and 0.734, respectively, since these two methods took into account the effects of the total area of the steel (A_s) of the CFST beam's cross-section. The above comparison confirmed that the lips of the C-sections used in the Slender CFST beams investigated in this study behaved as internal steel stiffeners, which led to sufficient improvements in their flexural strength capacities [19].

Table 4. Verification of the obtained M_u values of the tested specimens and analysed FE models.

Model Designations	λ (W/t)	λ_{st} (W_{eff}/t)	M_u (kN.m)	$M_{u\text{-}EC4}$ (kN.m)	$M_{u\text{-}EC4}$ /M_u	$M_{u\text{-}Han}$ (kN·m)	$M_{u\text{-}Han}$ /M_u	$M_{u\text{-}P1}$ (kN·m)	$M_{u\text{-}P1}$ /M_u	M_n (kN·m)	M_n /M_u
FB-RC0	98.0	48.0	57.7	39.1	0.678	42.3	0.732	42.6	0.738	48.9	0.848
FB-RC30	98.0	48.0	53.7	37.3	0.694	39.3	0.731	37.2	0.692	46.7	0.870
FB-RC50	98.0	48.0	52.6	37.2	0.706	39.2	0.745	36.8	0.700	46.6	0.886
FB-RC70	98.0	48.0	51.5	37.1	0.720	39.1	0.759	36.5	0.709	46.5	0.903
FB1-A	148.0	73.0	38.1	25.8	0.678	27.2	0.713	27.1	0.711	30.1	0.789
FB2-A	98.0	48.0	55.4	37.2	0.671	39.2	0.707	37.1	0.670	46.7	0.842
FB3-A	73.0	35.5	72.4	48.2	0.665	51.7	0.714	48.6	0.671	61.4	0.848
FB4-A	58.0	28.0	88.8	58.8	0.663	64.8	0.730	61.7	0.695	75.1	0.846
FB5-A	48.0	23.0	105.0	69.3	0.660	78.5	0.747	68.0	0.647	88.6	0.844
FB1-B	98.0	48.0	55.4	37.2	0.671	39.2	0.707	37.1	0.670	46.7	0.842
FB2-B	98.0	48.0	58.7	38.9	0.663	41.9	0.713	42.1	0.717	48.6	0.829
FB3-B	98.0	48.0	60.3	40.0	0.663	44.5	0.738	45.5	0.755	49.9	0.827
FB4-B	98.0	48.0	61.0	40.8	0.669	46.9	0.769	49.3	0.809	50.7	0.832
FB5-B	98.0	48.0	62.2	41.4	0.666	49.1	0.789	53.7	0.863	51.4	0.826
FB1-C	98.0	48.0	36.0	22.0	0.611	23.7	0.660	23.9	0.665	27.8	0.772
FB2-C	98.0	48.0	39.3	27.4	0.697	29.1	0.741	28.9	0.735	34.7	0.883
FB3-C	98.0	48.0	48.3	32.4	0.671	34.2	0.708	33.2	0.688	41.1	0.851
FB4-C	98.0	48.0	55.4	37.2	0.671	39.2	0.707	37.1	0.670	46.7	0.842
FB5-C	98.0	48.0	59.9	41.5	0.693	43.8	0.731	40.4	0.674	51.1	0.853
FB1-D	98.0	48.0	37.0	24.5	0.662	25.4	0.688	23.4	0.632	31.0	0.839
FB2-D	98.0	48.0	46.1	30.6	0.663	32.0	0.694	29.9	0.649	38.5	0.836
FB3-D	98.0	48.0	55.4	37.2	0.671	39.2	0.707	37.1	0.670	46.7	0.842
FB4-D	98.0	48.0	66.0	44.4	0.672	47.1	0.713	45.0	0.682	55.4	0.839
FB5-D	98.0	48.0	77.1	52.1	0.676	55.7	0.722	53.6	0.695	64.8	0.840
[#] SB2-SI (St1.5)	131.3	65.2	60.1	35.0	0.584	45.4	0.755	53.1	0.885	43.2	0.719
[#] SB2-SI (St3.0)	131.3	64.7	64.2	35.3	0.550	51.3	0.799	56.8	0.884	48.1	0.749
[#] SB2-SI (St4.5)	131.3	64.2	69.7	35.6	0.510	57.2	0.820	60.9	0.873	53.6	0.768
[#] SB3-DI (St1.5)	131.3	43.1	65.5	35.3	0.539	50.9	0.777	56.5	0.863	50.3	0.769
[#] SB3-DI (St3.0)	131.3	42.4	76.5	35.8	0.468	63.0	0.823	65.4	0.854	62.4	0.816
[#] SB3-DI (St4.5)	131.3	41.8	88.0	36.4	0.413	74.6	0.848	75.3	0.856	76.6	0.870
MV	-	-	-	-	0.643	-	0.740	-	0.734	-	0.831
COV	-	-	-	-	0.112	-	0.057	-	0.110	-	0.049

[#] FE CFST models stiffened with internal I-steel stiffeners analysed by Al Zand et al. [19].

5.3. Development of the New Analytical Method

Generally, the cross-sections of steel tubes are classified into Compact, Noncompact and Slender sections based on their ability to buckle under compression stress. The tube's effective-width (W_{eff})-to-thickness (t) ratio, usually known as the slenderness ratio (λ), is used as a limit for this classification in the majority of related standardised codes. In the current study, the slenderness limits that are specified in ANSI/AISC 360-10 [64] (Chapter I) were adopted for the classification of the rectangular steel tube beams that were filled with concrete (cross-sections of CFST members). Figure 13 presents the relationship between the nominal flexural strength (Nominal moment; M_n) and the slenderness ratio (λ) of the cross-section of CFST beam's tube. First, the tube's cross-section was classified as a "Compact section" if the λ value was within the limits of the compactness ratio (λ_p), which was equal to 2.26 $(E_s/F_y)^{0.5}$. Second, if the value of λ was larger than that of λ_p, but within the limits of the noncompactness ratio (λ_r), which was equal to 3.0 $(E_s/F_y)^{0.5}$, then the tube's cross-section was classified as a "Noncompact section". Third, when λ exceeded the limit of λ_r but was within the limits of the maximum ratio (λ_{limit}), which was equal to 5.0 $(E_s/F_y)^{0.5}$, then the cross-section of the CFST beam's tube was classified as a "Slender section". This was the case because the ANSI/AISC 360-10 code does not permit the use of Slender CFST beams if their λ values exceed the maximum limit (λ_{limit}).

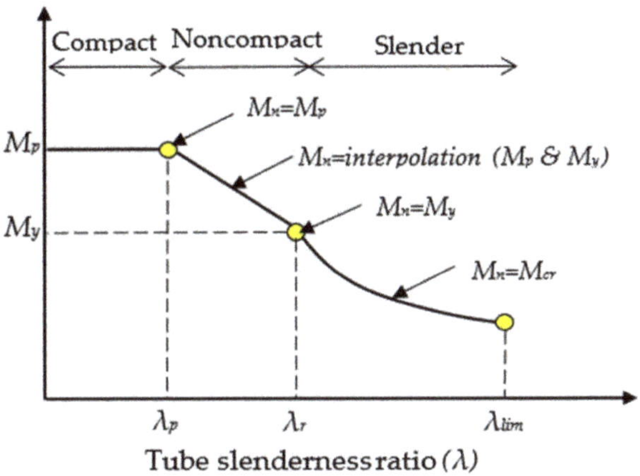

Figure 13. Moment vs. slenderness ratio relationship of the tube's cross-section.

In this study, a novel analytical method for the prediction of the nominal moment (M_n) was developed based on the fundamental theory of stress block diagrams of CFST beams that are internally stiffened with steel stiffeners, as shown in Figure 14. For this purpose, several assumptions were adopted, which are listed as follows:

i.   This method was limited to rectangular CFST beams with internal stiffeners under pure static bending.

ii.   The shear span-to-depth ratio was assumed to have no significant effects on the CFST beam's deflection behaviour [51,61].

iii.   Full interaction between the concrete infill and the steel tubes/stiffeners was assumed [19,51,61].

iv.   In the classification of the stiffened steel tube of the CFST beam, the effective flat width between the web and stiffener (w_{eff}) was used instead of the overall effective tube width (W_{eff}) for estimation of the stiffened slenderness ratio ($\lambda_{st} = w_{eff}/t$). In another words, ($\lambda_{st} = w_{eff}/t$) was used instead of ($\lambda = W_{eff}/t$) for the classification.

v.   A nominal concrete confinement, which varied considerably based on the slenderness limit, was assumed to have been generated by the steel tube.

vi.   The tension-related stress of the concrete, which occurred due to cracking failure, was ignored.

vii.   The variations in stress due to the stiffener's depth and the flange's thickness were ignored.

viii.   Compact section (see Figure 14a): this section was assumed to have rigid-plastic behaviour and the steel stress was assumed to remain within the yielding limit (F_y) at both the tension and compression zones. The concrete compression stress was assumed to be within the limits of the ultimate strength (f_{cu}) and distributed as a rectangular stress block to the N.A. position.

ix.   Noncompact section (see Figure 14b): this section was assumed to have elastic-plastic behaviour at the tension zone and elastic behaviour at the compression zone, and the steel stress was assumed to be within the limits of F_y [12,64]. The concrete compression stress was assumed to be within the limits of $0.9f_{cu}$ and distributed as a triangular stress block to the N.A. position.

x. Slender section (see Figure 14c): this section was assumed to have pure elastic behaviour, and the steel stress was assumed to be within the limits of F_y at the maximum tension face and within the limits of the buckling stress (F_{cr}) at the maximum compression face [12,64]. For this section, a lower concrete compression stress was assumed, which was taken to be within the limits of $0.8f_{cu}$.

xi. Finally, when the forces over the stiffened CFST beam's cross-section attained equilibrium (see Figure 14), the summarized forms of the new analytical formula for predicting the M_n for each section classification could be expressed as follows:

$$\text{For the Compact section } (\lambda_{st} \leq \lambda_p)$$

$$y_c = \left(2tDF_y + f_{cu}W_{eff}t\right) / \left(4tF_y + f_{cu}W_{eff}\right) \tag{2}$$

$$M_n = M_p = W_{eff}tF_y(D - t) + t_{st}d_{st}F_y(D - D_{st}) + tF_y[y_c^2 + (D - y_c)^2] + 0.5W_{eff}f_{cu}(y_c - t)^2 \tag{3}$$

$$\text{For the Noncompact section } (\lambda_p < \lambda_{st} \leq \lambda_r)$$

$$y_n = \left(2tDF_y + 0.45f_{cu}W_{eff}t\right) / \left(4tF_y + 0.45f_{cu}W_{eff}\right) \tag{4}$$

$$M_y = W_{eff}tF_y(D - t) + t_{st}D_{st}F_y(D - D_{st}) + tF_yD(D - 2y_n) + 4/3tF_yy_n^2 + 0.3W_{eff}f_{cu}(y_n - t)^2 \tag{5}$$

$$M_n = M_p - \left[(M_p - M_y) \cdot (\lambda - \lambda_p) / (\lambda_r - \lambda_p)\right] \tag{6}$$

$$\text{For the Slender section } (\lambda_r < \lambda_{st} \leq \lambda_{limit})$$

$$F_{cr} = 9E_s / (W_{eff}/t)^2 \tag{7}$$

$$y_s = \left[tDF_y + W_{eff}t\left(0.4f_{cu} + f_y - f_{cr}\right) + t_{st}d_{st}(F_y - f_{cr})\right] / \left(t(F_y + f_{cr}) + 0.4f_{cu}W_{eff}\right) \tag{8}$$

$$M_n = M_{cr} = W_{eff}tF_{cr}(y_s - t/2) + W_{eff}tF_y(D - y_s - t/2) + t_{st}d_{st}F_{cr}(y_s - d_{st}/2)$$
$$+ t_{st}d_{st}F_y(D - y_s - d_{st}/2) + 2/3tF_{cr}y_s^2 + 2/3tF_y(D - y_s)^2$$
$$+ 0.267W_{eff}f_{cu}(y_s - t)^2 \tag{9}$$

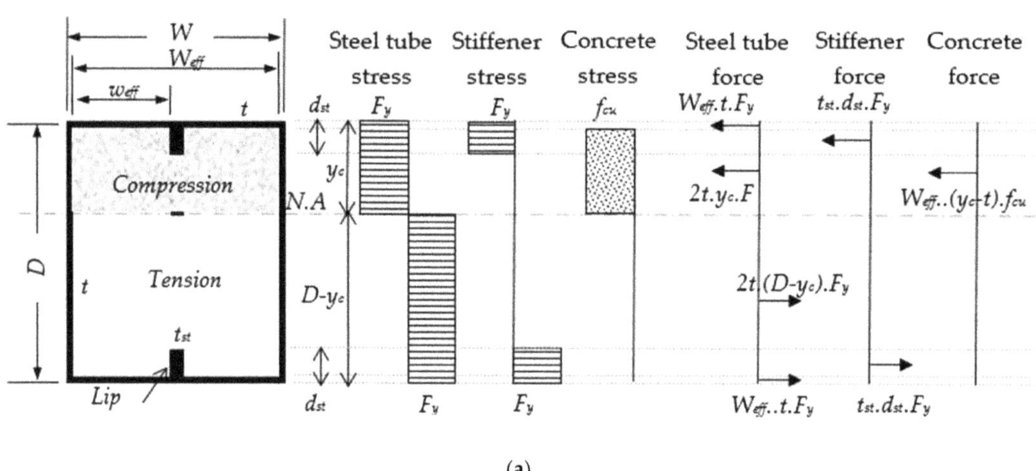

(a)

Figure 14. *Cont.*

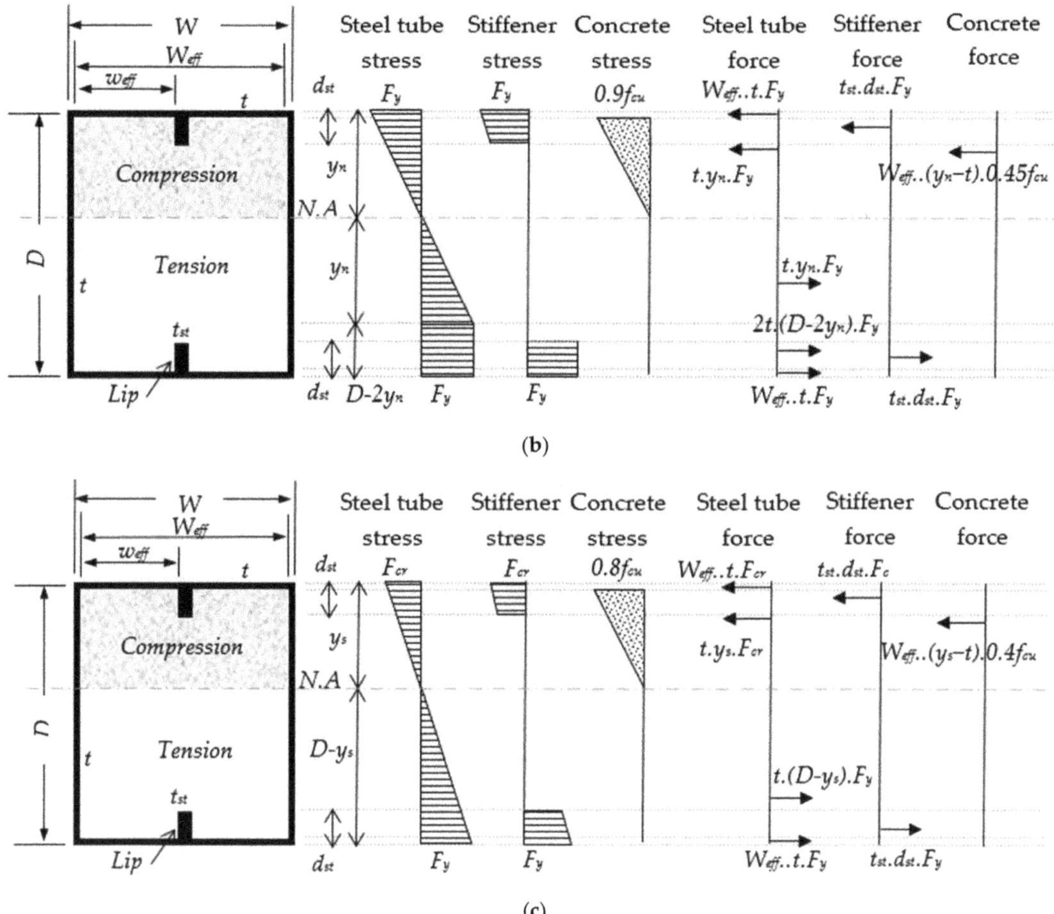

Figure 14. Stress block diagrams of stiffened CFST beam with internal stiffeners: (**a**) compact section; (**b**) non-compact section; (**c**) slender section.

Accordingly, the embedded steel stiffeners that were provided to stiffen the sections of the CFST beams/columns significantly reduced the flat distance of their tube's flanges/walls, which led to a delay in the tube's buckling failure, as discussed earlier in this paper and previously confirmed in the literature [19,22,47]. On that basis, the classification of the tube sections of the CFST beams could be changed from Slender to Noncompact and/or to Compact due to the influence of these stiffeners, as evidenced in the comparisons between the stiffened slenderness ratios (λ_{st}) and the λ values in Table 4. Furthermore, when compared to the M_u values obtained from the current study and the additional models analysed in [19], the new analytical method achieved the best prediction values (M_n), with MV and COV values of 0.831 and 0.049, respectively, as compared to the existing theoretical methods shown in Table 4.

6. Conclusions

The conclusions of the investigated CFST beams are summarized as follows:

> The experimental investigation confirmed that the bending capacity of the suggested prefabricated Slender CFST beams made from two pieces of C-sections was enhanced by about 3.7 times even when filled with 70% replacement recycled concrete material.

> Under the static bending load, the prefabricated tubular steel beams (double C-sections) filled with recycled concrete (0, 30%, 50%, and 70%) behaved very similarly to the conventional CFST beams. Additionally, the lips of these C-sections were adequately bonded to the concrete and acted as internal stiffeners for the Slender CFST beam's cross-section, which delayed the outward buckling failures at their top flanges. For example, the use of recycled aggregate to replace raw aggregate at a proportion of up to 70% resulted in slightly lower flexural stiffness and strength capacity values (-7.2% to -10.7%) compared to those obtained using normal concrete.

> Generally, it is worth highlighting that the suggested fabricated steel tube beams' self-weight was increased substantially due to the effect of concrete infill materials. In turn, the flexural strength capacities of these beams were remarkably enhanced, by approximately 409% and 363%, when using normal concrete and 70% recycled concrete mixtures, respectively. These findings are very important from a structural engineering point of view as, regardless the increment of the beams' self-weight, they can be used to prospectively determine the scenarios in which this composite system would be reliable for the targeted purpose of construction projects. Cost also played a vital role in the construction design process, which was demonstrated when we compared the cost of the concrete and steel sections in the local market.

> The flexural behaviour of tested CFST beam was accurately simulated using the ABAQUS software. The results obtained from the non-linear analyses of FE CFST models that were prepared for investigation of various parameters confirmed that slightly increasing the thickness of the tubes in these models had a major influence on their flexural strengths and stiffnesses as compared to the effects of other parameters. In contrast, a limited degree of influence was achieved when the compressive strengths of the concrete infill material and/or the yielding strengths of the steel tubes were increased.

> Lastly, the newly developed analytical method achieved the best prediction of the flexural strength capacities of the internally stiffened CFST beams that were tested and analysed in this study, since it was able to independently consider the influence of internal steel stiffeners along with the effects of the properties of steel tubes and concrete. However, this method was found to only be reliable for the internally stiffened rectangular CFST beams.

It is worth highlighting the main research limitation of the current investigation, namely that the uncertainties related to this model could be further examined as they are very significant in terms of the parameters of structural behavior. In addition, further experimental/numerical investigations, other than the rectangular cross-sections under static/dynamic loading scenarios, could be conducted on the stiffened CFST beams.

Author Contributions: Conceptualization, A.W.A.Z., R.A.-A. and W.H.W.B.; Data curation, A.W.A.Z., M.M.A. and E.H.; Formal analysis, A.W.A.Z., M.M.A. and Z.M.Y.; Funding acquisition, A.W.A.Z.; Investigation, M.M.A. and E.H.; Methodology, R.A.-A.; Project administration, A.W.A.Z., R.A.-A., W.H.W.B. and Z.M.Y.; Resources, W.H.W.B. and E.H.; Software, A.W.A.Z. and W.M.T.; Supervision, W.H.W.B.; Validation, A.W.A.Z., M.M.A., R.A.-A. and Z.M.Y.; Visualization, W.M.T., E.H. and Z.M.Y.; Writing—original draft, A.W.A.Z., M.M.A., W.M.T. and E.H.; Writing—review and editing, A.W.A.Z., W.H.W.B., W.M.T. and Z.M.Y. All authors have read and agreed to the published version of the manuscript.

Funding: This research was funded by Universiti Kebangsaan Malaysia, grant number GGPM-2020-001.

Institutional Review Board Statement: Not applicable.

Informed Consent Statement: Not applicable.

Data Availability Statement: Data are presented in the article.

Conflicts of Interest: The authors declare no conflict of interest.

Abbreviations

Area of concrete core cross-section (A_c), area of steel tube cross-section (*As*), coefficient of variation (COV), depth of rectangular steel tube (D), depth of steel stiffener (d_{st}), modulus of elasticity for concrete (E_c), modulus of elasticity for steel (E_s), compressive strength of concrete cube at 28 days (f_{cu}), characteristic concrete strength of $0.67f_{cu}$ (f_{ck}), ultimate strength of steel (f_u), yield strength of steel (f_y), steel tube buckling stress (F_{cr}), moment of inertia for concrete tube cross-section (I_c), moment of inertia for steel tube cross-section (I_s), initial flexural stiffness of composite section (K_i), serviceability level of the flexural stiffness of the composite section (K_s), effective length of the specimen (M), ultimate bending moment capacity (flexural strength capacity; M_u), nominal bending moment (M_n), plastic limit bending moment ($M_p = M_u$), yield limit bending moment (M_y), mean value (*MV*), steel tube/C-section thickness (t), steel stiffener thickness (t_{st}), width of rectangular steel tube (W), effective width of rectangular steel tube (W_{eff}), effective width/distance between two internal stiffeners and/or between the edge of the tube and the first internal stiffener (w_{eff}), distance of N.A. from the top flange of the tube for the Compact section (y_c), distance of N.A. from the top flange of the tube for the Noncompact section (y_n), distance of N.A. from the top flange of the tube for the Slender section (y_s), slenderness ratio of steel tube cross-section (λ), slenderness ratio at the compactness limit (λ_p), slenderness ratio at the noncompactness limit (λ_r), maximum limit of the slenderness ratio (λ_{limit}), stiffened slenderness ratio ($\lambda_{st} = w_{eff}/t$).

References

1. Zong, Z.; Jaishi, B.; Ge, J.-P.; Ren, W.-X. Dynamic analysis of a half-through concrete-filled steel tubular arch bridge. *Eng. Struct.* **2005**, *27*, 3–15. [CrossRef]
2. Han, L.-H.; Li, W.; Bjorhovde, R. Developments and advanced applications of concrete-filled steel tubular (CFST) structures: Members. *J. Constr. Steel Res.* **2014**, *100*, 211–228. [CrossRef]
3. Wu, D.; Gao, W.; Feng, J.; Luo, K. Structural behaviour evolution of composite steel-concrete curved structure with uncertain creep and shrinkage effects. *Compos. Part B Eng.* **2016**, *86*, 261–272. [CrossRef]
4. Gao, J.; Su, J.; Xia, Y.; Chen, B. Experimental Study of Concrete-Filled Steel Tubular Arches with Corrugated Steel Webs. *Adv. Steel Constr.* **2014**, *10*, 99–115. [CrossRef]
5. Al-Shaar, A.A.M.; Göğüş, M.T. Performance of Retrofitted Self-Compacting Concrete-Filled Steel Tube Beams Using External Steel Plates. *Adv. Mater. Sci. Eng.* **2018**, *2018*, 3284745. [CrossRef]
6. Yang, Y.-F.; Zhu, L.-T. Recycled aggregate concrete filled steel SHS beam-columns subjected to cyclic loading. *Steel Compos. Struct.* **2009**, *9*, 19–38. [CrossRef]
7. Zhang, Y.-B.; Han, L.-H.; Zhou, K.; Yang, S. Mechanical performance of hexagonal multi-cell concrete-filled steel tubular (CFST) stub columns under axial compression. *Thin-Walled Struct.* **2018**, *134*, 71–83. [CrossRef]
8. Luat, N.-V.; Lee, D.H.; Lee, J.; Lee, K. GS-MARS method for predicting the ultimate load-carrying capacity of rectangular CFST columns under eccentric loading. *Comput. Concr.* **2020**, *25*, 1–14. [CrossRef]
9. Qeshta, I.M.; Shafigh, P.; Jumaat, M.Z.; Abdulla, A.I.; Alengaram, U.J.; Ibrahim, Z. Flexural Behaviour of Concrete Beams Bonded with Wire Mesh-Epoxy Composite. *Appl. Mech. Mater.* **2014**, *567*, 411–416. [CrossRef]
10. Javed, M.F.; Sulong, N.R.; Memon, S.A.; Rehman, S.K.U.; Khan, N.B. FE modelling of the flexural behaviour of square and rectangular steel tubes filled with normal and high strength concrete. *Thin-Walled Struct.* **2017**, *119*, 470–481. [CrossRef]
11. Yang, Y.-F.; Ma, G.-L. Experimental behaviour of recycled aggregate concrete filled stainless steel tube stub columns and beams. *Thin-Walled Struct.* **2013**, *66*, 62–75. [CrossRef]
12. Lai, Z.; Varma, A.; Zhang, K. Noncompact and slender rectangular CFT members: Experimental database, analysis, and design. *J. Constr. Steel Res.* **2014**, *101*, 455–468. [CrossRef]
13. Wang, J.; Shen, Q.; Wang, F.; Wang, W. Experimental and analytical studies on CFRP strengthened circular thin-walled CFST stub columns under eccentric compression. *Thin-Walled Struct.* **2018**, *127*, 102–119. [CrossRef]
14. Wang, W.-H.; Han, L.-H.; Li, W.; Jia, Y.-H. Behavior of concrete-filled steel tubular stub columns and beams using dune sand as part of fine aggregate. *Constr. Build. Mater.* **2014**, *51*, 352–363. [CrossRef]
15. Al Zand, A.W.; Badaruzzaman, W.H.W.; Mutalib, A.A.; Hilo, S.J. Flexural Behavior of CFST Beams Partially Strengthened with Unidirectional CFRP Sheets: Experimental and Theoretical Study. *J. Compos. Constr.* **2018**, *22*, 04018018. [CrossRef]
16. Guler, S.; Yavuz, D. Post-cracking behavior of hybrid fiber-reinforced concrete-filled steel tube beams. *Constr. Build. Mater.* **2019**, *205*, 285–305. [CrossRef]
17. Al-Shaar, A.A.M.; Göğüş, M.T. Flexural behavior of lightweight concrete and self-compacting concrete-filled steel tube beams. *J. Constr. Steel Res.* **2018**, *149*, 153–164. [CrossRef]

18. Liu, J.-P.; Yang, J.; Chen, B.-C.; Zhou, Z.-Y. Mechanical performance of concrete-filled square steel tube stiffened with PBL subjected to eccentric compressive loads: Experimental study and numerical simulation. *Thin-Walled Struct.* **2020**, *149*, 106617. [CrossRef]

19. Al Zand, A.W.; Wan Badaruzzaman, W.H.; Ali, M.M.; Hasan, Q.A.; Al-Shaikhli, M.S. Flexural performance of cold-formed square CFST beams strengthened with internal stiffeners. *Steel Compos. Struct.* **2020**, *34*, 123–139.

20. Ansari, M.; Jeddi, M.; Badaruzzaman, W.; Tahir, M.; Osman, S.; Hosseinpour, E. A numerical investigation on the through rib stiffener beam to concrete-filled steel tube column connections subjected to cyclic loading. *Eng. Sci. Technol. Int. J.* **2020**, *24*, 728–735. [CrossRef]

21. Kwon, Y.B.; Park, S.W. Resistance of circular concrete-filled tubular sections to combined axial compression and bending. *Thin-Walled Struct.* **2017**, *111*, 93–102. [CrossRef]

22. Al Zand, A.W.; Badaruzzaman, W.H.W.; Al-Shaikhli, M.S.; Ali, M.M. Flexural performance of square concrete-filled steel tube beams stiffened with V-shaped grooves. *J. Constr. Steel Res.* **2020**, *166*, 105930. [CrossRef]

23. Lai, M.; Song, W.; Ou, X.; Chen, M.; Wang, Q.; Ho, J. A path dependent stress-strain model for concrete-filled-steel-tube column. *Eng. Struct.* **2020**, *211*, 110312. [CrossRef]

24. Liu, Z.; Lu, Y.; Li, S.; Zong, S.; Yi, S. Flexural behavior of steel fiber reinforced self-stressing recycled aggregate concrete-filled steel tube. *J. Clean. Prod.* **2020**, *274*, 122724. [CrossRef]

25. Nematzadeh, M.; Karimi, A.; Gholampour, A. Pre- and post-heating behavior of concrete-filled steel tube stub columns containing steel fiber and tire rubber. *Structures* **2020**, *27*, 2346–2364. [CrossRef]

26. Qiao, Q.-Y.; Zhang, W.-W.; Mou, B.; Cao, W.-L. Seismic behavior of exposed concrete filled steel tube column bases with embedded reinforcing bars: Experimental investigation. *Thin-Walled Struct.* **2019**, *136*, 367–381. [CrossRef]

27. Chen, Z.; Xu, J.; Chen, Y.; Lui, E. Recycling and reuse of construction and demolition waste in concrete-filled steel tubes: A review. *Constr. Build. Mater.* **2016**, *126*, 641–660. [CrossRef]

28. Ou, Z.J. The practice of concrete filled steel tube piers to bridges: A review. In *Applied Mechanics and Materials*; Trans Tech Publications Ltd.: Kapellweg, Switzerland, 2013; Volume 405, pp. 1602–1605.

29. Kozielova, M.; Marcalikova, Z.; Mateckova, P.; Sucharda, O. Numerical Analysis of Reinforced Concrete Slab with Subsoil. *Civ. Environ. Eng.* **2020**, *16*, 107–118. [CrossRef]

30. Montoya, E.; Vecchio, F.J.; Sheikh, S.A. Compression Field Modeling of Confined Concrete: Constitutive Models. *J. Mater. Civ. Eng.* **2006**, *18*, 510–517. [CrossRef]

31. Červenka, J.; Papanikolaou, V.K. Three dimensional combined fracture–plastic material model for concrete. *Int. J. Plast.* **2008**, *24*, 2192–2220. [CrossRef]

32. Ahmed, W.; Lim, C. Production of sustainable and structural fiber reinforced recycled aggregate concrete with improved fracture properties: A review. *J. Clean. Prod.* **2020**, *279*, 123832. [CrossRef]

33. Huang, L.; Yang, Z.; Li, Z.; Xu, Y.; Yu, L. Recycling of the end-of-life lightweight aggregate concrete (LWAC) with a novel approach. *J. Clean. Prod.* **2020**, *275*, 123099. [CrossRef]

34. Alsharie, H.; Masoud, T.; Abdulla, A.I.; Ghanem, A. Properties of lightweight cement mortar containing treated pumice and limestone. *J. Eng. Appl. Sci.* **2015**, *10*, 96–101.

35. Su, S.; Li, X.; Wang, T.; Zhu, Y. A comparative study of environmental performance between CFST and RC columns under combinations of compression and bending. *J. Clean. Prod.* **2016**, *137*, 10–20. [CrossRef]

36. Xu, J.; Wang, Y.; Ren, R.; Wu, Z.; Ozbakkaloglu, T. Performance evaluation of recycled aggregate concrete-filled steel tubes under different loading conditions: Database analysis and modelling. *J. Build. Eng.* **2020**, *30*, 101308. [CrossRef]

37. Yang, Y.-F.; Han, L.-H.; Zhu, L.-T. Experimental Performance of Recycled Aggregate Concrete-Filled Circular Steel Tubular Columns Subjected to Cyclic Flexural Loadings. *Adv. Struct. Eng.* **2009**, *12*, 183–194. [CrossRef]

38. Yang, Y.F.; Man, L.H. Compressive and flexural behaviour of recycled aggregate concrete filled steel tubes (RACFST) under short-term loadings. *Steel Compos. Struct.* **2006**, *6*, 257–284. [CrossRef]

39. Xu, J.J.; Zhao, X.Y.; Chen, Z.P.; Liu, J.C.; Xue, J.Y.; Elchalakani, M. Thin-Walled Structures Novel prediction models for composite elastic modulus of circular recycled aggregate concrete-filled steel tubes. *Thin-Walled Struct.* **2019**, *144*, 106317. [CrossRef]

40. Karimi, A.; Nematzadeh, M.; Mohammad-Ebrahimzadeh-Sepasgozar, S. Analytical post-heating behavior of concrete-filled steel tubular columns containing tire rubber. *Comput. Concr.* **2020**, *26*, 467–482.

41. Mouli, M.; Khelafi, H. Strength of short composite rectangular hollow section columns filled with lightweight aggregate concrete. *Eng. Struct.* **2007**, *29*, 1791–1797. [CrossRef]

42. Fu, Z.; Ji, B.; Wu, D.; Yu, Z. Behaviour of lightweight aggregate concrete-filled steel tube under horizontal cyclic load. *Steel Compos. Struct.* **2019**, *32*, 717–729.

43. Fu, Z.-Q.; Ji, B.-H.; Zhu, W.; Ge, H.-B. Bending behaviour of lightweight aggregate concrete-filled steel tube spatial truss beam. *J. Cent. South Univ.* **2016**, *23*, 2110–2117. [CrossRef]

44. D'Orazio, M.; Stipa, P.; Sabbatini, S.; Maracchini, G. Experimental investigation on the durability of a novel lightweight prefabricated reinforced-EPS based construction system. *Constr. Build. Mater.* **2020**, *252*, 119134. [CrossRef]

45. Mohammed, H.J.; Zain, M. Experimental application of EPS concrete in the new prototype design of the concrete barrier. *Constr. Build. Mater.* **2016**, *124*, 312–342. [CrossRef]

46. Sadrmomtazi, A.; Sobhani, J.; Mirgozar, M.; Najimi, M. Properties of multi-strength grade EPS concrete containing silica fume and rice husk ash. *Constr. Build. Mater.* **2012**, *35*, 211–219. [CrossRef]
47. Zhu, A.; Zhang, X.; Zhu, H.; Zhu, J.; Lu, Y. Experimental study of concrete filled cold-formed steel tubular stub columns. *J. Constr. Steel Res.* **2017**, *134*, 17–27. [CrossRef]
48. Javed, M.F.; Sulong, N.R.; Memon, S.A.; Rehman, S.K.-U.; Khan, N.B. Flexural behaviour of steel hollow sections filled with concrete that contains OPBC as coarse aggregate. *J. Constr. Steel Res.* **2018**, *148*, 287–294. [CrossRef]
49. Javed, M.F.; Sulong, N.H.R.; Memon, S.A.; Rehman, S.K.-U.; Khan, N.B. Experimental and numerical study of flexural behavior of novel oil palm concrete filled steel tube exposed to elevated temperature. *J. Clean. Prod.* **2018**, *205*, 95–114. [CrossRef]
50. Chen, Y.; Feng, R.; Gong, W. Flexural behavior of concrete-filled aluminum alloy circular hollow section tubes. *Constr. Build. Mater.* **2018**, *165*, 173–186. [CrossRef]
51. Han, L.-H.; Lu, H.; Yao, G.-H.; Liao, F.-Y. Further study on the flexural behaviour of concrete-filled steel tubes. *J. Constr. Steel Res.* **2006**, *62*, 554–565. [CrossRef]
52. Chen, Y.; Feng, R.; Xu, J. Flexural behaviour of CFRP strengthened concrete-filled aluminium alloy CHS tubes. *Constr. Build. Mater.* **2017**, *142*, 295–319. [CrossRef]
53. Sharafati, A.; Haghbin, M.; Aldlemy, M.S.; Mussa, M.H.; Al Zand, A.W.; Ali, M.; Bhagat, S.K.; Al-Ansari, N.; Yaseen, Z.M. Development of Advanced Computer Aid Model for Shear Strength of Concrete Slender Beam Prediction. *Appl. Sci.* **2020**, *10*, 3811. [CrossRef]
54. Bambach, M.; Jama, H.; Zhao, X.; Grzebieta, R. Hollow and concrete filled steel hollow sections under transverse impact loads. *Eng. Struct.* **2008**, *30*, 2859–2870. [CrossRef]
55. Al Zand, A.W.; Hosseinpour, E.; Badaruzzaman, W.H.W. The influence of strengthening the hollow steel tube and CFST beams using U-shaped CFRP wrapping scheme. *Struct. Eng. Mech.* **2018**, *66*, 229–235.
56. Al Zand, A.W.; Badaruzzaman, W.H.W.; Tawfeeq, W.M. New empirical methods for predicting flexural capacity and stiffness of CFST beam. *J. Constr. Steel Res.* **2019**, *164*, 105778. [CrossRef]
57. Moon, J.; Roeder, C.W.; Lehman, D.E.; Lee, H.-E. Analytical modeling of bending of circular concrete-filled steel tubes. *Eng. Struct.* **2012**, *42*, 349–361. [CrossRef]
58. Yang, Y.-F. Modelling of recycled aggregate concrete-filled steel tube (RACFST) beam-columns subjected to cyclic loading. *Steel Compos. Struct.* **2015**, *18*, 213–233. [CrossRef]
59. Ombres, L.; Verre, S. Experimental and Numerical Investigation on the Steel Reinforced Grout (SRG) Composite-to-Concrete Bond. *J. Compos. Sci.* **2020**, *4*, 182. [CrossRef]
60. Micelli, F.; Cascardi, A. Structural assessment and seismic analysis of a 14th century masonry tower. *Eng. Fail. Anal.* **2019**, *107*, 104198. [CrossRef]
61. Han, L.-H. Flexural behaviour of concrete-filled steel tubes. *J. Constr. Steel Res.* **2004**, *60*, 313–337. [CrossRef]
62. *AIJ-1997 Recommendations for Design and Construction of Concrete Filled Steel Tubular Structures*; Architectural Institute of Japan: Tokyo, Japan, 1997.
63. *EC4-2004 Design of Composite Steel and Concrete Structures. Part 1.1: General Rules and Rules for Buildings*; European Standard: Brussels, Belgium, 2004.
64. *ANSI/AISC 360-10, Specification for Structural Steel Buildings*; American Institute of Steel Construction (AISC): Chicago, IL, USA, 2010.

materials — MDPI

Evaluation and Numerical Investigations of the Cyclic Behavior of Smart Composite Steel–Concrete Shear Wall: Comprehensive Study of Finite Element Model

Hadee Mohammed Najm [1,*], Amer M. Ibrahim [2,*], Mohanad Muayad Sabri Sabri [3], Amer Hassan [1], Samadhan Morkhade [4], Nuha S. Mashaan [5], Moutaz Mustafa A. Eldirderi [6] and Khaled Mohamed Khedher [7,8]

1 Department of Civil Engineering, Aligarh Muslim University, Aligarh 202002, India; ameralburay@gmail.com
2 Department of Civil Engineering, College of Engineering, University of Diyala, Diyala 32001, Iraq
3 Peter the Great St. Petersburg Polytechnic University, 195251 St. Petersburg, Russia; mohanad.m.sabri@gmail.com
4 Department of Civil Engineering, VPKBIET, Baramati, Pune 413133, India; samadhanmorhade@gmail.com
5 Faculty of Science and Engineering, School of Civil and Mechanical Engineering, Curtin University, Bentley, WA 6102, Australia; nuhas.mashaan1@curtin.edu.au
6 Department of Chemical Engineering, College of Engineering, King Khalid University, Abha 61421, Saudi Arabia; maldrdery@kku.edu.sa
7 Department of Civil Engineering, College of Engineering, King Khalid University, Abha 61421, Saudi Arabia; kkhedher@kku.edu.sa
8 Department of Civil Engineering, High Institute of Technological Studies, Mrezgua University Campus, Nabeul 8000, Tunisia
* Correspondence: gk4071@myamu.ac.in (H.M.N.); amereng76@yahoo.com (A.M.I.)

Citation: Najm, H.M.; Ibrahim, A.M.; Sabri, M.M.S.; Hassan, A.; Morkhade, S.; Mashaan, N.S.; Eldirderi, M.M.A.; Khedher, K.M. Evaluation and Numerical Investigations of the Cyclic Behavior of Smart Composite Steel–Concrete Shear Wall: Comprehensive Study of Finite Element Model. *Materials* **2022**, *15*, 4496. https://doi.org/10.3390/ma15134496

Academic Editor: Stelios K. Georgantzinos

Received: 25 May 2022
Accepted: 22 June 2022
Published: 26 June 2022

Publisher's Note: MDPI stays neutral with regard to jurisdictional claims in published maps and institutional affiliations.

Abstract: The composite shear wall has various merits over the traditional reinforced concrete walls. Thus, several experimental studies have been reported in the literature in order to study the seismic behavior of composite shear walls. However, few numerical investigations were found in the previous literature because of difficulties in the interaction behavior of steel and concrete. This study aimed to present a numerical analysis of smart composite shear walls, which use an infilled steel plate and concrete. The study was carried out using the ANSYS software. The mechanical mechanisms between the web plate and concrete were investigated thoroughly. The results obtained from the finite element (FE) analysis show excellent agreement with the experimental test results in terms of the hysteresis curves, failure behavior, ultimate strength, initial stiffness, and ductility. The present numerical investigations were focused on the effects of the gap, thickness of infill steel plate, thickness of the concrete wall, and distance between shear studs on the composite steel plate shear wall (CSPSW) behavior. The results indicate that increasing the gap between steel plate and concrete wall from 0 mm to 40 mm improved the stiffness by 18% as compared to the reference model, which led to delay failures of this model. Expanding the infill steel plate thickness to 12 mm enhanced the stiffness and energy absorption with a ratio of 95% and 58%, respectively. This resulted in a gradual drop in the strength capacity of this model. Meanwhile, increasing concrete wall thickness to 150 mm enhanced the ductility and energy absorption with a ratio of 52% and 32%, respectively, which led to restricting the model and reduced lateral offset. Changing the distance between shear studs from 20% to 25% enhanced the ductility and energy absorption by about 66% and 32%, respectively.

Keywords: composite steel plate shear wall; hysteresis curves; ductility; energy absorption; finite element model

1. Introduction

The components of composite steel plate shear walls (CSPSWs) consist of web plates, infill steel plates, concrete, and shear studs. The composite steel plate shear wall proves to be excellent over reinforced concrete walls in terms of ultimate strength, stiffness, ductility, and energy dissipation [1].

Previous studies have demonstrated a high interest in using composite steel plate shear walls in buildings and construction [2–4]. The appropriate provisions for composite shear walls, such as various seismic behaviors of composite shear walls, are ASCE 7-10 and AISC 341-10 [5,6], which are prepared by allowing the use of CSPSW systems in earthquake zones. Scholars have conducted numerous experimental tests to examine the behavior of composite shear walls in the absence of boundary walls. Nie et al. [7], Mydin [8], Wright [9], and Wang [10] described that composite shear walls not only have high ultimate strength but also have excellent ductile behavior. The local buckling of the web and fracture failure of corners of the wall are the observed failure modes. Zhang et al. [11] and Zhang et al. [12] showed that more channels reduce the ultimate strength and stiffness of the wall but observed improvement in the ductile and energy dissipation behavior. Lastly, to calculate the ultimate strength, initial stiffness has been proposed. However, all the essential variables are not included in the proposed equation, which leads to moderate results. Therefore, this creates a strong demand to perform an exhaustive numerical analysis of composite shear walls [13,14].

Several studies have suggested the equation-based FE analysis for composite shear walls. Nguyen et al. [15], Epackachi et al. [16], and Rafiei et al. [17] developed finite element models and checked their accuracy. The parametric analysis of the connector in the wall shows that an increase in the number of connectors could improve the bearing capacity of the steel plate and that a change in the spacing of the connector could affect the failure mode of the steel plate. Wei et al. [18] studied the axial compression performance of composite shear walls. The effect of distance-to-thickness ratios on the failure mode was studied, and a formula to calculate the axial compression of a composite shear wall has been suggested. The higher axial compression ratio of the wall is beneficial to restrain the internal concrete and enhance the compressive strength of the concrete. Thus, the energy dissipation capacity of the composite shear wall is enhanced [19,20]. Increasing the thickness of the steel plate can increase the stiffness and ultimate bearing capacity of the wall, as the hysteretic curve of the wall is plumper [21–24]. Epackachi et al. [25] simulated shear walls with different aspect ratios. When the aspect ratio was between 0.6 and 3.0, the coupling effect of the moment and shear force was obviously achieved. The specifications [6,26–28] define the formula for the shear capacity of composite shear walls. The formulas for the shear and flexural capacity were given, but the formula for estimating the flexural–shear coupling was not supplied [29]. Kantaros et al. [30] reported a comparison of the mechanical properties of different scaffold designs that, however, featured the same porosity and similar dimensions. Compressive strength testing was conducted in three 3D-printed scaffold designs. Moreover, a finite element study was conducted, simulating the compressive strength testing. The results of the compression testing experiment were found to be in good agreement with the computational analysis results. Nedelcu and Cucu [31] studied the buckling modes identification by an FEA of thin-walled members using only GBT cross-sectional deformation modes. The authors presented the latest developments of an original method based on generalized beam theory (GBT) capable of identifying the fundamental deformation modes of global, distortional, or local nature in general buckling modes provided by the shell finite element analysis (FEA) of isotropic thin-walled members.

In summary, the seismic performance of the CSPSW is typically affected by various factors such as the thickness of the infill steel plate, thickness of the concrete wall, distance between the shear studs, ratio of reinforcement, concrete strength, steel yield strength, and layout of the shear stud [32]. Rahai, A. and Hatami, F. [33] investigated the performance of the composite shear wall and established the base for the research on seismic behavior. Although few analyses have suggested the formulas for calculation of lateral stiffness, ductility, and energy dissipation, they were all based on the test results. The predictive effect with other parameters is unknown. In addition, the previous studies mainly focused on the behavior of traditional CSPSW.

This study's objective was to examine the behavior of the smart CSPSW under cyclic load and to investigate the consequence of many parameters on this new type of CSPSW.

2. Smart Structure Technology

The research on how to maintain the safety of humans inside buildings is an important issue of concern in modern times, as buildings with a low security level pose a threat to human life. Smart structure technology is a modern building and structure control system that notes its own condition, detects impending failure, monitors the damage, and adapts to changing environments [34,35]. The smart technology of composite steel plate shear wall is installed by adding a gap between the steel frame and concrete wall; this gap is provided to improve the performance of CSPSW. The most important benefits of this gap are improvement in stiffness, ductility, and energy absorption of all the systems [36].

A smart composite shear wall system consists of an infill steel plate, boundary frame (beam and column), and concrete wall attached on one side of the infill steel plate or both sides (in this research, the steel plate was attached on one side only). The reinforced concrete wall is in mediating contact with the boundary steel frame because there is a gap in between [37]. The difference between the traditional and smart CSPSW is in the change in the behavior of the concrete wall under cyclic loading, whereas in the traditional CSPSW, the concrete wall works in conjunction with a steel plate. However, in smart CSPSW, due to the gap between the steel frame and concrete wall, RC will not work and resist lateral load until the inter-story drift has reached a certain value. Figure 1 shows the structural components of the newly developed shear walls [38–41].

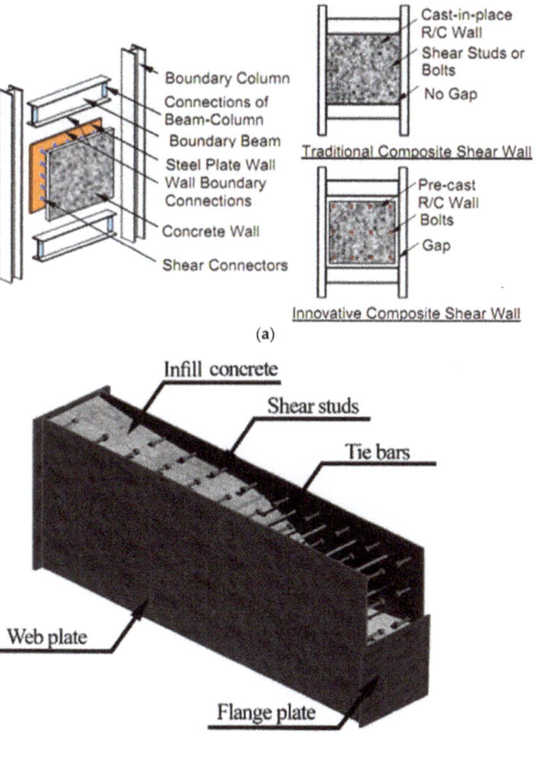

Figure 1. Traditional and innovative composite steel plate shear wall: (**a**) all compounds of the composite shear wall, (**b**) cross-section of composite shear wall.

3. Experimental Program

3.1. Sample Design

The experimental work conducted on CSPSW was carried out by Rahai and Hatami [33] to study the behavior of composite shear walls made of concrete and steel. The details of the experimental work of CSPSW are shown in Figure 2.

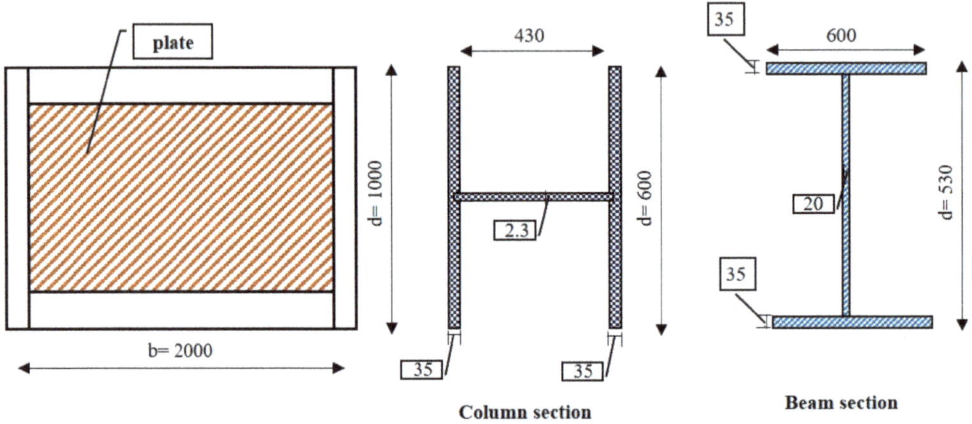

Figure 2. Experimental specimen dimensions.

The frame's parameters had a length (center to center of column) of 2000 mm and height (center to center of the beam) of 1000 mm. For this model, IPE2000 from ST37 was used for the flexural frame strengthened with 12 mm plates connected to both flanges. Steel of 3 mm thickness was used for the plates. The thickness of the concrete plate was 50 mm, reinforced with 1% of concrete volume.

There was a 30 mm gap between the concrete cover and the boundary elements. In order to connect the concrete cover to a steel plate, 7 mm diameter * 100 mm length bolts were used. Moreover, to reinforce the concrete, a 6.5 mm reinforcing bar diameter was used with a center-to-center distance of 65 mm [31]. Table 1 shows the characteristics and strength of steel that was used in all models St37 with yield stress of 240 MPa and ultimate stress of 370 MPa. The steel's behavior was considered a bilinear elastic–plastic curve for modeling. The compressive strength of concrete in the 28th day's cylindrical core sample is 45 MPa, and its tensile strength is equivalent to 3 MPa.

Table 1. Details of the experimental specimens and their mechanical properties.

Group No.	Group Name	SW	Gap	Thickness of Steel	Thickness of Concrete	Distance between Shear Studs	Ratio of Reinforcement	Compressive Strength	Yield Strength	Layout of Shear Stud (H*V)
1	Gap between steel frame and concrete wall	SW-G0mm	0	3	50	200	1%	45	240	3*8
		SW-G30mm	30	3	50	200	1%	45	240	3*8
		SW-G40mm (R)	40	3	50	200	1%	45	240	3*8
		SW-G50mm	50	3	50	200	1%	45	240	3*8
2	Thickness of infill steel plate	SW-TS3mm (R)	40	3	50	200	1%	45	240	3*8
		SW-TS6mm	40	6	50	200	1%	45	240	3*8
		SW-TS12mm	40	12	50	200	1%	45	240	3*8
3	Thickness of concrete wall	SW-TC50mm (R)	40	3	50	200	1%	45	240	3*8
		SW-TC75mm	40	3	75	200	1%	45	240	3*8
		SW-TC100mm	40	3	100	200	1%	45	240	3*8
		SW-TC150mm	40	3	150	200	1%	45	240	3*8
4	Distance between shear studs	SW-D200mm (R)	40	3	50	200	1%	45	240	3*8
		SW-D210mm	40	3	50	210	1%	45	240	3*8
		SW-D220mm	40	3	50	220	1%	45	240	3*8
		SW-D230mm	40	3	50	230	1%	45	240	3*8
		SW-D240mm	40	3	50	240	1%	45	240	3*8
		SW-D250mm	40	3	50	250	1%	45	240	3*8

3.2. Loading Program and Test Setup

Horizontal loading was controlled by force [33]. In the force loading phase, the horizontal forces were 600 kN, and loading was cyclically loaded with 1/60 Hz frequency. The experimental load characteristics are shown in Table 2. The loading history is illustrated in Figure 3.

Table 2. Cyclic loading time history.

Time (s)		Max. Load (KN)	Loading Shape	Frequencies (Hz.)
Start	End			
0.0	71	0.0	Cyclic	0.0
72	180	300	Cyclic	1/60
181	360	500	Cyclic	1/60
361	540	600	Cyclic	1/60

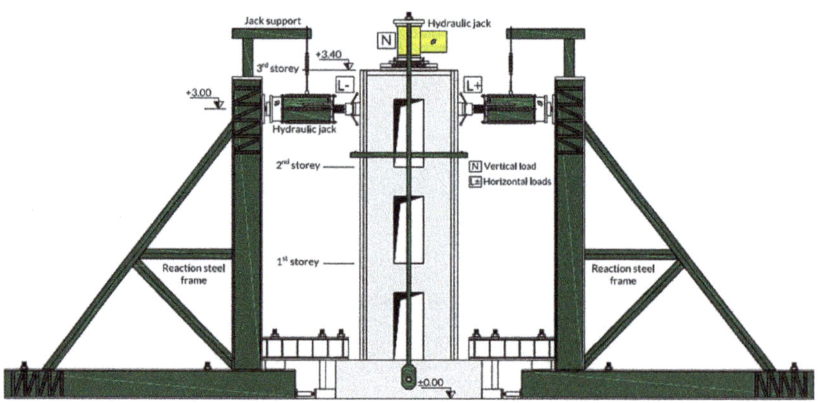

Figure 3. Cyclic loading arrangement.

4. Finite Element Model (FEM)

4.1. Model Overview

4.1.1. Part and Element of the FE Model

The numerical study was performed by finite element analysis using ANSYS. The finite element model consists of five parts: the infilled steel plate, shear studs, outer steel frame, reinforced concrete wall, and reinforcement. The outer steel frame consists of a web and flange plate.

The four nodded shell 181 elements from the ANSYS element library were used to model the outer steel frame, infilled plate, web, and flange plate. The element has six degrees of freedom at each node. The change in stress in the thickness direction cannot be ignored in shear stud and concrete walls because the sizes in the three directions have little difference. Therefore, the element choice to represent the shear stud and reinforcement was a link 180. The element used to represent the concrete wall was 3D solid65. In the test, the shear studs were welded on the web plate. Thus, the shear studs were tied to the steel plate in the FE model. In the test, the reinforcement and the steel studs were fixed in the concrete.

4.1.2. Contact of FE Mode

The friction contact model has been used between the steel plate and concrete. The tangential friction coefficient is 0.6 [37], as shown in Figure 4a. Interface surface between infill steel plate and concrete wall represented by targe170 and contact 174. Finally, the load slip action is represented by comb 39, as shown in Figure 4. The assembled FE model is shown in Figure 4e.

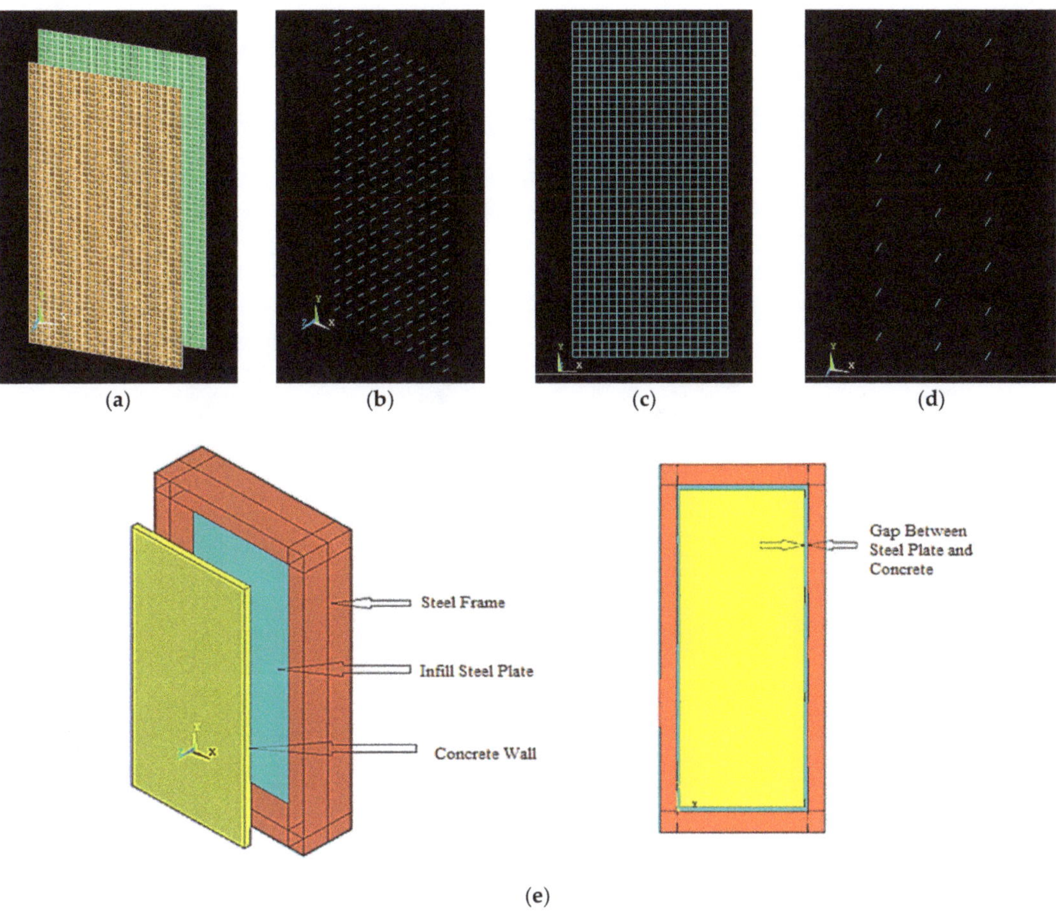

Figure 4. Contact between different elements: (**a**) interface surface; (**b**) load slip; (**c**) reinforcement; (**d**) shear stud; (**e**) assembled FEM.

4.1.3. Boundary Conditions

The bottom frame is a fixed-end constraint, and the boundary condition of the top beam is a sliding constraint. Therefore, six degrees of freedom are constrained at the bottom steel frame (i.e., U1 = U2 = U3 = UR1 = UR2 = UR3 = 0), and four degrees of freedom are constrained at the top steel frame (i.e., U3 = UR1 = UR2 = UR3 = 0).

4.1.4. Steel Constitutive Model

Steel plate is a major element in the composite shear wall. Preferably, this plate is chosen of steel with a low yield point. For example, an St37 steel plate is preferred for high-strength steel plates because an St37 steel plate, due to its low yield point, is preferred to encourage the yielding of steel plates.

4.1.5. Concrete Constitutive Model

The reinforced concrete cover on one side or both sides of a steel plate carries some of the story shears by improving the diagonal compression field and increasing strength and stiffness. Of course, the major role of the reinforced concrete cover is to prevent out-of-plane buckling of steel plate prior to reaching yielding. In some cases, shear studs not only are subjected to shear but also to a considerable tension due to local buckling of the steel plate. For cast-in-place concrete, welded shear studs are usually utilized; for pre-cast concrete walls, bolts can be used.

4.2. Validation of Finite Element Model

Before performing an actual parametric study, the validation of the FE model was performed. The results of hysteretic curves obtained from FE analysis and test results were compared, as shown in Figure 5. It is observed that predicted FE results are directly matching with actual test results. Therefore, it is concluded that the FE model is able to simulate the hysteretic curves of the composite steel plate shear wall in a significant way.

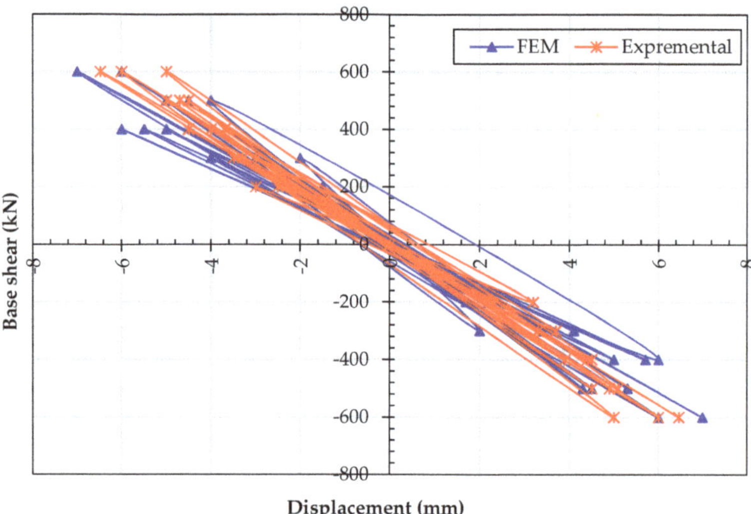

Figure 5. Comparison of deformation load in the numerical and experimental model for CSPSW.

In the test loading process, the steel plate experienced severe buckling at different positions with increasing horizontal displacement as shown in Figure 6. The FE model could simulate the local buckling phenomenon, as shown in Figure 7.

(**a**)

Figure 6. *Cont.*

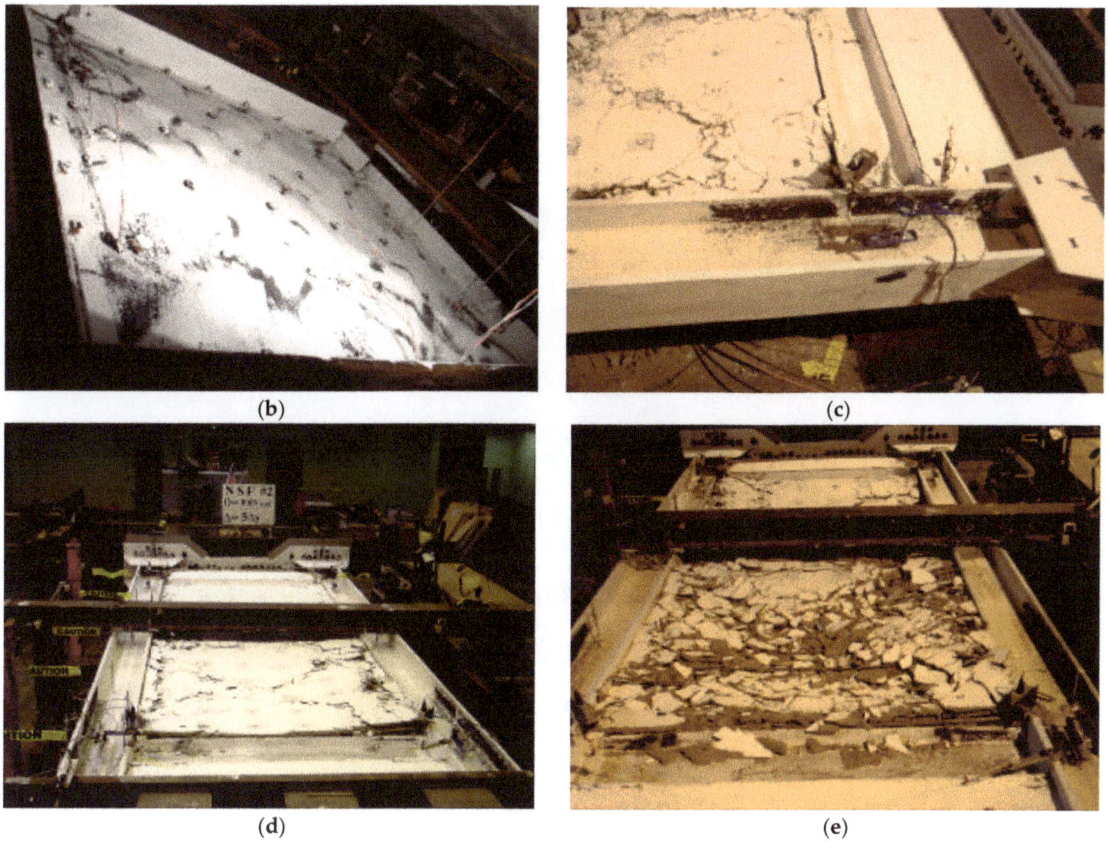

Figure 6 photos (b), (c), (d), (e)

Figure 6. Local buckling, out-of-plane and crack formation for experimental specimen. (**a**) A View of Test Specimen (**b**) steel plate buckling; (**c**) plastic hinge at the base of column; (**d**) concrete crack at mid of the test; (**e**) concrete crack at end of the test.

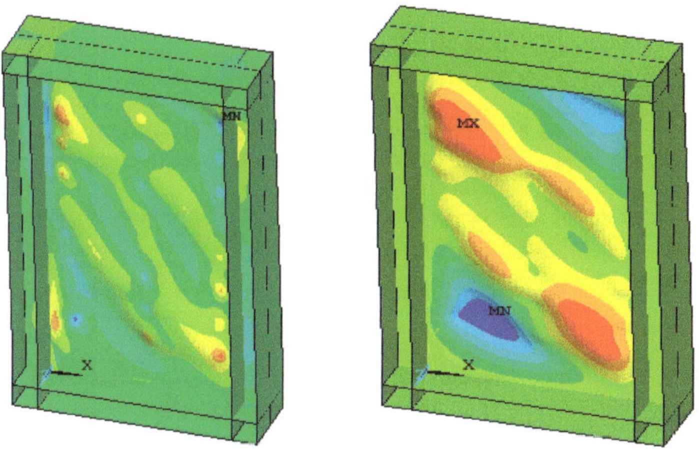

Figure 7. Local buckling, out-of-plane deflection for numerical specimen.

The results obtained during the last cyclic loading from the experimental test and the ANSYS output are presented in Figures 6–8, which show the comparison of out-of-plane deflection and crack formation in the composite shear wall in different experimental and numerical specimens. The lateral displacement was found to be 6.47 mm and 7 mm in the case of numerical and experimental tests, respectively.

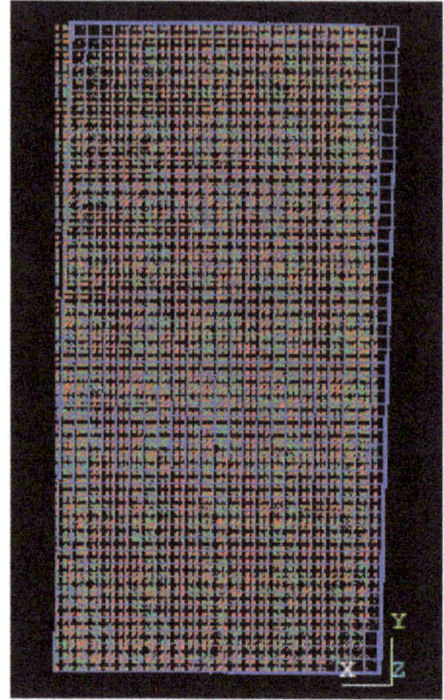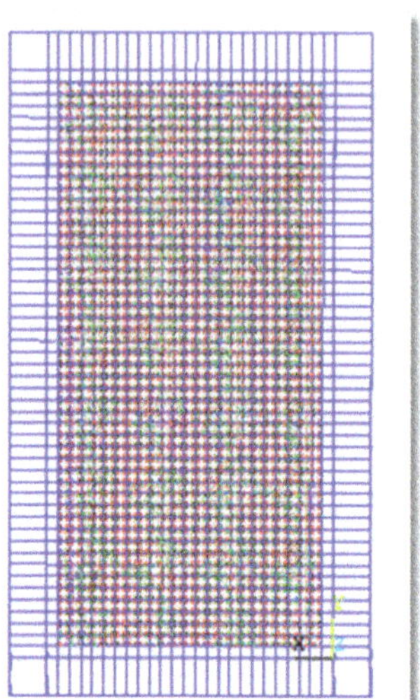

Figure 8. Concrete crack formation for numerical specimen.

5. Parametric Analysis

For optimization of shear wall parameters in tall structures, the parametric study was performed by changing the section size and design guidelines as suggested [33]. The consequences of numerous variables have been vitally studied on the stiffness, ductility, and energy dissipation of the composite shear wall. In order to study the hysteric behavior of the model in parametric analysis, the cyclic load was applied to the wall. The boundary conditions in the model were consistent with the test.

The variables consisted of the gap between the concrete wall and steel frame, the thickness of the infill steel plate, the concrete wall, the distance between the shear studs, the ratio of reinforcement, concrete strength, steel yield strength, and layout of a shear stud. The selected standard model parameters are different, and many were chosen as follows: the gap between the concrete wall and steel frame was 40 mm, concrete wall thickness was 50 mm, the steel ratio was 1%, and infill steel plate thickness was 3 mm. In addition, a distance of 200 mm between the shear studs was selected, the axial compressive strength of the concrete was 45 MPa, the yield strength of the infill steel plate was 240 MPa, and the layout of the shear stud (H*V) was 3*8.

In the parametric analysis, the gap had a range of 0–30–40–50 mm, the thickness of the infill steel plate had a range of 3–6–12 mm, the thickness of the concrete wall had a range of 50–75–100–150 mm, and the distance between the shear studs had a range of 200–210–220–230–240–250 mm.

5.1. Influence Rules of the Parameters

The influence rules of key design parameters are studied by including material displacement, stiffness, ductility, and energy dissipation. The structural behavior of the composite steel plate shear wall, when subjected to cyclic load, is characterized by four different stages when increasing the applied load; these stages are:

- The initial elastic stiffness phase;
- The shear yielding stiffness phase;
- The post-yielding stiffness phase;
- The pre-failure stiffness phase.

Meanwhile, the smart CSPSW, at first loading, displays a linear elastic response where the steel frame and infill steel plate, beams and columns, undergo inelastic deformations.

After that, the interaction between the infill steel plate and the reinforced concrete panel in the compression field is extra efficient. While the lateral loading is further raised, the infill steel plate responsible is immaterially and geometrically nonlinear. Moreover, the lateral shear stiffness of the wall decreases substantially owing to the shear yield of the infill steel plate.

During the third phase, with the excess of lateral unloading, the pure shear yield transpires pending the full shear yield appearing in the infill steel plate, and the lateral stiffness reduces gradually at this phase.

While the lateral load surpasses the shear yield capacity of the infill steel plate, the material and geometric nonlinearity responsible for steel frame and boundary elements is massive. At this phase, the frame supplies utmost lateral stiffness.

From the results, it can be seen that increasing the model thickness (infill steel plate and concrete wall) worked to increase the structural strength capacity and the model's ability to absorb and dissipate energy, which led to a delay in the model failure; at the same time, it prevents the rapid drop in the load-carrying capacity.

Similarly, increasing the distance and layout of the shear studs (certified number of shear studs) increased the structural strength capacity and enhanced the ability of the model to absorb energy and the model ductility.

What is more, if the properties of smart CSPSW were increased, the structural strength capacity and model ability to absorb and dissipate energy would be enhanced.

5.1.1. The First-Group Models (Influence of Gap between Concrete Wall and Steel Frame)

1. Lateral displacement

After loading the first-group models SW-G0mm, SW-G30mm, SW-40MM (R), and SW-G50mm gradually, it can be noted that the model passed through four phases as explained below. Furthermore, lateral displacement of group 1 at each phase can be seen in Figure 9.

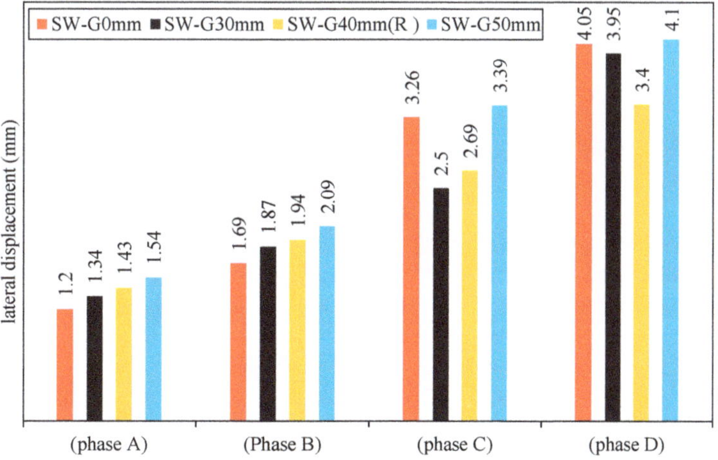

Figure 9. Lateral displacement of group 1.

- Phase A:

In this phase, the applied load causes linear relation between the horizontal unload and the resulting displacement. This relationship for the models SW-G0mm, SW-G30mm, SW-40MM (R), and SW-G50mm continued until yield deflection was reached.

In this phase, the lateral displacement of models SW-G30mm, SW-40mm (R), and SW-G50mm was larger by 12%, 17%, and 28%, respectively, as compared with the reference model SW-G0mm. By increasing the lateral load, the first yield will occur in the steel plate defined. The out-of-plane displacement of group 1 for 0 mm, 30 mm, 40 mm, and 40 mm gaps is shown in Figure 10.

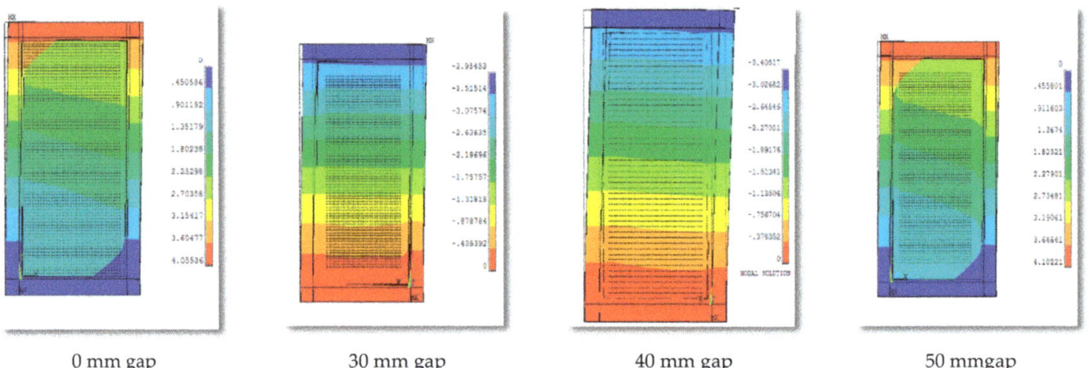

0 mm gap   30 mm gap   40 mm gap   50 mmgap

Figure 10. Out-of-plane displacement for the various gaps between the steel frame and concrete wall.

It is worth mentioning that, in this stage, all the models of traditional and smart composite steel plates with a gap of 30, 40, and 50 mm, respectively, were symmetric in their behavior until they reached the shear yield zone.

- Phase B:

This phase began at the shear yield zone, which represents the first point of the transformation curve to a flat line that has a high incline, as a result of the high increase in the specimen deflection.

When increasing the gap between the steel frame and concrete wall, the interaction between the reinforced concrete panel and infill steel plate in the compression zone became more active. Thus, the results caused a decrease in the lateral shear stiffness for this model. The shear yield zone of the models SW-G30mm, SW-40mm (R), and SW-G50mm was found to be larger by 11%, 15%, and 24% as compared to the reference model SW-G0mm.

- Phase C:

During this phase, when lateral load increases, the shear yield spreads until the full shear yield occurs in the infill steel plate; the models with gaps gave good results under increased lateral load.

For models with a gap of 30 and 40 mm, the lateral displacement was lower than that of the model without a gap of SW-G0mm by 23% and 20%. However, this displacement for models with a gap (50) mm was larger than that of reference model SW-G0mm by 4%.

- Phase D:

This phase refers to the pre-failure of the models. In CSPSW, the infill steel plate of CSPSW resists lateral load by pure shear yield, as a reinforced concrete panel prevents inelastic buckling.

In this phase and in the same cycle loading (600 KN), the deflection of SW-G50mm is larger by 14% compared with the reference model SW-G0mm. Meanwhile, for the other models (SW-G30mm and SW-G40mm (R)) at the same cycle loading, the deflection was lower by 2% and 16%, respectively.

In this phase, all the models show a nonlinear inelastic reaction of the steel frame, and the lateral stiffness is supplied by steel boundary elements. It is worth mentioning at this stage that the symmetry between the behavior of SW-G50mm and the behavior of SPSW shows the large gap between the steel frame and concrete wall, which leads to the formation of cracks in the concrete before contact between them. Furthermore, the system loses the idea of a smart composite steel plate shear wall, which depends essentially on concrete contribution delay to work with steel frame.

2. Stiffness

Stiffness is the ratio of load vs. deformation and can be used to describe either the elastic or plastic (after yield) range. It can be seen from the load-deflection curve that the stiffness refers to the slope of the curve at any point along the curve. Figure 11 presents the initial elastic stiffness (K_e) and the proportion of the initial elastic stiffness of SPSW to CSPSW (K_{ec}/K_{es}) (Rahai and Hatami, 2009).

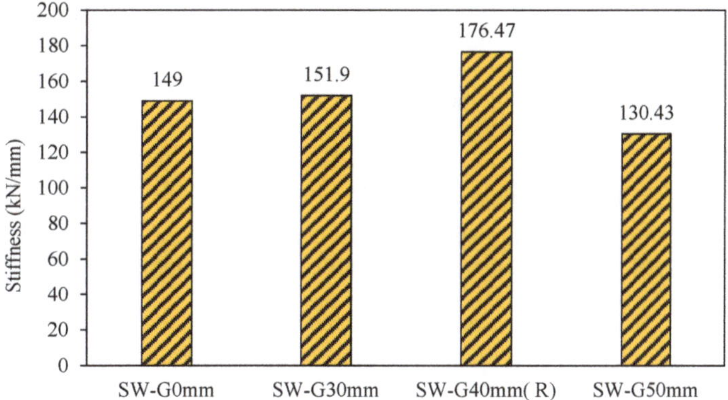

Figure 11. Stiffness of group 1.

Figure 11 shows the stiffness for all group one models. From this result, it can be noted that when increasing the gap to 30 and 40 mm for SW-G30mm and SW-40mm (R), the stiffness was increased by 2% and 18%, respectively, in comparison with the reference model SW-G0mm. Consequently, the model resistance deformations increased. However, in the model of gap 50 mm, its stiffness decreased significantly by 13% as compared to the reference model SW-G0mm, where the specimen was resistant to lateral load. SW-40mm (R) has terrific stiffness among all the models of the first group, and stiffness is equal to 176.47 kN/mm.

3. Ductility

Ductility refers to the deformation that a material can undergo after it has yielded or exceeded its elastic range. From the load-deflection curve, it is concluded that ductility refers to the length of the curve after the yield point to failure. The ductility ratio is calculated as the ratio of the maximum displacement to the yield displacement ($\mu = \delta_{max}/\delta_y$). The yield point displacement (δ_y) is calculated through the notion of equal plastic energy. Hence, the area bounded by the perfect elastic–plastic curve is equal to that of the actual push-over curve [30] (Shafaei et al., 2016).

The results of Figure 12 show that the model SW-G50mm has minimal ductility up to 1.71 as a result of the high value of the deflection at the ultimate load, and this caused sudden buckling and rapid drop in the load-carrying capacity when the ultimate load was achieved.

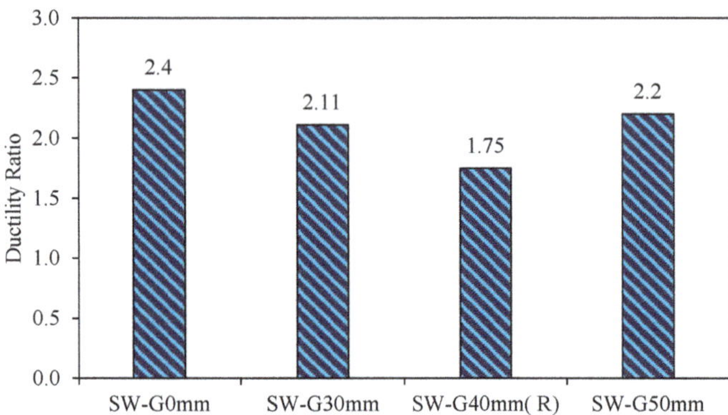

Figure 12. Ductility ratio of group 1.

The great ductility of models with a small gap was because of the high estimation of the lateral displacement at an ultimate load contrasted and the lateral displacement value at the yield load and, in this manner, brought on a gradual failure in the model load capacity.

From the above results, it appears that there was a proportional relationship between the ductility and the gap between the steel frame and concrete wall; therefore, increasing the gap resulted in a decrease in the ductility due to the decreased moment of inertia when the gap was increased. Consequently, this affects the ability of shear wall to resist the lateral load.

4.   Energy Absorption

Energy absorption ability is a critical indicator of the model's resilience to loading. Models with high energy absorption ability are typically found to have high imperviousness to impact and crash loading and hence are valuable for high-performance structures.

The absorbed energy by a shear wall in a half-cycle can be objectively accepted as the zone under shear load displacement, from which the region of recoverable elastic is not subtracted. It is assumed that the unloading and the elastic moduli are approximately the same. For instance, when the max cycle load of the reference model is equal to 600 kN, the displacement value will be about 4.01 mm, and its curve will take a parabola shape. Therefore, energy absorption is equal to (1/3*displacement*load), where the former law represents the area of the parabola shape [30] (Shafaei et al., 2016).

Figure 13 shows the energy absorption of each model through each phase. Through phase B, it can be noticed that an increased gap between the steel frame and concrete wall could increase the energy absorption by 2%, 34%, and 35% for SW-G30mm, SW-G40mm (R), and SW-G50mm, respectively, as compared to reference model SW-G0mm.

Models with the gap (SW-G30mm, SW-G40mm (R), and SW-G50mm) had good energy absorption, which was due to the high area under the curve of load deformation, and it referred to the increase in the resistance of the model to the deformation.

In phase D, it can be seen that the energy absorption of the models SW-G30mm, SW-G40mm (R), and SW-G50mm was larger by 10%, 6%, and 12% as compared to model SW-G0mm.

From the results of the phases above and the calculation of stiffness, ductility, and energy absorption, it is noticed that the gap between the steel frame and concrete wall should be limited by a specific value of 4% of the width because this value gives a good result, which results in a delay in the failures of the model, and this model is economical in the amount of concrete. Therefore, the other groups of smart CSPSW can use SW-G40mm (R) as a reference model.

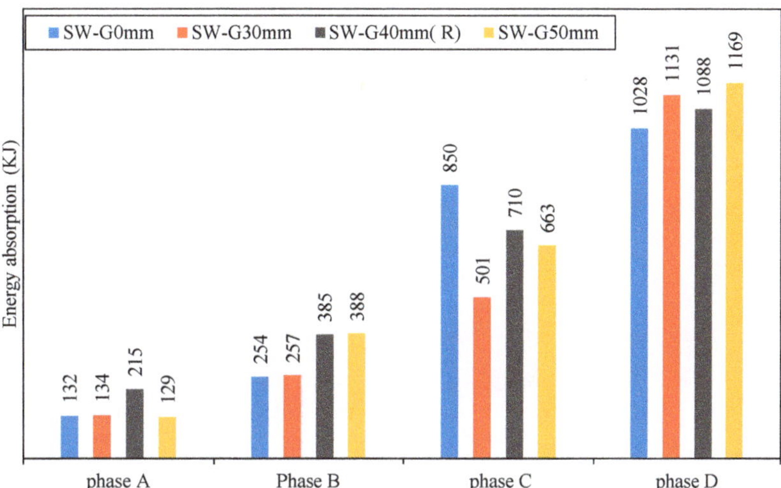

Figure 13. Energy absorption of group 1.

5.1.2. The Second-Group Models (Influence of Infill Steel Plate Thickness)

1. Lateral displacement

The second group, models SW-TS6mm and SW-TS12mm, was loading gradually; the lateral displacement of group 2 at each phase can be seen in Figure 14. Moreover, it can be noted that they passed through four phases depending on the applied load, as discussed below.

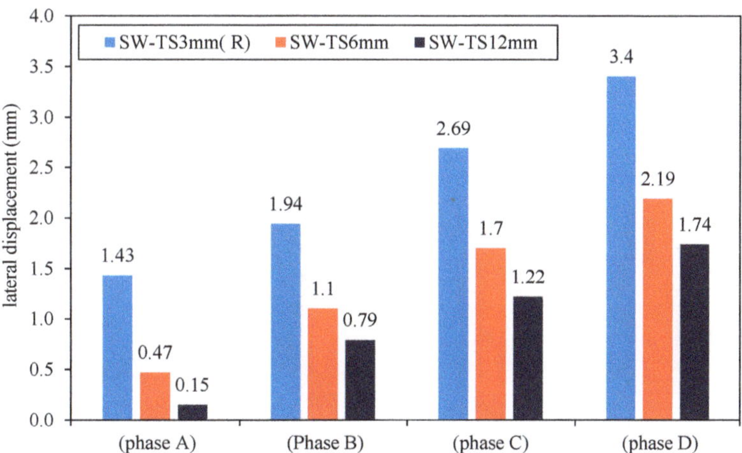

Figure 14. Lateral displacement of group 2.

- Phase A:

The applied loads cause a linear relationship between the lateral load and the resulting displacement. This relationship for models SW-TS6mm and SW-TS12mm continued until yield displacement was achieved. In this stage, the lateral displacement value of the models SW-TS6mm and SW-TS3mm (R) was lower by 67% and 89% as compared with the reference model SW-TS3mm (R), as a result of an increase in the smart CSPSW thickness of infill steel plate which caused a decrease in the yield displacement and increase in the yield load values which led to an increase in the elastic stage for the models SW-TS6mm and SW-TS12mm as compared to the reference model SW-TS3mm (R).

- Phase B:

In this phase, increasing the thickness of the infill steel plate caused an increase in the strain hardening capacity for these models and led to an increase in the stress redistribution significantly until ultimate displacement was achieved. For SW-TS6mm and SW-TS12mm, the lateral displacement was lower by about 43–59% as compared to the reference model SW-TS3mm (R). At the same load, as a result of the increased thickness of the steel plate of smart CSPSW, there was a decrease in the yield displacement and increase in yield load values that led to an increase in the elastic stage for all the models, as compared to reference model SW-TS3mm (R).

- Phase C:

During this phase, when increasing the lateral load, the shear yield propagated until the full shear yield occurred in the infill steel plate, therefore increasing the thickness of the infill steel plate from 3 to 6 and 12 mm. The lateral displacement was lower by 36% and 54%, respectively, as compared to the reference model SW-TS3mm (R).

- Phase D:

The collapse of the reference model SW-TS3mm (R) began before the SW-TS6mm and SW-TS12mm. When comparing the result in this phase at the same cycle loading, it can be noticed that the lateral displacement of SW-TS6mm and SW-TS12mm is lower by 35% and 48%, respectively, as compared with the reference model SW-TS3mm (R). At this stage for SW-TS12MM, the frame provides the most lateral stiffness. Furthermore, by increasing the load, the frame reaches its collapse mechanism, and lateral stiffness declines gradually to a near-zero value.

Generally, it can be pointed out that the thickness of steel has a substantial effect on the lateral displacement, as shown in Figure 15.

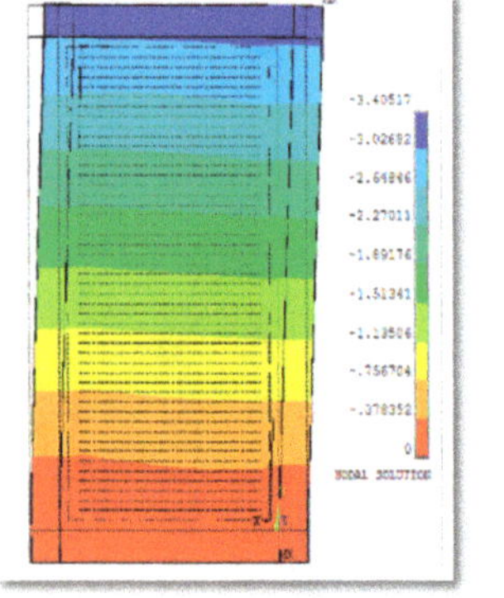
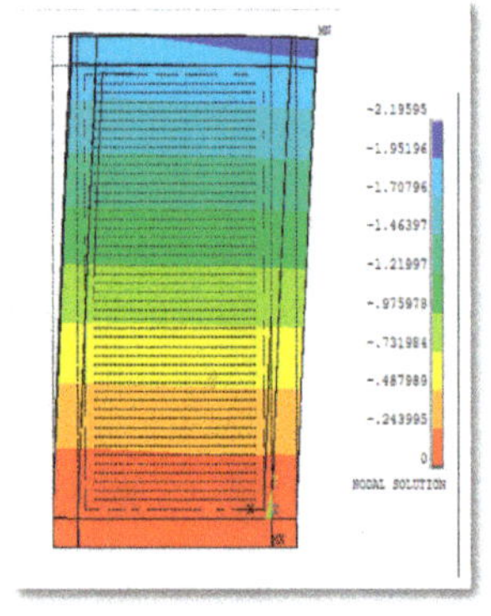
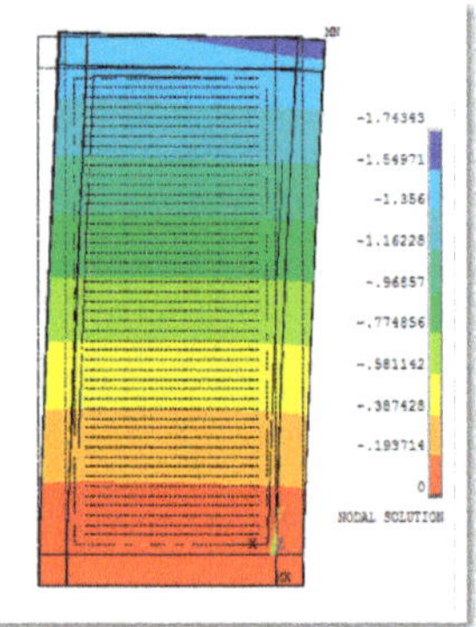

<table>
<tr><td>3 mm steel thickness</td><td>6 mm steel thickness</td><td>12 mm steel thickness</td></tr>
</table>

Figure 15. Out-of-plane displacement of group 2 for various steel thicknesses.

2. Stiffness

Figure 16 demonstrates the stiffness values for the second-group models. From this result, it can be noted that increasing the thickness of the infill steel plates from 3 to 6 and 12 mm led to an increase in their stiffness by about 55% and 95%, respectively, as compared to the reference model SW-TS3mm (R). Thus, it can be concluded that the stiffness of the models was directly compared to the infill steel plate thickness.

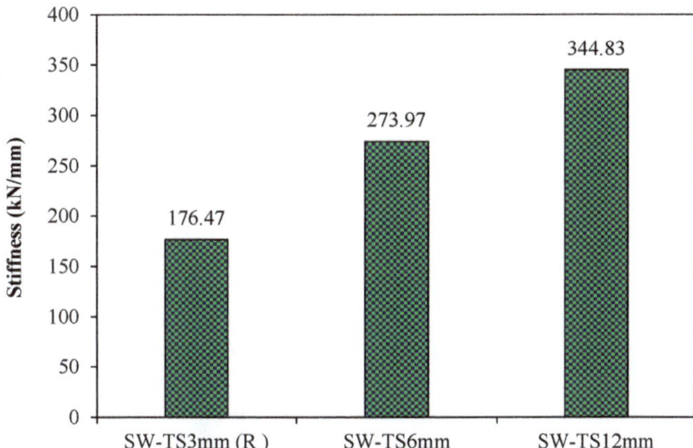

Figure 16. Stiffness of group 2 (thickness of infill steel plate).

3. Ductility

Figure 17 shows the ductility ratio of all the models when increasing the thickness of the infill steel plate. From Figure 17, it can be found that the models SW-TS6mm and SW-TS12mm have larger ductility, up to 12% and 21%, respectively, as compared to the reference model SW-TS3mm (R). It was because of the high estimation of the deflection at an ultimate load and the deflection value at the yield load, and in this manner, it brought on a continuous failure in the model failure limit.

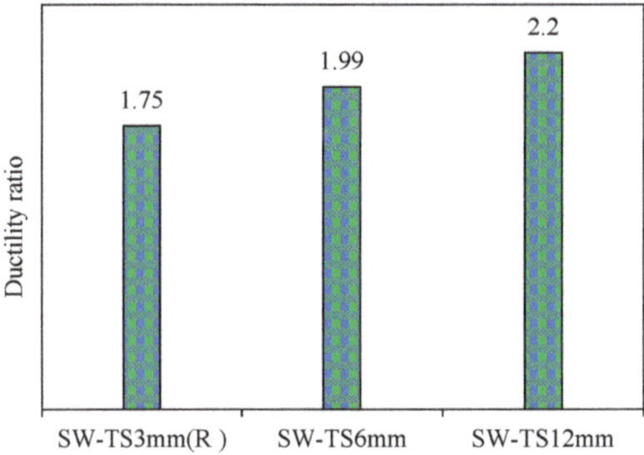

Figure 17. Ductility ratio of group 2.

Therefore, it seems that there is a proportional relationship between the ductility and the infill steel plate thickness; consequently, increasing the thickness causes an increase in the ductility.

4. Energy Absorption

Figure 18 shows the energy absorption of each model through each phase. For SW-TS6mm and SW-TS12mm in phase C, when there was an increase in the thickness of the infill steel plate, the energy absorption increased by 5% and 14% as compared to the reference model SW-TS3mm (R). Meanwhile, through phase D, it is noticed that an increase in the thickness of the infill steel plate resulted in an increase in the energy absorption by 53% and 57% for SW-TS6mm and SW-TS12mm as compared to the reference model SW-TS3mm (R).

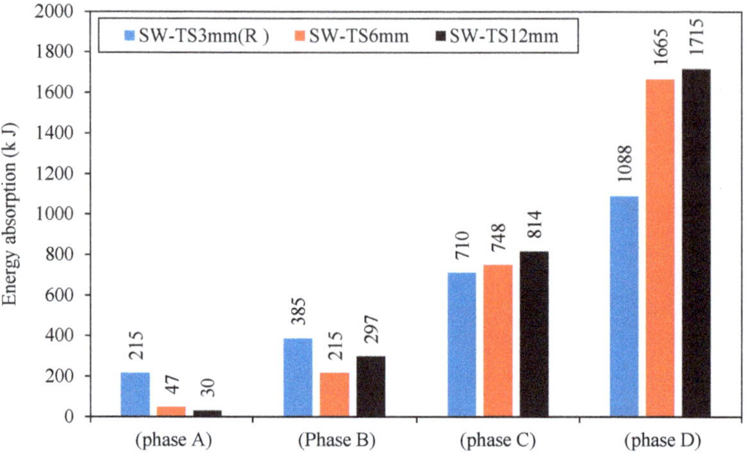

Figure 18. Energy absorption of group 2.

Models with a large thickness (SW-TS12mm) had great energy absorption, and it was due to the high area under the curve of load deflection. It refers to the increased resistance of the model to the deformation. From the results of the phases above and calculation of stiffness, ductility, and energy absorption, it can be noticed that the thickness of infill steel plate for (2000*1000) mm (length*width) specimen dimensions should be limited by a specific value (min 3 mm) due to the trail thickness of 1 mm of the infill steel plate which results in a quick failure in the model. The type of failure was expressed as an opening in the steel plate. As a result, the best range for using the thickness of the infill steel plate was between 3 and 12 mm. The best value in terms of cost economy was 6 mm.

5.1.3. The Third-Group Models (Influence of Concrete Wall Thickness)

1. Lateral displacement

The third-group models SW-TC75mm, SW-TC100mm, and SW-TC150MM were loading gradually. From the result, it is noted that they passed through four phases depending on the applied load, and the lateral displacement of each phase is shown in Figure 19:

- Phase A:

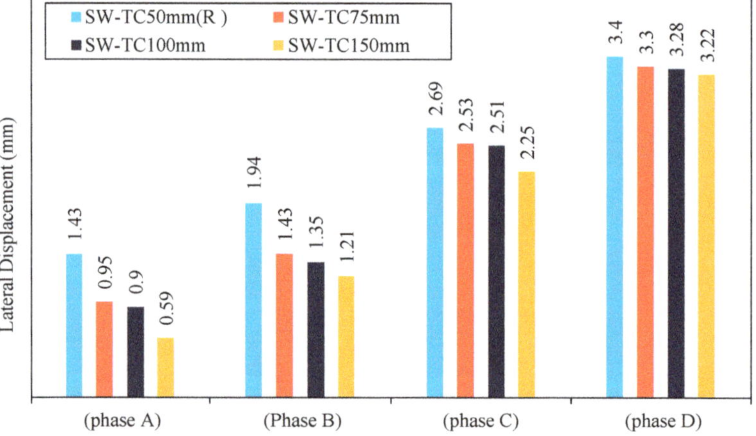

Figure 19. Lateral displacement of group 3.

The applied load causes a linear relationship between the lateral load and the resulting lateral displacement. This relationship for the models SW-TC75mm, SW-TC100mm, and SW-TC150mm continued until yield displacement was achieved. In this phase, the lateral displacement value of the models SW-TC75mm, SW-TC100mm, and SW-TC150mm was lower by 33%, 37%, and 58% as compared with the reference model SW-TC50mm (R). Increasing the smart CSPSW thickness of the concrete wall caused a decrease in the yield displacement and increase in yield load values and thus an increase in the elastic stage for the specimens SW-TC75mm, SW-TC100mm, and SW-TC150mm as compared to the reference model SW-TC50mm (R).

- Phase B:

After increasing the thickness of the concrete wall, a comparison of the result was performed, and it was noticed that lateral displacement of the models SW-TC75mm, SW-TC100mm, and SW-TC150mm at the same load was lower than the reference model SW-TS3mm (R) by 26%, 30%, and 37%. This is because of increasing the thickness of the concrete wall of this model, which increases the strain hardening capacity for the models, and that led to an increase in the stress redistribution significantly until data achieved the ultimate displacement.

- Phase C:

When increasing the lateral load, the shear yield propagates until the full shear yield occurs in the infill steel plate. In this phase, after comparison of the result, when increasing the thickness of concrete wall from 50 to 75, 100, and 150 mm, the lateral displacement was lower by 5%, 6%, and 16%, respectively, as compared to the reference model SW-TC50mm (R).

- Phase D:

The collapse of the reference model SW-TS3mm (R) began before the SW-TC75mm, SW-TC100mm, and SW-TC150mm because of the increased thickness of the concrete wall. Thus, the failure possibility of this model under lateral load was lower than under other loads. Consequently, when comparing the result in this phase at the same cycle of loading, it can be noticed that the lateral displacement of SW-TC75mm, SW-TC100mm, and SW-TC150mm is lower by 3%, 4%, and 5%, respectively.

In general, it was observed that the thickness of concrete has an influence on the term lateral displacement, as shown in Figure 20.

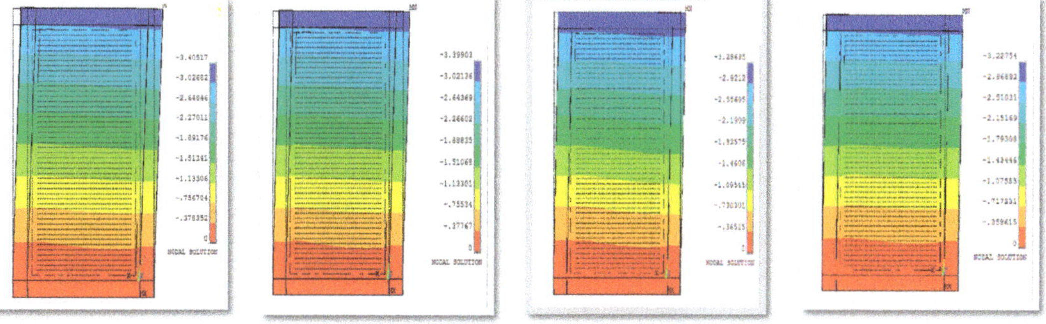

50 mm concrete thickness   75 mm concrete thickness   100 mm concrete thickness   150 mm concrete thickness

Figure 20. Out-of-plane displacement of group 3 for different concrete thicknesses.

2. Stiffness

Figure 21 shows the stiffness values for third-group models. From these results, it can be concluded that increasing the thickness of concrete wall from 50 to 75, 100, and 150 mm leads to an increase in its stiffness by 3%, 4%, and 5%, respectively, as compared to the reference model SW-TC50mm (R). Thus, it can be noted that the thickness of concrete wall has a slight effect on the model's stiffness.

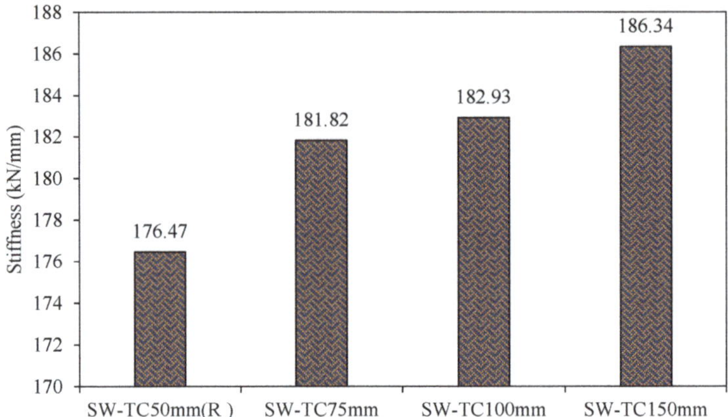

Figure 21. Stiffness of group 3.

3. Ductility

The values of ductility of the third-group models when increasing the thickness of the concrete walls are shown in Figure 22. After comparison of the result, it can be found that the models (SW-TC75mm, SW-TC100mm, and SW-TC150mm) have large ductility of 32%, 38%, and 52%, respectively, as compared to the reference model SW-TC50mm (R). This result was due to a high value of lateral displacement at the ultimate load compared to the deflection value at the yield load and thus caused a gradual failure in the model load capacity.

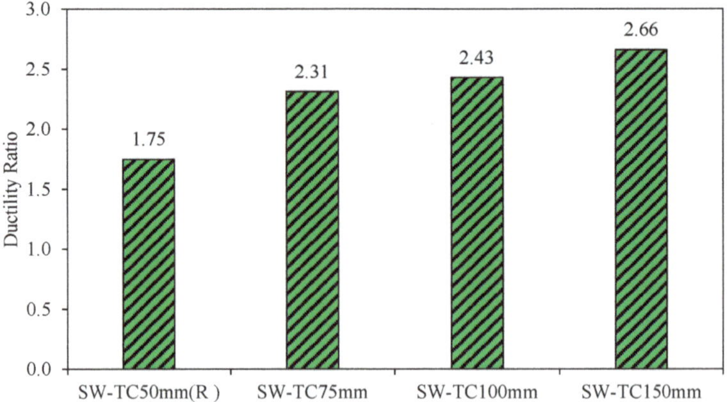

Figure 22. Ductility ratio of group 3.

Thus, it appears that there is a proportional relationship between the ductility and the concrete wall thickness; therefore, increasing the thickness causes an increase in the ductility and ultimately gives a gradual drop in the load-carrying capacity until the failure load is reached.

4.   Energy Absorption

The energy absorption of each model through each phase is shown in Figure 23. For SW-TC75mm, SW-TC100mm, and SW-TC150mm in phase C, when there is an increase in the thickness of the concrete wall, the energy absorption increases by 15%, 23%, and 24%, respectively, as compared to reference model SW-TC50mm (R).

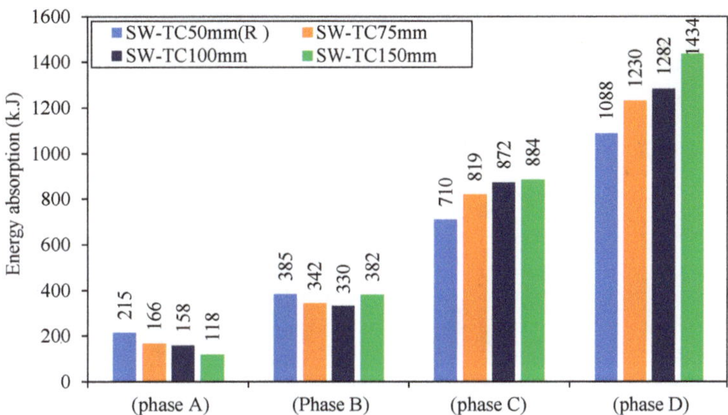

Figure 23. Energy absorption of group 3.

Meanwhile, through phase D, it can be noticed that an increased thickness of concrete wall increased the energy absorption by 13%, 18%, and 32%, respectively, for SW-TC75 mm, SW-TC100 mm, and SW-TC150 mm as compared with reference model SW-TC50mm (R). The models with the large thickness (SW-TC150mm) had good energy absorption, and it was due to the high area under the curve of load deformation. It refers to the increased resistance of the model to the deformation.

From the calculation of stiffness, ductility, and energy absorption, it can be noticed that concrete wall has a large effect on the behavior of smart CSPSW because increasing the thickness of concrete wall leads to an increase in the contribution of concrete in force transfer; therefore, the influence of lateral load on the infill steel plate becomes low. Moreover, increased thickness leads to restricting the frame and reducing the lateral offset.

From the results of the phases above, the thickness of concrete wall for (2000 * 1000) mm (length * width) specimen dimensions should be limited by a specific value (max 150 mm) because the behavior of smart CSPSW remains the same beyond that thickness. Therefore, the best range for using the thickness of the concrete wall was 50–100 mm.

5.1.4. The Fourth-Group Models (Influence of Distance between Shear Studs)

1.   Lateral displacement

From Figure 24, it can be seen that the models SW-D210mm, SW-D220mm, SW-D230mm, SW-D240mm, and SW-D250mm go through the same phases as the reference model SW-D200mm (R) when loaded gradually, which are as follows:

- Phase A:

For the models SW-D210mm, SW-D220mm, SW-D230mm, SW-D240mm, and SW-D250mm, the elastic phase starts from the beginning of loading to the yield displacement. The lateral displacement values of the models SW-D210mm, SW-D220mm,

SW-D230mm, SW-D240mm, and SW-D250mm were lower by 20%, 30%, 58%, 59%, and 60% as compared to the reference model SW-D200mm (R). At the same load, decreasing the distance between the shear studs of smart CSPSW, which causes a decrease in the yield displacement and increase in the yield load values, leads to an increase in the elastic stage for all the models as compared to the reference model SW-D200mm (R). Therefore, it can be noted that a decreased distance between shear studs in all the models has a large effect on the elastic stage for these models.

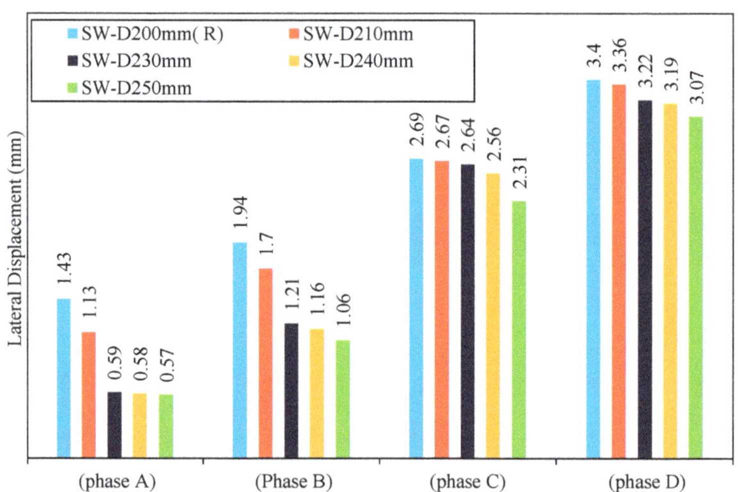

Figure 24. Lateral displacement of group 4.

- Phase B:

The strength capacity of the models SW-D210mm, SW-D220mm, SW-D230mm, SW-D240mm, and SW-D250mm was governed by the plastic deformation, which occurred because of the moment capacity and the shear force at the shear studs. This moment capacity of the models decreases due to a decrease in distance between shear studs (increased number of shear studs) because of the occurrence of a high reduction in the moment contribution of these models. Consequently, the lateral displacement of the models SW-D210mm, SW-D220mm, SW-D230mm, SW-D240mm, and SW-D250mm was lower by 12%, 24%, 37%, 40%, and 45% as compared to the reference model SW-D200mm (R). From this, it can be noted that the presence of increased distance between shear studs affected significantly the shear yield phase through the escalation of strain hardening capacity and led to a significant change in the escalation models' stress redistribution compared with the reference model SW-D200mm (R).

- Phase C:

In this phase, all models of the fourth group had very similar behaviors with very close values of the yield load and yield displacement. It is also observed that the increased distance between shear studs in the CSPSW models did not have a large effect on the post shear yielding phase for these models. In this phase, when there was a variation in the distance between shear studs of 200–250 mm for SW-D210mm, SW-D220mm, SW-D230mm, SW-D240mm, and SW-D250mm, the lateral displacement was lower by 0.74%, 0.74%, 1.86%, 4%, and 14% as compared to the reference model SW-D200mm (R).

- Phase D:

This phase began when the model reached the ultimate load by exposure of all model elements that are situated above and below the shear stud to high stresses. The collapse of the reference model SW-D200mm (R) began with the occurrence of buckling in the infill steel plate because of using many shear studs. In other words, the small distance between the shear studs leads to high plastification around shear studs because of exposure to high compression, which leads to a global buckling failure mode.

After the comparison of the result in this phase at the same cycle of loading, it is noticed that the lateral displacement of SW-D210mm, SW-D220mm, SW-D230mm, SW-D240mm, and SW-D250mm is lower by 1%, 2%, 5%, 6%, and 9% as compared to the reference model SW-D200mm (R).

Overall, it was observed that the distance of the shear stud has a slight effect on the term of lateral displacement, as shown in Figure 25.

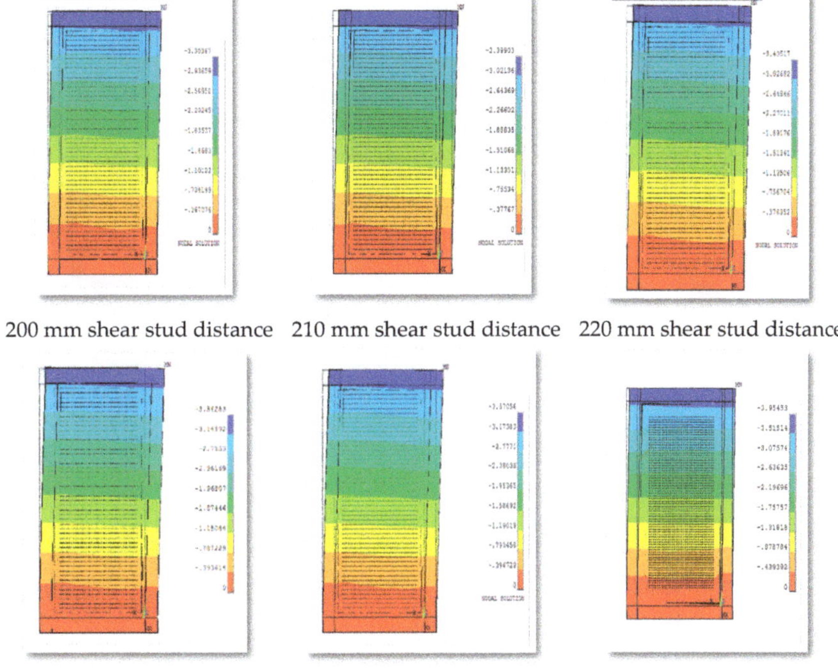

200 mm shear stud distance   210 mm shear stud distance   220 mm shear stud distance

230 mm shear stud distance   240 mm shear stud distance   250 mm shear stud distance

Figure 25. Out-of-plane displacement of group 4 for various shear stud distances.

2. Stiffness

Figure 26 gives the values of the stiffness for the fourth-group models. Figure 26 shows that increased distance between shear studs in the models SW-D210mm, SW-D220mm, SW-D230mm, SW-D240mm, and SW-D250mm enhances their stiffness by 1%, 3%, 5%, 6%, and 10%, respectively, compared to the reference model SW-D200mm as a result of the small lateral displacement of these models. From the results, it can be seen that the distance between shear studs has very little effect on the model stiffness.

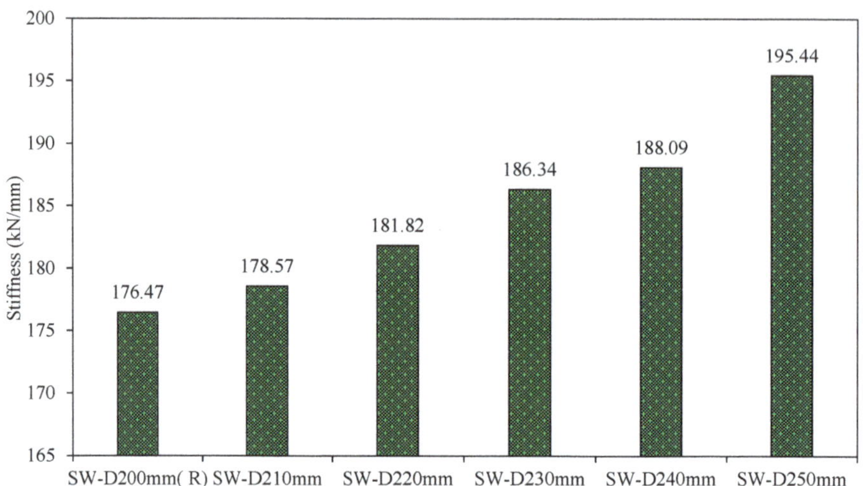

Figure 26. Stiffness of group 4.

3.   Ductility

Figure 27 shows the ductility values of the fourth-group models SW-D210mm, SW-D220mm, SW-D230mm, SW-D240mm, and SW-D250mm. From Figure 27, it can be concluded that the increased distance between shear studs in the models SW-D210mm, SW-D220mm, SW-D230mm, SW-D240mm, and SW-D250mm enhanced their ductility substantially by 12%, 22%, 34%, 37%, and 40%, respectively.

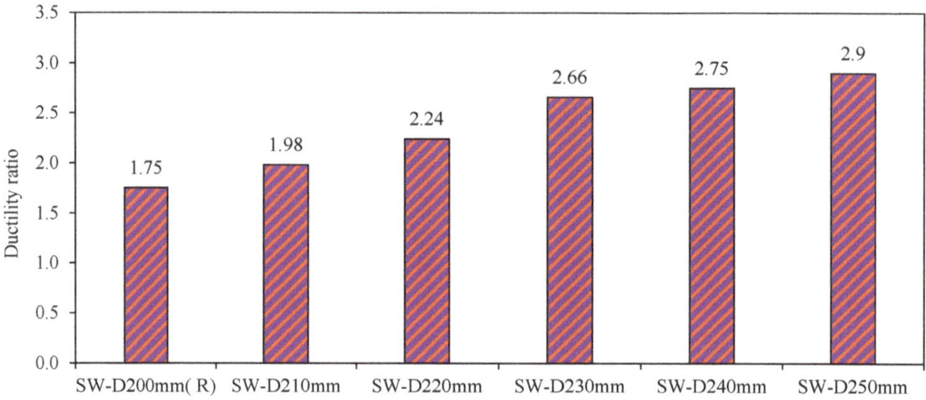

Figure 27. Ductility ratio of group 4.

A gradual drop in the load-carrying capacity of these models was observed when they reached the ultimate load compared with the sudden and rapid drop of the reference model SW-D200mm (R).

4.   Energy Absorption

Figure 28 shows the energy absorption of each model through each phase. For SW-D210mm, SW-D220mm, SW-D230mm, SW-D240mm, and SW-D250mm in phase C, when increasing the distance between shear studs, the energy absorption increases by 14%, 23%, 32%, 33%, and 30% as compared to reference model SW-D200mm (R). While through phase D, it can be seen that increased distance between shear studs increased the energy absorption by 9%, 15%, 20%, 22%, and 24% for SW-D210mm, SW-D220mm, SW-D230mm, SW-D240mm, and SW-D250mm as compared to reference model SW-D200mm (R). The models with a large distance (SW-D250mm) had good energy absorption, and it was due to the high area under the curve of load deflection. This refers to the improved resistance of the model to the deformation.

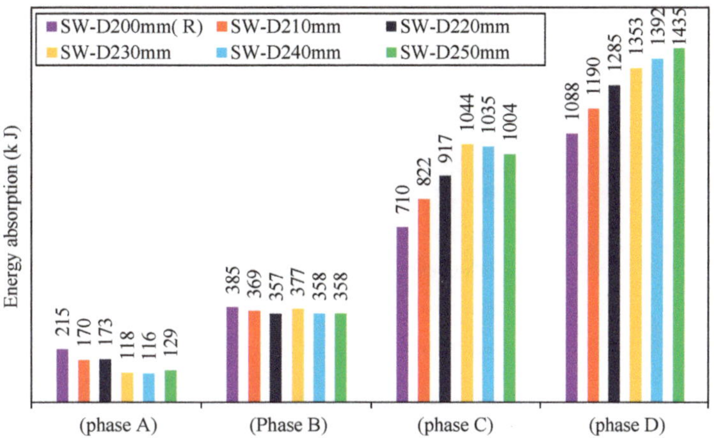

Figure 28. Energy absorption of group 4.

From the results of stiffness, ductility, and energy absorption, it is noticed that the distance between shear studs for (2000 * 1000) mm (length * width) specimen dimensions should be limited by a specific value (250 mm). This is because large distances will cause widespread buckling of the steel plate in free sub-panels between the shear stud and thus will result in no improvement. Therefore, the ideal range for the distance between the shear studs was 200–250 mm.

6. Conclusions

Based on the numerical results conducted in this study, the conclusions were drawn as follows:

- Increasing the gap between the steel frame and concrete wall influences the sequences of the yielding of components, where yielding shows in the beam and infill steel plate first. At the end of the test, the columns showed yielding at the base but did not buckle. The gap between the steel frame and the concrete wall should be limited by a specific value of 4% of the width, as this value has a considerable effect on delaying failures of the model. Moreover, this model is economical in terms of the volume of concrete.
- The thickness of infill steel plate for 2000*1000 mm (length*width) specimen dimensions should be limited by a specific value (min 3 mm) because using 1 mm of infill steel plate resulted in a quick failure in the model; the type of failure was expressed as an opening in the steel plate. Therefore, the ideal range of infill steel plate thickness was 3–12 mm. The best value in terms of cost economy is 6 mm.
- The thickness of the concrete wall for (2000*1000) mm (length*width) specimen dimensions should be limited by a specific value (max 150 mm) because the behavior of smart CSPSW remains the same beyond that thickness. Therefore, the best range for using the thickness of the concrete wall was 50–100 mm.
- The distance between shear studs for 2000*1000 mm (length*width) specimen dimensions should be limited by a specific value of 250 mm because large distances will cause widespread buckling of the steel plate and will result in no enhancement. Therefore, the best range for the distance between the shear stud was 200–250 mm.

Author Contributions: Conceptualization, H.M.N., A.M.I., M.M.S.S., A.H., S.M., N.S.M., M.M.A.E. and K.M.K.; methodology, H.M.N., A.M.I., M.M.S.S., A.H., S.M., N.S.M., M.M.A.E. and K.M.K.; software, H.M.N., A.M.I., M.M.S.S., A.H., S.M., N.S.M., M.M.A.E. and K.M.K.; validation, H.M.N., A.M.I., M.M.S.S., A.H., S.M., N.S.M., M.M.A.E. and K.M.K.; formal analysis, H.M.N., A.M.I., M.M.S.S., A.H., S.M., N.S.M., M.M.A.E. and K.M.K.; investigation, H.M.N.; resources, M.M.S.S.; data curation, H.M.N., A.M.I., M.M.S.S., A.H., S.M., N.S.M., M.M.A.E. and K.M.K.; writing—original draft preparation, H.M.N. and A.M.I.; writing—review and editing, H.M.N., A.M.I., M.M.S.S., A.H., S.M., N.S.M., M.M.A.E. and K.M.K.; visualization, H.M.N., A.M.I., M.M.S.S., A.H., S.M., N.S.M., M.M.A.E. and K.M.K.; supervision, H.M.N. and A.M.I.; project administration, M.M.S.S.; funding acquisition, M.M.S.S. All authors have read and agreed to the published version of the manuscript.

Funding: The research is partially funded by the Ministry of Science and Higher Education of the Russian Federation as part of the World-class Research Center program: Advanced Digital Technologies (contract No. 075-15-2022-311 dated 20 April 2022). This research work was supported by the Deanship of Scientific Research at King Khalid University, Abha, Saudi Arabia, under grant number RGP. 2/246/43.

Institutional Review Board Statement: Not applicable.

Informed Consent Statement: Not applicable.

Data Availability Statement: Not applicable.

Acknowledgments: The authors extend their thanks to the Deanship of Scientific Research at King Khalid University, Abha, Saudi Arabia, for supporting this work. The authors also extend their thanks to the Ministry of Science and Higher Education of the Russian Federation for funding this work.

Conflicts of Interest: The authors declare no conflict of interest.

References

1. Ramesh, S. *Behavior and Design of Earthquake-Resistant Dual-Plate Composite Shear Wall Systems*; Purdue University: West Lafayette, IN, USA, 2013.
2. Bruneau, M.; Alzeni, Y.; Fouché, P. Seismic Behavior of Concrete-Filled Steel Sandwich Walls and Concrete-Filled Steel Tube Columns. In Proceedings of the Steel Innovations Conference 2013, Christchurch, New Zealand, 21–22 February 2013.
3. Yan, J.B.; Yan, Y.Y.; Wang, T. Cyclic Tests on Novel Steel-Concrete-Steel Sandwich Shear Walls with Boundary CFST Columns. *J. Constr. Steel Res.* **2020**, *164*, 105760. [CrossRef]
4. Liu, Y.; Zhou, Z.; Cao, W.; Wei, L. Study on Seismic Performance of the Masonry Wall with Lightweight Concrete Filled Steel Tube Core Column and Thermal-Insulation Sandwich. *J. Earthq. Eng. Eng. Vib.* **2016**, *15*, 850. [CrossRef]
5. ASCE 7. *Minimum Design Loads for Buildings and Other Structures*; American Society of Civil Engineers: Reston, VA, USA, 2022.
6. AISC American Institute of Steel Construction-Seismic Provisions for Structural Steel Buildings. *Seismic Provisions for Structural Steel Buildings*; American Institute of Steel Construction: Chicago, IL, USA, 2010.
7. Nie, X.; Wang, J.J.; Tao, M.X.; Fan, J.S.; Bu, F.M. Experimental Study of Flexural Critical Reinforced Concrete Filled Composite Plate Shear Walls. *Eng. Struct.* **2019**, *197*, 109439. [CrossRef]
8. Mydin, M.A.O.; Wang, Y.C. Structural Performance of Lightweight Steel-Foamed Concretesteel Composite Walling System under Compression. *Thin Walled Struct.* **2011**, *49*, 66–76. [CrossRef]
9. Wright, H. Axial and Bending Behavior of Composite Walls. *J. Struct. Eng.* **1998**, *124*, 758–764. [CrossRef]
10. Wang, M.Z.; Guo, Y.L.; Zhu, J.S.; Yang, X. Flexural-Torsional Buckling and Design Recommendations of Axially Loaded Concrete-Infilled Double Steel Corrugated-Plate Walls with T-Section. *Eng. Struct.* **2020**, *208*, 110345. [CrossRef]
11. Zhang, X.; Qin, Y.; Chen, Z. Experimental Seismic Behavior of Innovative Composite Shear Walls. *J. Constr. Steel Res.* **2016**, *116*, 218–232. [CrossRef]
12. Zhang, W.; Wang, K.; Chen, Y.; Ding, Y. Experimental Study on the Seismic Behaviour of Composite Shear Walls with Stiffened Steel Plates and Infilled concrete. *Thin Walled Struct.* **2019**, *144*, 106279. [CrossRef]
13. Najem, H.M. Influence of Concrete Strength on the Cycle Performance of Composite Steel Plate Shear Walls. *Diyala J. Eng. Sci.* **2018**, *11*, 1–7. [CrossRef]
14. Najem, H.M. The Effect of Infill Steel Plate Thickness on the Cycle Behavior of Steel Plate Shear Walls. *Diyala J. Eng. Sci.* **2018**, *11*, 1–6. [CrossRef]
15. Nguyen, N.H.; Whittaker, A.S. Numerical Modelling of Steel-Plate Concrete Composite Shear Walls. *Eng. Struct.* **2017**, *150*, 1–11. [CrossRef]
16. Epackachi, S.; Whittaker, A.S.; Varma, A.H.; Kurt, E.G. Finite Element Modeling of Steel-Plate Concrete Composite Wall Piers. *Eng. Struct.* **2015**, *100*, 369–384. [CrossRef]

17. Rafiei, S.; Hossain, K.M.A.; Lachemi, M.; Behdinan, K.; Anwar, M.S. Finite Element Modeling of Double Skin Profiled Composite Shear Wall System under In-Plane Loadings. *Eng. Struct.* **2013**, *56*, 46–57. [CrossRef]
18. Wei, F.F.; Zheng, Z.J.; Yu, J.; Wang, Y.Q. Computational Method for Axial Compression Capacity of Double Steel-Concrete Composite Shear Walls with Consideration of Buckling. *Gongcheng Lixue Eng. Mech.* **2019**, *36*, 154–164. [CrossRef]
19. Guo, L.; Wang, Y.; Zhang, S. Experimental Study of Rectangular Multi-Partition Steel-Concrete Composite Shear Walls. *Thin-Walled Struct.* **2018**, *130*, 577–592. [CrossRef]
20. Wang, M.Z.; Guo, Y.L.; Zhu, J.S.; Yang, X. Flexural Buckling of Axially Loaded Concrete-Infilled Double Steel Corrugated-Plate Walls with T-Section. *J. Constr. Steel Res.* **2020**, *166*, 105940. [CrossRef]
21. Shafaei, S.; Ayazi, A.; Farahbod, F. The Effect of Concrete Panel Thickness upon Composite Steel Plate Shear Walls. *J. Constr. Steel Res.* **2016**, *117*, 81–90. [CrossRef]
22. Qin, Y.; Shu, G.P.; Zhou, G.G.; Han, J.H. Compressive Behavior of Double Skin Composite Wall with Different Plate Thicknesses. *J. Constr. Steel Res.* **2019**, *157*, 297–313. [CrossRef]
23. Huang, Z.; Liew, J.Y.R. Structural Behaviour of Steel-Concrete-Steel Sandwich Composite Wall Subjected to Compression and End Moment. *Thin-Walled Struct.* **2016**, *98*, 592–606. [CrossRef]
24. Li, X.; Li, X. Steel Plates and Concrete Filled Composite Shear Walls Related Nuclear Structural Engineering: Experimental Study for out-of-Plane Cyclic Loading. *Nucl. Eng. Des.* **2017**, *315*, 144–154. [CrossRef]
25. Epackachi, S.; Nguyen, N.H.; Kurt, E.G.; Whittaker, A.S.; Varma, A.H. In-Plane Seismic Behavior of Rectangular Steel-Plate Composite Wall Piers. *J. Struct. Eng.* **2015**, *141*, 04014176. [CrossRef]
26. *JGJ/T 380*; Technical Specification for Steel Plate Shear Walls. Ministry of Housing and Urban-Rural Development: Beijing, China, 2015.
27. *ACI Committee 318*; Building Code Requirements for Structural Concrete and Commentary (ACI 318M-11). American Concrete Institute: Indianapolis, IN, USA, 2011.
28. *ANSI/AISC N690-18*; Specification for Safety-Related Steel Structures for Nuclear Facilities. American Institute of Steel Construction: Chicago, IL, USA, 2018.
29. Epackachi, S.; Nguyen, N.H.; Kurt, E.G.; Whittaker, A.S.; Varma, A.H. An Experimental Study of the In-Plane Response of Steel-Concrete Composite Walls. In Proceedings of the 22nd International Conference on Structural Mechanics in Reactor Technology (SMiRT-22), San Francisco, CA, USA, 18–23 August 2013.
30. Kantaros, A.; Chatzidai, N.; Karalekas, D. 3D Printing-Assisted Design of Scaffold Structures. *Int. J. Adv. Manuf. Technol.* **2016**, *82*, 559–571. [CrossRef]
31. Nedelcu, M.; Cucu, H.L. Buckling Modes Identification from FEA of Thin-Walled Members Using Only GBT Cross-Sectional Deformation Modes. *Thin Walled Struct.* **2014**, *81*, 150–158. [CrossRef]
32. Yan, J.B.; Liew, J.Y.R.; Zhang, M.H.; Sohel, K.M.A. Experimental and Analytical Study on Ultimate Strength Behavior of Steel–Concrete–Steel Sandwich Composite Beam Structures. *Mater. Struct. Mater. Constr.* **2015**, *48*, 1523–1544. [CrossRef]
33. Rahai, A.; Hatami, F. Evaluation of Composite Shear Wall Behavior under Cyclic Loadings. *J. Constr. Steel Res.* **2009**, *65*, 1528–1537. [CrossRef]
34. Li, Q.; He, Y.; Zhou, K.; Han, X.; He, Y.; Shu, Z. Structural Health Monitoring for a 600 m High Skyscraper. *Struct. Des. Tall Spec. Build.* **2018**, *27*, e1490. [CrossRef]
35. Wang, R.; Cao, W.L.; Yin, F.; Dong, H.Y. Experimental and Numerical Study Regarding a Fabricated CFST Frame Composite Wall Structure. *J. Constr. Steel Res.* **2019**, *162*, 105718. [CrossRef]
36. Tong, J.Z.; Yu, C.Q.; Zhang, L. Sectional Strength and Design of Double-Skin Composite Walls Withre-Entrant Profiled Faceplates. *Thin-Walled Struct.* **2021**, *158*, 107196. [CrossRef]
37. Zhao, Q.; Astaneh-Asl, A. Cyclic Behavior of Traditional and Innovative Composite Shear Walls. *J. Struct. Eng.* **2004**, *130*, 271–284. [CrossRef]
38. Luo, Y.; Guo, X.; Li, J.; Xiong, Z.; Meng, L.; Dong, N.; Zhang, J. Experimental Research on Seismic Behaviour of the Concrete-Filled Double-Steel-Plate Composite Wall. *Adv. Struct. Eng.* **2015**, *18*, 1845–1858. [CrossRef]
39. Huang, S.T.; Huang, Y.S.; He, A.; Tang, X.L.; Chen, Q.J.; Liu, X.; Cai, J. Experimental Study on Seismic Behaviour of an Innovative Composite Shear Wall. *J. Constr. Steel Res.* **2018**, *148*, 165–179. [CrossRef]
40. Nie, J.G.; Hu, H.S.; Fan, J.S.; Tao, M.X.; Li, S.Y.; Liu, F.J. Experimental Study on Seismic Behavior of High-Strength Concrete Filled Double-Steel-Plate Composite Walls. *J. Constr. Steel Res.* **2013**, *88*, 206–219. [CrossRef]
41. Najm, H.M.; Ibrahim, A.M.; Sabri, M.M.; Hassan, A.; Morkhade, S.; Mashaan, N.S.; Eldirderi, M.M.A.; Khedher, K.M. Modelling of Cyclic Load Behaviour of Smart Composite Steel-Concrete Shear Wall Using Finite Element Analysis. *Buildings* **2022**, *12*, 850. [CrossRef]

Article

Thermomechanical Behavior of Bone-Shaped SWCNT/Polyethylene Nanocomposites via Molecular Dynamics

Georgios I. Giannopoulos [1,*] and Stylianos K. Georgantzinos [2]

[1] Department of Mechanical Engineering, School of Engineering, University of Peloponnese, 1 Megalou Alexandrou Street, GR-26334 Patras, Greece
[2] Department of Aerospace Science and Technology, National and Kapodistrian University of Athens, GR-34400 Psachna Evias, Greece; sgeor@uoa.gr
* Correspondence: ggiannopoulos@uop.gr

Abstract: In the present study, the thermomechanical effects of adding a newly proposed nanoparticle within a polymer matrix such as polyethylene are being investigated. The nanoparticle is formed by a typical single-walled carbon nanotube (SWCNT) and two equivalent giant carbon fullerenes that are attached with the nanotube edges through covalent bonds. In this way, a bone-shaped nanofiber is developed that may offer enhanced thermomechanical characteristics when used as a polymer filler, due to each unique shape and chemical nature. The investigation is based on molecular dynamics simulations of the tensile stress–strain response of polymer nanocomposites under a variety of temperatures. The thermomechanical behavior of the bone-shaped nanofiber-reinforced polyethylene is compared with that of an equivalent nanocomposite filled with ordinary capped single-walled carbon nanotubes, in order to reach some coherent fundamental conclusions. The study focuses on the evaluation of some basic, temperature-dependent properties of the nanocomposite reinforced with these innovative bone-shaped allotropes of carbon.

Keywords: bone-shaped; fullerene; nanotube; polymer; nanocomposite; stress-strain

Citation: Giannopoulos, G.I.; Georgantzinos, S.K. Thermomechanical Behavior of Bone-Shaped SWCNT/Polyethylene Nanocomposites via Molecular Dynamics. *Materials* **2021**, *14*, 2192. https://doi.org/10.3390/ma14092192

Academic Editor: Franz Saija

Received: 8 April 2021
Accepted: 23 April 2021
Published: 24 April 2021

Publisher's Note: MDPI stays neutral with regard to jurisdictional claims in published maps and institutional affiliations.

1. Introduction

The majority of the nanocomposite (NC) problems and applications are typically related to combinations of loads instead of single types of loads. Perhaps the most common problem is the study of an NC under the simultaneous action of mechanical as well as thermal loadings. Nowadays, the research on nanomaterial reinforced composites that are subjected to thermomechanical loads is of great interest since it may provide valuable practical and efficient solutions in a variety of novel applications. Today, intensive research is carried for the production of polymer-based nanocomposites with special and enhanced thermal conductivity properties for use in thermal management systems [1,2] or, on the contrary, for the development of nanofilled polymers for thermal insulation applications from energy storage to power delivery [3]. Additional attention is paid in the field of structural applications where nanoreinforced polymers seem to be ideal candidates for high-temperature operation devices [4,5]. The accurate prediction of the thermomechanical properties especially of polymer-based NCs, which provide enhanced mechanical characteristics such as high strength-to-weight ratio, is of high importance. In this context, Burgaz [6] has investigated the current status of thermomechanical properties of polymer NCs containing nanofillers in the form of nanocylinders, nanospheres, and nanoplatelets, using case studies from the literature to highlight significant innovations and potential applications. In another interesting attempt, Reddy et al. [7] have discussed some of the recent developments in multiscale modeling of the thermal and mechanical properties of advanced NC systems by including relevant works from the literature to improve the theoretical background.

Theoretical approximations for analyzing the thermomechanical properties of NCs include molecular dynamics (MD) [8–16], molecular mechanics (MM) [17,18], and continuum mechanics (CM) [19] based methods. In addition, multi-scale numerical schemes have recently been proposed, which combine atomistic simulations such as MD or MM with other CM methods such as FEM in an effort to provide reliable predictions with low-computational cost [17,18,20]. Despite the fact that the MM and the CM formulations require significantly smaller computational efforts, the MD approaches seem to be more versatile and provide more accurate and reliable numerical solutions when investigating multiphase nanomaterial components in the nanoscale. This is due to the variety of potential models, force fields, algorithm choices, and simulation modulus that are available in most of the relevant commercial codes [21].

There are several recent attention-grabbing works associated with the MD simulation of the thermomechanical behavior of carbon nanomaterial reinforced polymers. In a relatively early effort, Cho and Yang [8] performed a parametric study to investigate the effects of composition variables on the thermic and mechanical properties of carbon nanotube (CNT)-reinforced NCs using MD simulations. Aiming at a different outcome, Liu et al. [9] adopted classical MD simulations to investigate the absorption and diffusion behavior of polyethylene (PE) chains on the surface of the side-wall of a CNT at different temperatures. Much later, Herasati et al. [10] investigated the effects of polymer chain branches, crystallinity, and CNT additives on the glass transition temperature of PE. In a characteristic attempt, Jeyranpour et al. [11] adopted MD to carry out a comparative study regarding the effects of fullerenes on the thermo-mechanical properties of a specialized resin epoxy. An extended study was performed by Pandey et al. [12] who focused on the study of viscoelastic, thermal, electrical, and mechanical properties of graphite flake-reinforced high-density PE composites. Adopting a different matrix material, Zhou et al. [13] conducted a comparative study to determine the effects of graphene and CNTs on the thermomechanical properties of asphalt binder using MD. On the other hand, Park et al. [14] investigated the thermomechanical characteristics of silica-mineralized nitrogen-doped CNT reinforced poly (methyl methacrylate) (PMMA) NCs for the first time by MD simulations. An interesting study was presented by Singh and Kumar [15] in which they examined the interfacial behavior of functionalized CNT/PE NC at different temperatures using MD simulations, utilizing the second-generation polymer consistent force field (PCFF). Finally, in a more recent attempt, Zhang et al. [16] investigated via MD simulations the thermomechanical properties of a NC consisting of weaved PE and CNT junctions.

At least at the micro-scale, it is well established that bone-shaped (BS) fibers may carry the load more effectively and provide higher fiber pull-out resistance because of the mechanical interlocking between the enlarged fiber ends and the matrix [22]. On the other hand, in a notable attempt, Xu et al. [23] presented a novel approach for the template synthesis of BS CNT nanomaterials. Taking into consideration the aforementioned facts, in the present work, the reinforcing ability of a recently presented nanofiber (NFB) [24], when used as filler in a polymeric matrix made of PE, is numerically analyzed via MD simulations at various temperature levels. Typically, the circular edges of single-walled carbon nanotubes (SWCNTs) are capped with fullerene hemispheres of the same diameter [25]. However, here, the SWCNT edges are enlarged by attaching to them giant spherical molecular formations, which are based on the atomistic structure of stable high-order carbon fullerenes [26]. The NC is tested by using a periodic unit cell that contains at its center this special carbonic single-walled molecular structure as reinforcement. The proposed BS NFB is surrounded by a number of PE chains composing the polymeric matrix phase. A uniform and periodic NFB dispersion is assumed at a rather high mass fraction of 20% in order to better unveil all the temperature-dependent reinforcing effects. The thermomechanical behavior of the NC is examined via the presentations of various temperature-dependent diagrams regarding its axial stiffness coefficients, tensile strength, and linear coefficient of thermal expansion. The influence of the temperature rise on the longitudinal and transverse tensile stress–strain behavior is also illustrated. At all times, for comparison reasons, the

BS NFB/PE NC under major investigation is set into contrast with an equivalent PE NC reinforced by an ordinary capped (OC) SWCNT of the same tubular diameter and total length. To the author's best knowledge, this is the first time that the effects of this BS NBF on the thermomechanical behavior of a polymer are being examined via MD or any other theoretical approach.

2. Primary Geometry and Density Assumptions

2.1. Structure of Single Molecules

Typically, the PE matrix phase is assumed to consist of polymeric chains of 100 monomers. The repeat unit of the PE chains is illustrated in Figure 1a. Figure 1b depicts the molecular structure and basic geometric characteristics of the investigated BS SWCNT, while Figure 1c shows the atomistic formation of the OC SWCNT, which is also tested for comparison reasons.

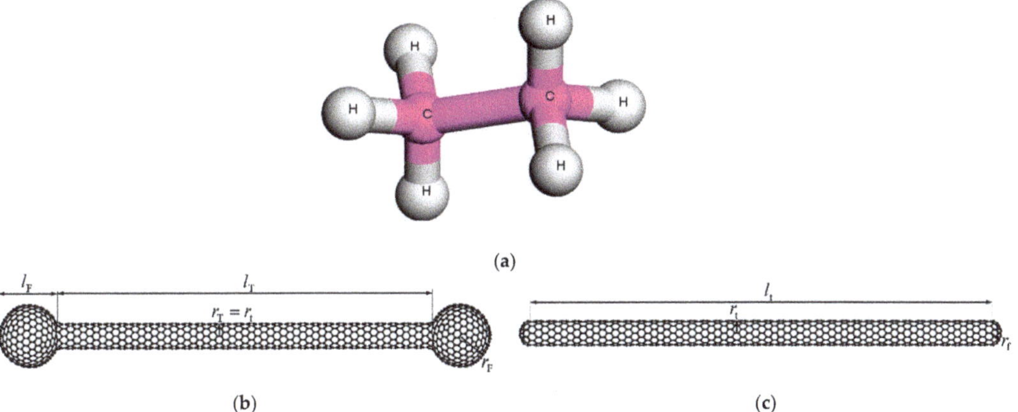

Figure 1. Atomistic structure of the: (**a**) repeat unit of the PE chain, (**b**) BS SWCNT, and (**c**) OC SWCNT.

The tubular shape of both NFBs is achieved by using the molecular structure of the zigzag (10,0) SWCNT, the radius of which is $r_t = 0.397$ nm [25]. The edges of the BS SWCNT are capped by using enlarged spherical segments based on the molecular structure of C_{500} fullerene [26]. The radius and the length of the spherical C_{500} fullerene-like segment are $r_F = 0.997$ nm [26] and $l_F = 1.912$ nm $< 2\,r_F$, respectively. The length of the (10,0) SWCNT-like tubular shape is $l_T = 12.58$ nm, leading to a total BS NFB length of 16.40 nm.

On the other hand, the edges of the OC SWCNT are formed by using the hemispherical molecular structure of C_{60} fullerene [25]. The radius of C_{60} hemisphere is obviously equal to $r_f = r_t = 0.397$ nm [25]. By selecting the specific nano-dimensions, the total length of the OC NFB becomes 16.30 nm, which is almost equal to the BS NFB total length. Thus, a comparison of the reinforcing ability between these two types of NFBs may be enabled.

It may be proved that the lattice area A_{NFB} of each NFB is given by:

$$A_{NFB} \approx \begin{cases} 2\pi r_t l_T + 2[(2\pi r_F^2 + 2\pi r_F(l_F - r_F)], & NFB = BS\ SWCNT \\ 2\pi r_t l_t + 2(2\pi r_f^2), & NFB = OC\ SWCNT \end{cases} \tag{1}$$

The total number of atoms N_{NFB} of the BS and OC NFB is 2130 and 1520, respectively. In addition, the wall thickness of both NFBs is assumed to be equal to the usual distance between two successive carbon layers in graphite, i.e., $t = 0.335$ nm. Given the specific wall thickness, the density of each NFB may be approximated by the following equation:

$$\rho_{NFB} = \frac{m_{NFB}}{A_{NFB}t} \tag{2}$$

where m_{NFB} is the mass of the NFB, which may easily be calculated by the relationship:

$$m_{\text{NFB}} = N_{\text{NFB}} m_{\text{C}} \tag{3}$$

where $m_{\text{C}} = 1.9927 \times 10^{-23}$ g is the mass of a carbon atom.

2.2. Unit Cells

The initial model domains are analyzed according to a global Cartesian coordinate system (x,y,z). In order to comprehensively examine the two-phase NC models, the pure PE amorphous material should be analyzed beforehand for each tested temperature level $T = 300, 325, 350, 375, 400$ K. It should be mentioned that all simulations are performed for temperatures higher than the glass transition temperature of polyethylene [10]. For all cases, it is assumed that the PE has an initial density equal to $^{\text{in}}\rho_{\text{PE}} = 0.6$ g/cm^3. According to this PE density value, by utilizing 10 PE polymer chains and by taking into account the molecular weight of each PE chain, a cubic unit cell of equal initial side lengths of $^{\text{in}}L_{\text{PE}x}$, $^{\text{in}}L_{\text{PE}y}$, and $^{\text{in}}L_{\text{PE}z}$ along the x-, y-, and z-axis is constructed [21]. It should be noticed that the use of more than 10 chains inside the unit cell had a negligible effect on the overall numerical outcome regarding the thermomechanical behavior of PE. After conducting the full MD procedure described in the following section, the final converged values of the PE unit cell density $^{\text{fi}}\rho_{\text{PE}}(T)$, the side lengths $^{\text{fi}}L_{\text{PE}x}(T) = {}^{\text{fi}}L_{\text{PE}y}(T) = {}^{\text{fi}}L_{\text{PE}z}(T)$, as well as the thermomechanical behavior of the PE at a given temperature T are estimated. The final equilibrated formation of an amorphous unit cell of the pure PE at 300 K is illustrated in Figure 2. The depicted vectors σ_{xx}, σ_{yy}, and σ_{zz} correspond to the normal stresses in the x, y, and z direction, while the vectors σ_{xy}, σ_{yz}, and σ_{zx} denote the shear stresses in the x-y, y-z, and z-x plane, respectively, required for the thermomechanical characterization of a given unit cell.

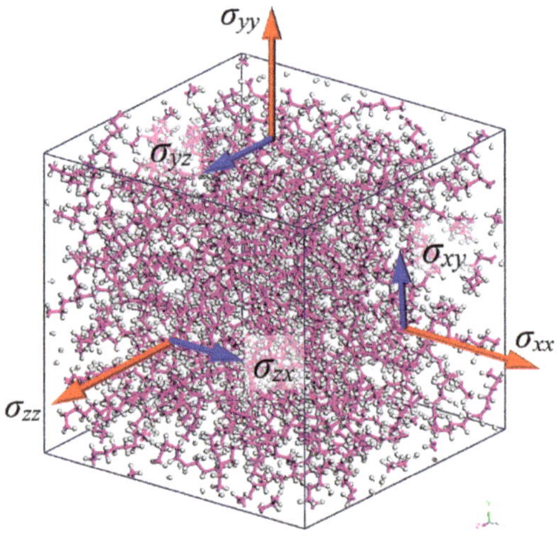

Figure 2. The equilibrated unit cell of the pure PE at $T = 300$ K.

The initial geometry of the NC unit cells is defined in a more complicated manner. First, both the BS and the OC SWCNT of Figure 1a,b, respectively, are kept constantly aligned with the x-axis. Moreover, their centroid is maintained at the center of the unit cells at all times. To assure an effective distribution of the reinforcements within the polymeric material, it is assumed that the longitudinal length of the NC unit cell $^{\text{in}}L_{\text{NC}x}(T)$ is six times higher than its transverse lengths $^{\text{in}}L_{\text{NC}y}(T)$ and $^{\text{in}}L_{\text{NC}z}(T)$, i.e., $^{\text{in}}L_{\text{NC}x}(T) = 6 \times {}^{\text{in}}L_{\text{NC}y}(T) = 6 \times {}^{\text{in}}L_{\text{NC}z}(T)$. This aspect ratio is kept stable at all times until the molecular structure of the final equilibrated unit cell is achieved, which means that in the final stage of the analysis there is $^{fi}L_{NCx}(T) = 6 \times {}^{fi}L_{NCy}(T) = 6 \times {}^{fi}L_{NCz}(T)$.

In order to enable packing [21] of the PE chains into each unit cell, an initial NC density $^{in}\rho_{NC}$ should be predefined. It is convenient to assume that the density of both NFBs ρ_{NFB} is negligibly affected within the temperature range from 300–400 K. Then, the initial density of the NC may be estimated by the following relationship:

$$^{in}\rho_{NC} = \frac{m_{NFB} + m_{PE}}{\dfrac{m_{NFB}}{\rho_{NFB}} + \dfrac{m_{PE}}{{}^{in}\rho_{PE}}} \tag{4}$$

where m_{PE} is the mass of the PE inside the unit cell.

The specific mass of the PE may be estimated via the following equation:

$$m_{PE} = m_{NFB}\frac{(100 - M_{NFB})}{M_{NFB}} \tag{5}$$

where M_{NFB} is the mass fraction of the NFB taken equal to 20% for all cases under consideration.

Evidently, in Equation (4), the initial density of the polymeric matrix component is taken equal to $^{in}\rho_{PE} = 0.6 \text{ g/cm}^3$, i.e., the initially assumed density for the construction of the pure PE unit cell.

It is easy to prove that the initial longitudinal and transverse lengths of the NC unit cell may be calculated by:

$$^{in}L_{NCx} = 6 \times {}^{in}L_{NCy} = 6 \times {}^{in}L_{NCz} = \sqrt[3]{36\left(\frac{m_{NFB}}{\rho_{NFB}} + \frac{m_{PE}}{{}^{in}\rho_{PE}}\right)} \tag{6}$$

Having the initial geometry of the NC fully defined, the MD formulation may be carried out in order to compute the final unit cell shape expressed by the lengths $^{fi}L_{NCx}(T)$, $^{fi}L_{NCy}(T)$, and $^{fi}L_{NCz}(T)$; the density $^{fi}\rho_{NC}(T)$; and the temperature-dependent mechanical behavior characterized by the corresponding stress–strain curves. A representative final equilibrated unit cell of the BS and OC SWCNT reinforced polymer is shown in Figure 3a,b, respectively.

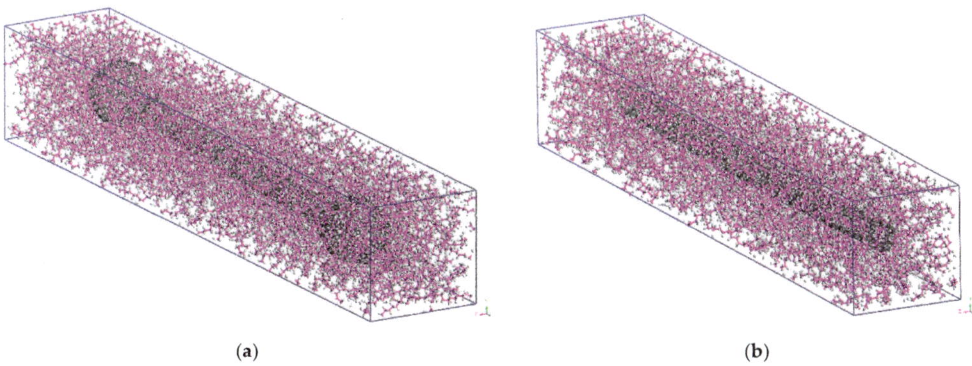

(a) (b)

Figure 3. The equilibrated unit cell of the (a) BS SWCNT and (b) OC SWCNT reinforced PE at a mass fraction of 20% and temperature $T = 300$ K.

3. MD Simulation

The full MD simulation procedure that is proposed here is divided into the following described stages, realized by using the "Materials Studio" software package (Version 2017).

3.1. Geometry Optimization of Single Molecular Structures

In the first stage, geometric optimization (GO) [21] is performed for each initially assumed molecular structure, i.e., the main PE chain as well as both NFBs, which are depicted in Figure 1. During the GO, energy minimization is achieved by using the steepest descent algorithm [21]. It is assumed that convergence is accomplished when the absolute difference of the computed system energy and force between two subsequent iterations becomes less than 0.001 Kcal/mol and 5 Kcal/mol/nm, respectively. The required numerical calculations are based on the Dreiding potential that contains four contributing terms for representing bond stretching, changes in bond angle, changes in dihedral rotation, and van der Waals non-bonded interactions. The total energy according to the Dreiding generic force field may be expressed as [27]:

$$
\begin{aligned}
U_{\text{total}} &= \sum_{\text{bond}} \left[\frac{1}{2} k_b (b - b_0)^2 \right] + \sum_{\text{angle}} \left[\frac{1}{2} k_\theta (\theta - \theta_0)^2 \right] \\
&+ \sum_{\text{dihedral}} \left[\sum_{n=1}^{4} k_n (\cos \varphi)^{n-1} \right] + \sum_{\text{nonbond}} \left\{ 4\varepsilon_0 \left[\left(\frac{\delta}{r_{ij}} \right)^{12} - \left(\frac{\delta}{r_{ij}} \right)^{6} \right] \right\}
\end{aligned}
\tag{7}
$$

In the last equation, the first three sums denote the energies required to stretch bonds from their equilibrium length b_0 to b, change bend angles from their equilibrium value θ_0 to θ, and twist atoms about their bond axis by an angle φ. The final sum, which contains functions of the atom pair distance r_{ij} denotes the Lennard-Jones-based van der Waals (vdW) non-bond interactions. The constant ε_0 and δ is the energy well depth and the zero-energy spacing of the Lennard-Jones potential, respectively. Depending on the atom type combinations, the Dreiding force-field predefines the stiffness-like parameters k_b, k_θ, and k_n [21,27]. It should be mentioned that, here, the vdW contributions are computed according to the atom-based summation method using a cut-off radius of 1 nm and long-range corrections [28].

After conducting the GO of the single BS SWCNT, some negligible cross-sectional asymmetries are revealed on the molecular structure of its edges. Characteristically, Figure 4 demonstrates the molecular configuration of the BS SWCNT after being geometrically optimized in the first stage of the MD analysis. Overall, it may be observed that the enlarged spherical edges obtain a 3d hexagon-like shape which, however, does not influence that provided by the NFB mechanical interlocking phenomena.

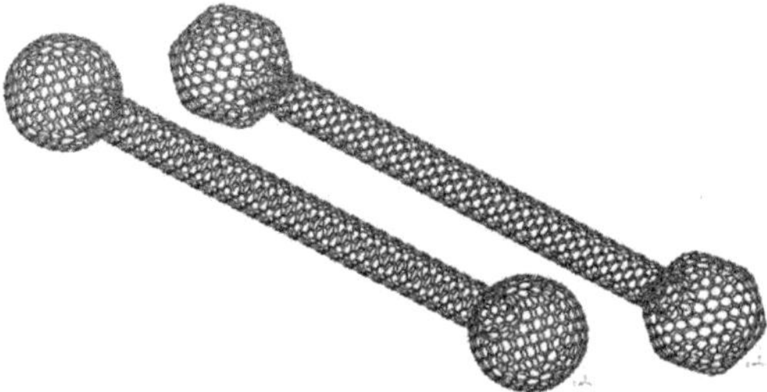

Figure 4. The BS SWCNT molecular structure before and after the GO with respect to the global Cartesian coordinate system (x,y,z).

3.2. Construction and Geometry Optimization of Unit Cells

In the second stage, the periodic unit cell representing the problem under consideration is constructed using standard packing algorithms available by commercial software packages [21] and by following the procedure that is analytically described in the previous section. After defining the three-dimensional (3d) unit cell box for the tested temperature T, a number of PE chains are inserted into it while the packing algorithm evenly increases their population until the initial unit cell density is achieved, i.e., $^{in}\rho_{PE} = 0.6 \, g/cm^3$ or $^{in}\rho_{NC}$ of Equation (4) when the pure matrix or the NC is to be analyzed, respectively. Evidently, the relevant positioning of the molecules is performed after computing the interactions between neighbor atoms via the Dreiding force field whereas the single-chain conformations, ring spearing, and close contacts are constantly monitored. To achieve a minimized initial unit cell state, low energy sites are preferred over high energy sites for each molecular structure. A GO process, like the one described in the first stage, is executed to additionally reduce the overall potential energy of the 3d problem domain.

3.3. Dynamic Analysis of Unit Cells

In the third stage, a three-phase dynamic analysis is performed for each investigated temperature T by using a time step of 1 fs in all cases. The followed MD numerical scheme including the utilized force field, the equilibrium algorithms, and the dynamic analysis computational process, is similar to the one proposed and validated by Bao et al. [29], in an effort to investigate amorphous PE under cyclic tensile-compressive loading below the glass transition temperature. Due to the dynamic nature of the simulation, in order to keep the molecular systems under a specific temperature and pressure level, the Andersen thermostat and Berendsen barostat are utilized, respectively [28]. Initially, an MD simulation takes place for a 50 ps time period under the NVT ensemble, assuming a temperature of 500 K. Then, the NPT ensemble is utilized for a 250 ps time interval to keep the temperature and the external pressure of the unit cell at 500 K and 1 atm, respectively. Finally, an NPT dynamic analysis is carried out to drop slowly the temperature from 500 K to $T = 300, 325, 350, 375,$ and 400 K by adopting 2 ps simulation time per 1 K of temperature decrease. After the finalization of the procedure, the relaxed equilibrium state, true final density, and side lengths of the unit cell are obtained. Performing an NPT dynamic analysis using even higher time intervals or a time step lower than 1 fs had no observable effect on the final numerical solutions.

3.4. Thermomechanical Properties Calculation

In the fourth and final stage of the process, the tensile stress–strain curves at a given temperature T are computed by applying to the 3d unit cell a set of quasi-static uniaxial tension and shear deformations. To avoid an extraordinary computational cost, the maximum chosen amplitude of strains is +0.1. The normal stresses, i.e., $\sigma_{xx}, \sigma_{yy},$ and σ_{zz}, and shear stresses, i.e., $\sigma_{xy}, \sigma_{yz},$ and σ_{zx}, (see Figure 2) at each strain level may be estimated through the following average virial stress of a system of particles [28]:

$$\overline{\sigma} = \frac{1}{2V} \sum_{j(\neq i)} (\mathbf{m}_{ij} \otimes \mathbf{u}_{ij} + \mathbf{r}_{ij} \otimes \mathbf{f}_{ij}) \tag{8}$$

where V is the volume of the system, i and j denote two particles at positions $\mathbf{r}_i$ and $\mathbf{r}_j$, respectively, $\mathbf{r}_{ij}$ is equal to $\mathbf{r}_i - \mathbf{r}_j$, $\mathbf{f}_{ij}$ is the inter-particle force applied on particle i by particle i, and $\mathbf{m}_{ij}$ and $\mathbf{u}_{ij}$ are the corresponding mass and velocity contributions.

In order to estimate the axial stiffness coefficients $E_{xx}, E_{yy},$ and E_{zz}, Hooke's law may be utilized in the three axial directions as:

$$\sigma_{xx} = E_{xx}\varepsilon_{xx}; \quad \sigma_{yy} = E_{yy}\varepsilon_{yy}; \quad \sigma_{zz} = E_{zz}\varepsilon_{zz} \tag{9}$$

where $\sigma_{xx}, \sigma_{yy},$ and σ_{zz} and $\varepsilon_{xx}, \varepsilon_{yy},$ and ε_{zz} are the tensional normal stresses and strains in the x-, y-, and z-axis, respectively.

Note that despite the amorphous nanostructure of the pure PE unit cell, it may be assumed that it has an almost isotropic behavior due to the high length accompanied by the random distribution of the polymer chains in the simulation box. Contrary, regarding the NC unit cells, significant anisotropy is present due to the NFB reinforcements.

Finally, note that the computation of the final unit cell size at the reference temperature T_0 and an arbitrary temperature T_1, permits the calculation of the coefficient of linear thermal expansion a_L. The specific thermal coefficient along the x-axis, with which the NFBs are aligned, can be approximated for the temperature range $[T_0,T_1]$ via the following equation:

$$a_{Lx} = \frac{{}^{\text{fi}}L_{\text{UC}x}(T_1) - {}^{\text{fi}}L_{\text{UC}x}(T_0)}{T_1 - T_0} \frac{1}{{}^{\text{fi}}L_{\text{UC}x}(T_0)} \tag{10}$$

where ${}^{\text{fi}}L_{\text{UC}x}$ is the final unit cell length along the x-axis, defined after the MD process is finalized.

4. Results and Discussion

For all the material cases under investigation, i.e., the BS SWCNT reinforced PE, the OC SWCNT reinforced PE, and, last but not least, the pure PE, simulations are conducted for five different temperatures, i.e., $T = 300, 325, 350, 375$, and 400 K. Furthermore, all the numerical tests regarding the NC materials correspond to a reinforcement (BS SWCNT of OC SWCNT) mass fraction M_{NFB} of 0.2 (20%). Finally, in order to reduce the computational cost without, however, excluding any key information from the numerical solutions, the maximum applied tensile strain is 10% for all cases.

To show the density variations of the BS SWCNT/PE NC, the OC SWCNT/PE NC, and the pure PE during the three-phase dynamic analysis, Figure 5a–c are given, respectively. Each figure shows the density variation of the corresponding unit cell with time, at all the tested temperatures. Generally, the analysis shows that convergence of the density values is achieved after a total MD simulation time of 450 ps. The final density values ${}^{\text{fi}}\rho(T)$, obtained after the three-phase dynamic analysis, correspond to the final points of the curves. These right endpoints of the curves in Figure 5, which define the final converged density at each temperature level, show that the pure PE density values, as expected, are lower than those of the NFB/PE NCs due to the absence of interphase interactions. Note that the use of larger unit cells, i.e., lower mass fractions, would not be so helpful in revealing the effects of the investigated BS NFB and, thus, is avoided here.

Figure 6 depicts the numerically computed final densities for all the tested materials and temperatures. The density of the pure PE varies between 0.83–0.80 as the temperature increases from 300 to 400 K, a prediction that is in good agreement with other computational and experimental estimations [30,31]. As it can be seen, all the density-temperature variations present an almost linear drop as the temperature increases. The OC SWCNT/PE NC seems to have slightly higher final density values in comparison with the BS SWCNT/PE NC at all temperature levels. However, the linear density decrease of both NC materials presents almost the same slope of decrease. Possibly, the lower density of the BS SWCNT reinforced PE is due to the larger lattice area that the BS NFB presents. The larger the NFB external area, the greater the interface region between the NFB and the matrix, which is characterized by the interlayer distance $t = 0.335$ nm.

Figure 7a,b illustrates the tensile and shear stress–strain behavior, respectively, of the pure amorphous PE material at various temperature levels. Note that the thermomechanical behavior of the PE, in the absence of an NFB reinforcement, is practically isotropic, and thus the same tensile and shear curve applies to all directions. The first peak in the tensile stress–strain curves corresponds to the tensile yield stress of the material. The tensile curve is characterized by a stress-softening region after the yield point. For even higher tensile deformations, PE presents a stress-hardening response [32,33] which, however, may not be illustrated for the rather small strains up to 10% that are investigated here. The tensile yield stress found here for $T = 300$ K is about 78.4 MPa and is in good agreement with the corresponding value of about 76.8 MPa found in an MD computational study based on Dreiding potential model [32]. A higher value of about 108.6 MPa is reported for a room temperature in another similar MD simulation [33]. In addition, an elastic modulus of 1.63 GPa and a Poisson's ratio of 0.37 are estimated here for the pure PE at 300 K, which are rather lower than the corresponding computed values of 1.32 GPa and 0.32, respectively, found elsewhere [33]. In another MD formulation in which the COMPASS force field has been used instead [34], an elastic modulus of 1.22 GPa and a Poisson's ratio of 0.37 have been proposed for a temperature of 298 K. For all the tested temperatures, the computed shear stress–strain curves are almost linear for stains up to 10% while their slop decreases in a linear manner as the temperature increases.

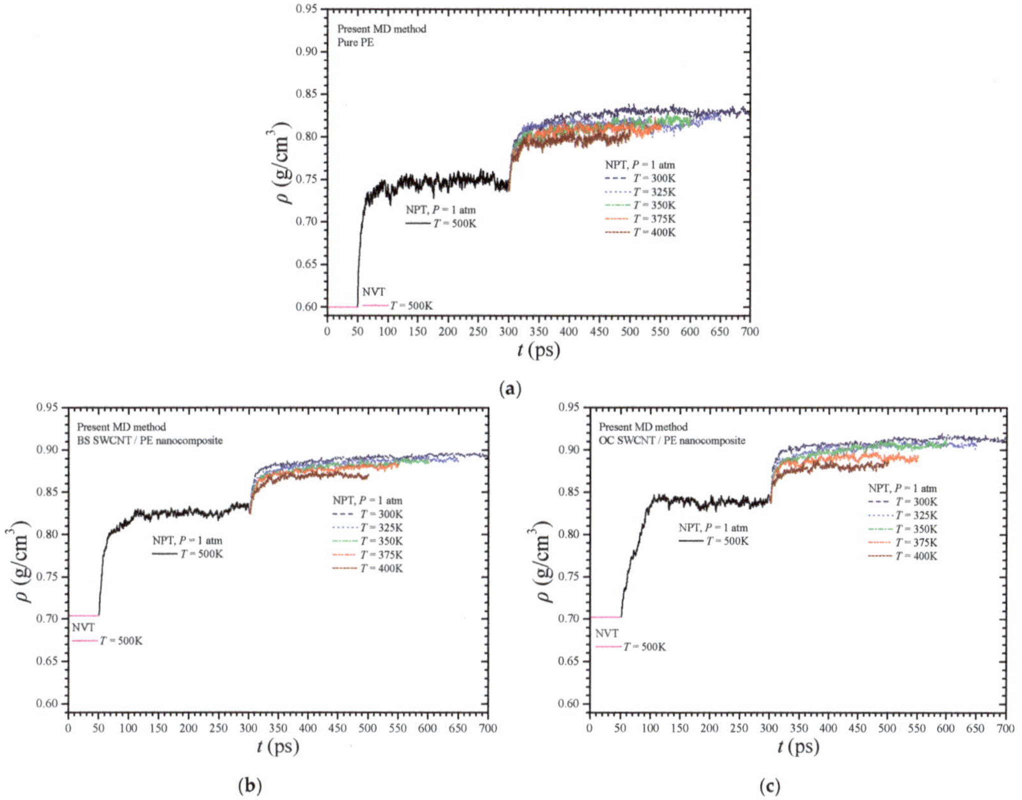

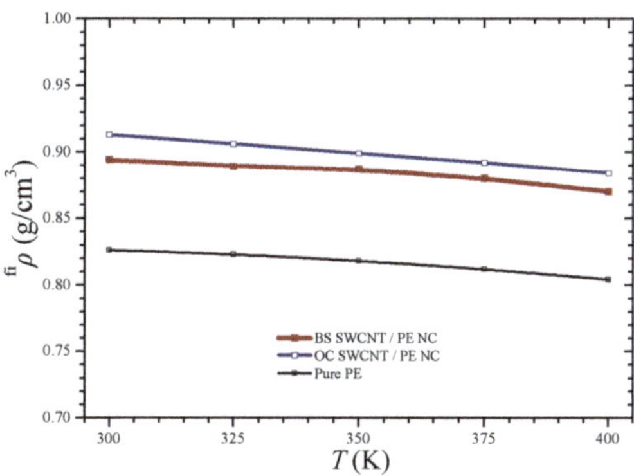

Figure 5. The density of the simulated (**a**) pure PE, (**b**) BS SWCNT-reinforced PE, and (**c**) OC SWCNT-reinforced PE versus time during the dynamic analysis and at various temperatures.

Figure 6. The final density of the simulated materials versus the temperature.

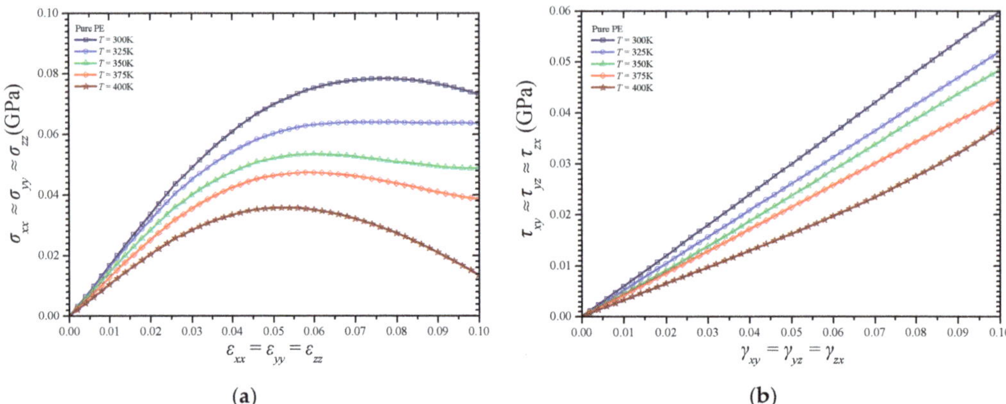

Figure 7. Stress–strain curves in (**a**) tension and (**b**) shear of the pure PE for various temperatures.

On the other hand, the computations showed that the NC materials present a distinct behavior along the longitudinal, i.e., effective, direction x in which the NFB is oriented, while they demonstrate an almost identical tensional response along the two transverse directions y and z because of the transverse cross-sectional symmetry of both NFBs. The tensile stress–strain temperature-dependent response of the BS and the OC SWCNT reinforced PE along the longitudinal x-axis and transverse y- or z-axis is illustrated in Figure 8a,b, respectively. In addition, Figure 9 depicts the variation of the zx shear stress versus the zx shear strain and temperature.

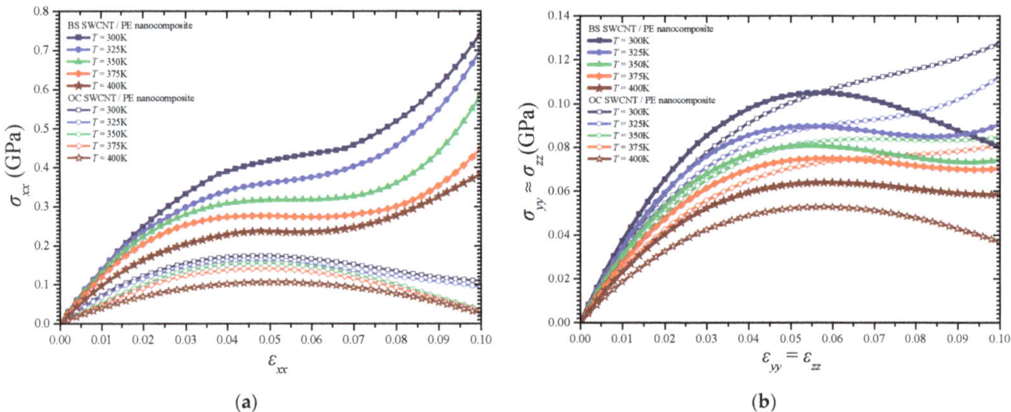

Figure 8. The tensile stress–strain curves of the simulated NCs in the (**a**) longitudinal x- and (**b**) transverse y- or z-axis for various temperatures.

The longitudinal direction x presents an upgraded temperature-dependent mechanical response compared with the other two directions. Both NFBs significantly improve the effective mechanical characteristics in the whole temperature range. However, it becomes obvious that the BS SWCNT/PE NC may carry a significantly higher maximum stress than the OC SWCNT/PE one, while it presents an advanced axial stiffness coefficient (tangent of the slope of the linear part of the tensile curves) in the longitudinal direction. This is due to the advanced geometric interlocking and load transfer mechanisms provided by the BS NFB mainly in the longitudinal direction. In addition, the special edge shape of the BS NFB seems to lead to an intense tensile stress hardening in the x-axis soon after the yield point is reached. Instead, there is no notable thermomechanical behavior enhancement provided by the BS SWCNT over the OC one, regarding the tensional yield stress in the transverse directions y and z. Furthermore, almost the same transverse elastic moduli increase may be observed by using both fibers in the PE matrix. Finally, according to the shear stresses–strain variations depicted in Figure 9, the BS NFB seems to offer a rather improved shear stiffness in the z-x plane for the whole temperature range under investigation.

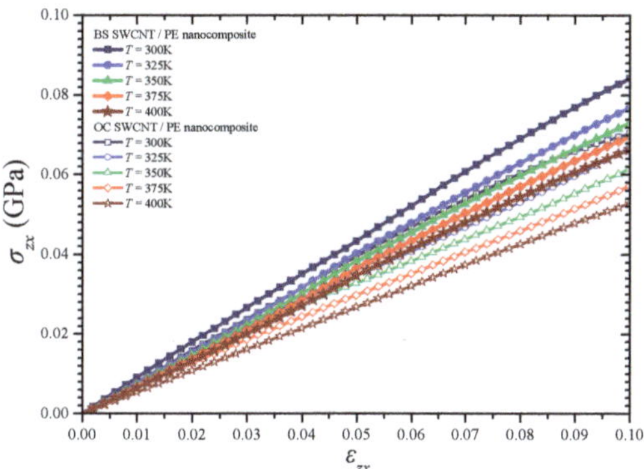

Figure 9. Shear stress–strain curves of the simulated NCs in the z-x plane for various temperatures.

To better demonstrate the enhancing ability of the BS proposed NFB, the key temperature-dependent axial properties arisen from Figures 7a and 8a are summarized and better analyzed in Figure 10.

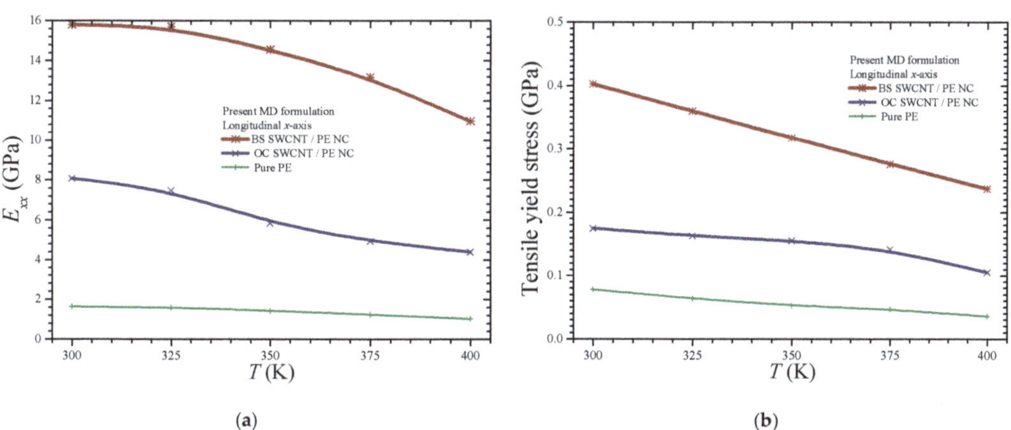

(a)

(b)

Figure 10. The temperature dependence of the longitudinal (**a**) stiffness coefficient and (**b**) tensile yield stress provided by the three materials.

Specifically, Figure 10a,b presents the effective axial stiffness coefficients and the tensile yield stresses of the three investigated materials, i.e., the pure PE, the BS, and OC SWCNT/PE NC, at a variety of temperature levels, respectively. A steady drop in the mechanical performance of all the materials as the temperature rises is observed in both figures. The positive influence of the BS NFB, due to the better 3d interlocking and stress transfer that is provided by its enlarged edges, on both the elastic and yield region, may be concluded by Figure 10a,b, respectively. Figure 10a proves that the BS NFB improves the longitudinal stiffness of the PE more effectively than the OC NFB in the whole investigated temperature range. A significantly higher longitudinal tensile yield stress is observed when the BS SWCNT is used as a reinforcing agent. Specifically, Figure 10b reveals that the BS SWCNT/PE NC may carry at least two times higher axial load than the OC SWCNT/PE NC independently of the temperature level.

The present numerical tests are carried out considering several simplifications regarding the NFBs such as uniform dispersion, perfect alignment, single-walled molecular structure, straight shape, and specific type and length for the CNT reinforcement. Thus, straightforward comparisons between the present results and corresponding experimental ones using the same NC design parameters may not be provided. Therefore, only a qualitative comparison is attempted with an experimental measurement regarding the elastic modulus of high-density PE (HDPE) reinforced with multi-walled CNTs (MWCNTs), having lengths of 10–30 μm and diameters of 5–15 nm, at a 10% volume fraction [35]. The specific experimentally tested volume fraction is comparable with the one investigated here. It should be mentioned at this point that by using the computed density of the pure PE at 300 K presented in Figure 5a and combining Equations (3) and (5), it may easily be proved that the OC SWCNT mass fraction of 20% corresponds to a volume fraction of about 8.6%. The reported experimental elastic modulus value at room temperature for the above-described MWCNT/HDPE NC is 7.86 GPa [35], which is in good agreement with the present numerical prediction of 8.09 GPa concerning the OC SWCNT/PE NC case.

Finally, the computed values of the linear coefficient of thermal expansion a_{Lx} along the x-axis for the three investigated materials are included in Table 1. The relevant average calculations for the three materials are based on Equation (10). For all cases, a linear increase of the longitudinal length of the unit cells is observed in the temperature range from 300–400 K. The computed linear coefficient of thermal expansion for the pure PE is in excellent agreement with corresponding reported experimental values that typically vary from 1.2×10^{-4} to 1.5×10^{-4} (1/K) [36]. According to the computed data, the linear coefficient of thermal expansion of the pure PE exhibits a notable increase when filled with both NFBs. However, the influence of the OC SWCNT is more significant, perhaps due to the fact that it leads to denser NC unit cell structures (Figure 6).

Table 1. The computed values of the linear coefficient of thermal expansion a_{Lx} for the three materials and the temperature range from 300–400 K.

Simulated Material	Average Computed Linear Coefficient of Thermal Expansion for x-axis a_{Lx} (1/K)
Pure PE	1.431×10^{-4}
OC SWCNT/PE nanocomposite	1.257×10^{-4}
BS SWCNT/PE nanocomposite	1.056×10^{-4}

Evidently, the extent of accuracy of the presented results is rather affected by the adopted theoretical assumptions as well as the inherent numerical restrictions of the adopted atomistic technique. For example, in the present work, the possible cross-linking phenomena are not considered while a rather high NC mass fraction is investigated in order to minimize the unit cell size and the complexity of molecular interactions and, thus, advance the convergence and the overall computational process. In addition, since the effect of the reinforcement on the NC material yielding becomes apparent by just applying strains up to 10%, the investigation of very high strains up to fracture is avoided. In this way, the computational cost is simultaneously reduced. In addition, the molecular simulations and their outcome are strongly dependent on the adopted potential field, cut-off distance, size of time-step, and equilibrium/convergence criteria. Nevertheless, despite the abovementioned limitations, the reliability of the research conclusions demonstrating the superiority of the proposed BS over the OC NFB, may be considered certain, given that the comparison between the two different reinforcements is realized by using the same theoretical fundamentals and computational options for the whole temperature range under investigation.

5. Conclusions

The present study has focused on the numerical prediction of the load-carrying capability of BS NBFs when utilized as reinforcements in polyethylene, under a variety of temperature levels. The computations have been grounded on the MD technique utilizing the Dreiding force field. The molecular structure of the proposed BS NFB has been based on the combination of a typical SWCNT and two giant carbonic fullerenes appropriately attached at the open SWCNT edges. An OC SWCNT of equivalent length and tubular diameter has been also tested in order to reach some distinct conclusions about the superiority of the BS NFB reinforcement regarding the provided temperature-dependent mechanical interlocking at the interphase. The numerical results and comparisons with the standard SWCNT NFB have shown that the proposed NFB increases more considerably both the axial stiffness as well as the tensile yield stress of the NC, especially in the longitudinal direction of the fiber for all the tested temperatures. In the near future, relevant research and relevant MD simulations at temperatures around the glass transition point of the polymeric matrix phase may reveal even more special features of the proposed BS carbonic nanomaterial.

Author Contributions: Conceptualization, G.I.G.; methodology, G.I.G. and S.K.G.; software, G.I.G. and S.K.G.; formal analysis, G.I.G. and S.K.G.; investigation, G.I.G. and S.K.G.; resources, G.I.G. and S.K.G.; data curation, G.I.G. and S.K.G.; writing—original draft preparation, G.I.G.; writing—review and editing, G.I.G. and S.K.G.; visualization, G.I.G. and S.K.G.; supervision, G.I.G. and S.K.G. All authors have read and agreed to the published version of the manuscript.

Funding: This research received no external funding.

Institutional Review Board Statement: Not applicable.

Informed Consent Statement: Not applicable.

Data Availability Statement: The data presented in this study are available on request from the corresponding author.

Conflicts of Interest: The authors declare no conflict of interest.

References

1. Zhang, X.; Zheng, J.; Du, Y.Q.; Wang, Z.W.; Wu, Y.D.; Luo, G. Graphene-polymer nanocomposites for thermal conductive applications. *IOP Conf. Ser. Mater. Sci. Eng.* **2020**, *770*, 012015. [CrossRef]
2. Chen, J.; Wei, H.; Bao, H.; Jiang, P.; Huang, X. Millefeuille-Inspired Thermally Conductive Polymer Nanocomposites with Overlapping BN Nanosheets for Thermal Management Applications. *ACS Appl. Mater. Interfaces* **2019**, *11*, 31402–31410. [CrossRef]
3. Ogbonna, V.E.; Popoola, A.P.I.; Popoola, O.M.; Adeosun, S.O. A review on polyimide reinforced nanocomposites for mechanical, thermal, and electrical insulation application: Challenges and recommendations for future improvement. *Polym. Bull.* **2020**, 1–33. [CrossRef]
4. Kumar, A.; Sharma, K.; Dixit, A.R. A review of the mechanical and thermal properties of graphene and its hybrid polymer nanocomposites for structural applications. *J. Mater. Sci.* **2019**, *54*, 5992–6026. [CrossRef]
5. Kianfar, A.; Seyyed Fakhrabadi, M.M.; Mashhadi, M.M. Prediction of mechanical and thermal properties of polymer nanocomposites reinforced by coiled carbon nanotubes for possible application as impact absorbent. *Proc. Inst. Mech. Eng. Part C J. Mech. Eng. Sci.* **2020**, *234*, 882–902. [CrossRef]
6. Burgaz, E. Thermomechanical analysis of polymer nanocomposites. In *Polymer Nanocomposites: Electrical and Thermal Properties*, 1st ed.; Huang, X., Zhi, C., Eds.; Springer: Cham, Switzerland, 2016; pp. 191–242.
7. Reddy, J.N.; Unnikrishnan, V.U.; Unnikrishnan, G.U. Recent developments in multiscale thermomechanical analysis of nanocomposites. In *Advances in Nanocomposites: Modeling, Characterization and Applications*, 1st ed.; Meguid, S.A., Ed.; Springer: Cham, Switzerland, 2016; pp. 177–188.
8. Cho, M.; Yang, S. Atomistic simulations for the thermal and mechanical properties of CNT-polymer nanocomposites. *Collect. Tech. Pap. AIAA/ASME/ASCE/AHS/ASC Struct. Struct. Dyn. Mater. Conf.* **2007**, *5*, 4718–4726.
9. Liu, J.; Wang, X.-L.; Zhao, L.; Zhang, G.; Lu, Z.-Y.; Li, Z.-S. The absorption and diffusion of polyethylene chains on the carbon nanotube: The molecular dynamics study. *J. Polym. Sci. Part B Polym. Phys.* **2007**, *46*, 272–280. [CrossRef]
10. Herasati, S.; Ruan, H.; Zhang, L.C. Effect of Chain Morphology and Carbon-Nanotube Additives on the Glass Transition Temperature of Polyethylene. *J. Nano Res.* **2013**, *23*, 16–23. [CrossRef]

11. Jeyranpour, F.; Alahyarizadeh, G.; Minuchehr, A. The thermo-mechanical properties estimation of fullerene-reinforced resin epoxy composites by molecular dynamics simulation—A comparative study. *Polymer* **2016**, *88*, 9–18. [CrossRef]
12. Pandey, A.K.; Singh, K.; Kar, K.K. Thermo-mechanical properties of graphite-reinforced high-density polyethylene composites and its structure–property corelationship. *J. Compos. Mater.* **2017**, *51*, 1769–1782. [CrossRef]
13. Zhou, X.; Zhang, X.; Xu, S.; Wu, S.; Liu, Q.; Fan, Z. Evaluation of thermo-mechanical properties of graphene/carbon-nanotubes modified asphalt with molecular simulation. *Mol. Simul.* **2016**, *43*, 312–319. [CrossRef]
14. Park, C.; Jung, J.; Yun, G.J. Thermomechanical properties of mineralized nitrogen-doped carbon nanotube/polymer nanocomposites by molecular dynamics simulations. *Compos. Part B Eng.* **2019**, *161*, 639–650. [CrossRef]
15. Singh, A.; Kumar, D. Temperature effects on the interfacial behavior of functionalized carbon nanotube–polyethylene nanocomposite using molecular dynamics simulation. *Proc. Inst. Mech. Eng. Part N J. Nanomater. Nanoeng. Nanosyst.* **2019**, *233*, 3–15. [CrossRef]
16. Zhang, B.; Li, J.; Gao, S.; Liu, W.; Liu, Z. Comparison of Thermomechanical Properties for Weaved Polyethylene and Its Nanocomposite Based on the CNT Junction by Molecular Dynamics Simulation. *J. Phys. Chem. C* **2019**, *123*, 19412–19420. [CrossRef]
17. Montazeri, A.; Rafii-Tabar, H. Multiscale modeling of graphene- and nanotube-based reinforced polymer nanocomposites. *Phys. Lett. Sect. A Gen. At. Solid State Phys.* **2011**, *375*, 4034–4040. [CrossRef]
18. Tsiamaki, A.; Anifantis, N. Simulation of the thermomechanical behavior of graphene/PMMA nanocomposites via continuum mechanics. *Int. J. Struct. Integr.* **2019**, *11*, 655–669. [CrossRef]
19. Fornes, T.D.; Paul, D.R. Modeling properties of nylon 6/clay nanocomposites using composite theories. *Polymer* **2003**, *44*, 4993–5013. [CrossRef]
20. Giannopoulos, G.I. Linking MD and FEM to predict the mechanical behaviour of fullerene reinforced nylon-12. *Compos. Part B Eng.* **2019**, *161*, 455–463. [CrossRef]
21. BIOVIA Materials Studio 2017. Available online: https://www.3ds.com/products-services/biovia/products/molecular-modeling-simulation/biovia-materials-studio/ (accessed on 29 March 2021).
22. Zhu, Y.T.; Beyerlein, I.J. Issues on the design of bone-shaped short fiber composites. *J. Adv. Mater.* **2003**, *35*, 51–60.
23. Xu, T.T.; Fisher, F.T.; Brinson, L.C.; Ruoff, R.S. Bone-shaped nanomaterials for nanocomposite applications. *Nano Lett.* **2003**, *3*, 1135–1139. [CrossRef]
24. Giannopoulos, G.I. Introducing bone-shaped carbon nanotubes to reinforce polymer nanocomposites: A molecular dynamics investigation. *Mater. Today Commun.* **2019**, *20*, 100570. [CrossRef]
25. Robinson, M.; Marks, N.A. NanoCap: A framework for generating capped carbon nanotubes and fullerenes. *Comput. Phys. Commun.* **2014**, *185*, 2519–2526. [CrossRef]
26. Terrones, M.; Terrones, G.; Terrones, H. Structure, chirality, and formation of giant icosahedral fullerenes and spherical graphitic onions. *Struct. Chem.* **2002**, *13*, 373–384. [CrossRef]
27. Mayo, S.L.; Olafson, B.D.; Goddard III, W.A. DREIDING: A generic force field for molecular simulations. *J. Phys. Chem.* **1990**, *94*, 8897–8909. [CrossRef]
28. Marx, D.; Hutter, J. *Ab Initio Molecular Dynamics: Basic Theory and Advanced Methods*, 1st ed.; Cambridge University Press: New York, NY, USA, 2009.
29. Bao, Q.; Yang, Z.; Lu, Z. Molecular dynamics simulation of amorphous polyethylene (PE) under cyclic tensile-compressive loading below the glass transition temperature. *Polymer* **2020**, *186*, 121968. [CrossRef]
30. Zhuang, X.; Zhou, S. Molecular dynamics study of an amorphous polyethylene/silica interface with shear tests. *Materials* **2018**, *11*, 929. [CrossRef]
31. Yi, P.; Locker, C.R.; Rutledge, G.C. Molecular dynamics simulation of homogeneous crystal nucleation in polyethylene. *Macromolecules* **2013**, *46*, 4723–4733. [CrossRef]
32. Tschopp, M.A.; Bouvard, J.L.; Ward, D.K.; Bammann, D.J.; Horstemeyer, M.F. Influence of Ensemble Boundary Conditions (Thermostat and Barostat) on the Deformation of Amorphous Polyethylene by Molecular Dynamics. *arXiv* **2013**, arXiv:1310.0728.
33. Vu-Bac, N.; Zhuang, X.; Rabczuk, T. Uncertainty quantification for mechanical properties of polyethylene based on fully atomistic model. *Materials* **2019**, *12*, 3613. [CrossRef] [PubMed]
34. Al-Ostaz, A.; Pal, G.; Mantena, P.R.; Cheng, A. Molecular dynamics simulation of SWCNT-polymer nanocomposite and its constituents. *J. Mater. Sci.* **2008**, *43*, 164–173. [CrossRef]
35. Arora, G.; Pathak, H. Experimental and numerical approach to study mechanical and fracture properties of high-density polyethylene carbon nanotubes composite. *Mater. Today Commun.* **2020**, *22*, 100829. [CrossRef]
36. Mwanang'onze, H.; Moore, I.D.; Green, M. Coefficient of Thermal Expansion Characterization for Plain Polyethylene Pipe. *Proc. ASCE Int. Conf. Pipeline Eng. Constr. N. Pipeline Technol. Secur. Saf.* **2003**, *2*, 1302–1311.

Article

Vibration Analysis of Carbon Fiber-Graphene-Reinforced Hybrid Polymer Composites Using Finite Element Techniques

Stelios K. Georgantzinos [1,*], **Georgios I. Giannopoulos** [2] and **Stylianos I. Markolefas** [3]

[1] Laboratory for Advanced Structures and Smart Systems, General Department, National and Kapodistrian University of Athens, 34400 Psachna, Greece

[2] Department of Mechanical Engineering, University of Peloponnese, 1 Megalou Alexandrou Street, 26334 Patras, Greece; ggiannopoulos@uop.gr

[3] General Department, National and Kapodistrian University of Athens, 34400 Psachna, Greece; stelmarkol@uoa.gr

* Correspondence: sgeor@uoa.gr

Received: 30 August 2020; Accepted: 21 September 2020; Published: 23 September 2020

Abstract: In this study, a computational procedure for the investigation of the vibration behavior of laminated composite structures, including graphene inclusions in the matrix, is developed. Concerning the size-dependent behavior of graphene, its mechanical properties are derived using nanoscopic empiric equations. Using the appropriate Halpin-Tsai models, the equivalent elastic constants of the graphene reinforced matrix are obtained. Then, the orthotropic mechanical properties of a composite lamina of carbon fibers and hybrid matrix can be evaluated. Considering a specific stacking sequence and various geometric configurations, carbon fiber-graphene-reinforced hybrid composite plates are modeled using conventional finite element techniques. Applying simply support or clamped boundary conditions, the vibrational behavior of the composite structures are finally extracted. Specifically, the modes of vibration for every configuration are derived, as well as the effect of graphene inclusions in the natural frequencies, is calculated. The higher the volume fraction of graphene in the matrix, the higher the natural frequency for every mode. Comparisons with other methods, where it is possible, are performed for the validation of the proposed method.

Keywords: composite structures; graphene; carbon fiber; hybrid matrix; vibrations; finite element analysis

---

1. Introduction

Composites are made up of two or more constituent materials that, when combined, generate a material with physical and chemical characteristics different from the individual components. Composite structures have presented outstanding advances in the last decades. The recent applications of fiber-reinforced composite structures to marine [1], civil [2], aerospace [3] and other industries [4–6] have already been reviewed in the open literature. Fast progress in manufacturing has resulted to the necessity for the improvement of composites in terms of strength, stiffness, density, and lower cost.

Graphene is a carbon allotrope, the thinnest known material, i.e., one carbon atom thick, which additionally is exceptionally strong—almost 200 times stronger than steel [7–9]. Furthermore, graphene presents outstanding electric [10,11], heat [12] and optical [13] properties. The excellent properties of graphene make it an excellent candidate as reinforcing material in composites. Consequently, many studies can be found in the literature investigating experimentally [14–17], computationally [18–20], or analytically [21,22] the effect of graphene in nanocomposites. The superb behavior of graphene nanocomposites has driven the research and technology for finding advanced applications. Chang and

Wu [23] have reported recent energy-related development of graphene nanocomposites in solar energy conversion, such as photovoltaic and photoelectrochemical devices and artificial photosynthesis, as well as electrochemical energy devices and improvements in environmental applications of functionalized graphene nanocomposites for the detection and removal of heavy metal ions, organic pollutants, gas and bacteria. Kumar et al. [24] have systematically examined the recent progress in illustrating the complexities of mechanical and thermal behavior of graphene- and modified graphene-based polymer nanocomposites. They also reported the potential applications, existing challenges, and potential perspectives regarding the multi-scale capabilities and promising advancements of the graphene family-based nanocomposites materials.

The fiber and matrix-dominant behavior of fiber-based polymer composites are crucial in many mechanical, aerospace, or structural applications. Pre-mixing the polymer matrix with nanoparticles could enhance the through-thickness or matrix-dominant properties. Rafiee et al. [25] employed several graphene-based nanomaterials, including graphene oxide and its thermally reduced version (rGO), graphene nanoplatelets, and multi-walled carbon nanotubes to modify the epoxy matrix and the surface of glass fibers. Unmodified and modified epoxy and fibers were used for fabricating multiscale glass fiber-reinforced composites, following the vacuum-assisted resin transfer molding process. The composites obtained combined improvements in both the fiber and matrix-dominant properties, resulting in superior composites. Kostagiannakopoulou et al. [26] developed a new class of carbon fiber-reinforced polymers with a nano-modified matrix, based on graphene nano-species. Their results have shown that nano-doped composites present a significant improvement of the interlaminar strain energy release rate of the order of 50% for the case of graphene nano-platelets. Taş and Soykok [27] have determined engineering constants of carbon nanotube based unidirectional carbon fiber-reinforced composite lamina theoretically with two different approaches. They have modeled using the finite element method and analyzed the behavior of a specific stacking sequence and configuration of a composite plate under static loadings. Their results have shown a considerable improvement of the plate stiffness due to the presence of nanotubes.

Although some works have already been done on vibrations of composite structures reinforced by nanoparticles, there are only a few works, which focused on vibration behavior of carbon fiber-based laminate composites with pure graphene inclusions. Hulun et al. [28] have investigated free vibration of graphene nanoplatelet (GPL) reinforced laminated composite quadrilateral plates using the element-free IMLS-Ritz method. They concluded that increasing GPLs weight fraction can increase the natural frequency of quadrilateral plate. Shen et al. have investigated nonlinear vibration of functionally graded graphene-reinforced composite laminated plates [29], functionally graded graphene-reinforced composite laminated cylindrical panels resting on elastic foundations [30], and functionally graded graphene-reinforced composite laminated cylindrical shells [31] in thermal environments. Wang et al. [32] have proposed a two-dimensional elasticity model of laminated graphene-reinforced composite (GRC) beams, where the graphene disperses uniformly in each layer, but the graphene volume fraction may vary from layer to layer. It is found that the laminated GRC beam with graphene distribution pattern X has the least deflection and highest fundamental frequency at extreme length-to-thickness ratios, but it has the highest deflection and least fundamental frequency at a very low length-to-thickness ratio due to its reduced transverse shear stiffness.

In the present paper, a computational approach is proposed for the estimation of vibration characteristics of carbon fiber laminate composite plates with graphene inclusions. The method firstly computes the size-dependent mechanical properties of graphene used in the composite. In the second phase, the orthotopic properties of the hybrid matrix are calculated adopting Halpin-Tsai models for specific volume fractions of graphene in the matrix. Knowing the elastic constants of hybrid matrix, the elastic mechanical properties of the composite lamina can be evaluated. In the next step, finite element models are developed for various configurations of laminated composite plates using conventional shell elements. Applying different boundary conditions and solving the free vibration problem, modes of vibration and corresponding natural frequencies are finally extracted. The effect of graphene volume fraction on vibration characteristics of various composite plates is determined. To the authors' best knowledge, it is the first time where this holistic and relatively simple approach (four-stage technique), taking into account the size-dependent behavior of graphene inclusions, is proposed for the vibration analysis of graphene-carbon fiber-reinforced composite structures.

2. Computational Approach

2.1. Problem Definition

Figure 1 depicts the composite plate considered to be analyzed. In Figure 1a, the finite element model of the graphene in nanoscale is illustrated. The information about the size-dependent mechanical performance of graphene derived from nanoscale is used for the determination of the elastic constants of hybrid matrix (Figure 1b) using Halpin-Tsai homogenization model. The elastic constants of the hybrid matrix are utilized in order to compute the orthotropic mechanical properties of the unidirectional composite lamina (Figure 1c). Then, the finite element modeling of a composite plate considering a specific stacking sequence (Figure 1d) can be performed for its vibration analysis.

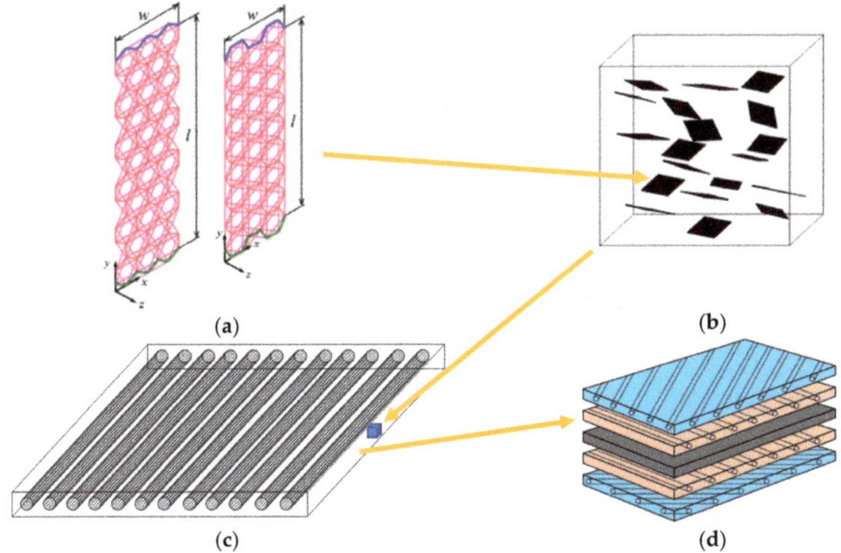

Figure 1. Modeling procedure (four-stage technique) of a laminated composite plate with graphene inclusions (**a**) finite element model of graphene in nanoscale; (**b**) homogenized model of hybrid matrix; (**c**) composite lamina with hybrid matrix; (**d**) composite plate with specific stacking sequence.

2.2. Graphene Mechanical Properties in Nanoscale

The elastic mechanical behaviour of different-sized graphene can be numerically examined and predicted using a spring-based finite element approach [7,33]. According to those approaches, three-dimensional, two-nodded, spring-based, finite elements of three degrees of freedom per node are combined in order to model effectively the interatomic interactions presented inside the graphene. The calculated variations of mechanical elastic constants have been approached by appropriate size-dependent non-linear functions of two independent variables, i.e., length and width for graphene, in order to express the analytical rules governing the elastic behaviour. The numerical results have been already validated through comparisons with corresponding data given in the open literature.

The Young's modulus in TPa of the graphene can be predicted using the following rational function [33]

$$E_{gr}\left(w_{gr}, l_{gr}\right) = \frac{P_0 + a_{01}w_{gr} + b_{01}l_{gr} + b_{02}l_{gr}^2 + c_{02}w_{gr}l_{gr}}{1 + a_1 w_{gr} + b_1 l_{gr} + a_2 w_{gr}^2 + b_2 l_{gr}^2 + c_2 w_{gr}l_{gr}}, 0 < w_{gr}, l_{gr} \leq 10 \text{ nm} \tag{1}$$

where $P_0, a_{01}, b_{01}, b_{02}, c_{02}, a_1, b_1, a_2, b_2,$ and c_2 are constants, which are determined by non-linear fitting, while w_{gr} and l_{gr} are the graphene width and length, respectively. The values of the constants of Equation (1), concerning both the two chirality directions of graphene are presented in Table 1. The accuracy of this equation compared to the finite element predictions and expressed by the coefficient of determination is calculated to be over than 99.5%.

Table 1. Fitting parameters of Equation (1) for the prediction of Young's modulus variation in TPa.

Direction	P_0	a_{01}	b_{01}	b_{02}	c_{02}	a_1	b_1	a_2	b_2	c_2
Zig zag	676	13640	−2532	58250	19170	11760	-4.7×10^{-4}	63.72	85900	17870
Armchair	2.98	3.03	13.55	0.893	−1.197	3.042	19.53	1.1487×10^{-1}	1.225	−1.634

2.3. Graphene/Polymer Matrix Physical Properties

The three-dimensional (3D) Halpin-Tsai model for randomly oriented discontinuous rectangular reinforcements is utilized to compute the Young's modulus of the hybrid graphene/polymer matrix. According to this model, the Young's modulus of the graphene/polymer matrix can be predicted by the following equation [34]

$$E_{m-gr} = \frac{1}{5}E_L + \frac{4}{5}E_T \tag{2}$$

where

$$E_L = \frac{1 + \xi_L \eta_L V_{gr}}{1 - \eta_L V_{gr}} E_m \tag{3}$$

$$E_T = \frac{1 + \xi_T \eta_T V_{gr}}{1 - \eta_T V_{gr}} E_m \tag{4}$$

and

$$\eta_L = \frac{\frac{E_{gr}}{E_m} - 1}{\frac{E_{gr}}{E_m} + \xi_L}, \quad \eta_T = \frac{\frac{E_{gr}}{E_m} - 1}{\frac{E_{gr}}{E_m} + \xi_T} \tag{5}$$

$$\xi_L = 2\frac{l_{gr}}{t_{gr}} + 40V_{gr}^{10}, \quad \xi_T = 2\frac{w_{gr}}{t_{gr}} + 40V_{gr}^{10} \tag{6}$$

where V_{gr} and t_{gr} stand for volume fraction and thickness of graphene, respectively. Furthermore, E_m is the Young's modulus of the polymer matrix, while $\xi_T, \xi_L, \eta_L,$ and η_T are Halpin-Tsai parameters, depending on the geometry of the reinforcement.

Assuming that Poisson's ratio v_m for both pure polymer matrix and hybrid matrix are approximately similar, it is calculated that Poisson's ratio of graphene/polymer matrix is

$$v_{m-gr} = v_m \tag{7}$$

Consequently, the shear modulus of the hybrid matrix exhibiting quasi-isotropic behaviour may be evaluated from the next equation

$$G_{m-gr} = \frac{E_{m-gr}}{2(v_{m-gr} + 1)} \tag{8}$$

Considering the inertia effect for the dynamic problem, the equivalent mass density of the hybrid matrix can be determined using rule of mixtures as

$$\rho_{m-gr} = V_{gr}\rho_{gr} + V_m\rho_m \tag{9}$$

where ρ_{gr}, ρ_m, and V_m are the mass density of graphene, mass density of matrix, and the volume fraction of the matrix, respectively. It is also known that

$$V_{gr} + V_m = 1 \tag{10}$$

2.4. Physical Properties of Composite Lamina

Considering a unidirectional lamina, its Young's modulus in longitudinal direction, E_1, and Poisson's ratio, v_{12} can be computed by employing the standard rule of mixture [35] as follows:

$$E_1 = E_{f11}\, V_f + E_{m-gr}V_{m-gr} \tag{11}$$

$$v_{12} = v_f V_f + v_{m-gr}V_{m-gr} \tag{12}$$

where E_{f11}, V_f, v_f and V_{m-gr} are Young's modulus of fiber along in longitudinal direction, volume fraction of fiber, Poisson's ratio of fiber and volume fraction of the hybrid matrix, respectively. The remaining orthotropic material properties, i.e., Young's modulus in transverse direction, E_2, Poisson's ratio, v_{23}, and shear moduli, G_{12} and G_{23}, can be computed by the next general Halpin-Tsai [36] expression

$$\frac{P}{P_{m-gr}} = \frac{1 + \xi\eta V_f}{1 - \eta V_f} \tag{13}$$

where P_{m-gr} are the related properties of hybrid matrix. Moreover, η is an experimental factor, which is computed by using the next equation,

$$\eta = \frac{\dfrac{P_f}{P_{m-gr}} - 1}{\dfrac{P_f}{P_{m-gr}} + \xi} \tag{14}$$

where P_f are the related properties of fiber. Notice that ξ is a measure of reinforcement geometry, which affected by loading conditions. Here, ξ is chosen as 2 for computation of E_2 and as 1 for computations of v_{23}, G_{12} and G_{23}.

Considering inertia effects, the equivalent mass density of the composite structure can be determined using rule of mixtures as

$$\rho_c = V_f\rho_f + V_{m-gr}\rho_{m-gr} \tag{15}$$

where ρ_f, ρ_{m-gr} and V_{m-gr} are the mass density of the fiber, mass density of the hybrid matrix, and volume fraction of the hybrid matrix, respectively. It is also known that

$$V_f + V_{m-gr} = 1 \tag{16}$$

2.5. Finite Element Modelling of the Composite Plate

In the present study, the finite element method is used for conducting free vibration analysis of the composite plates with graphene inclusion. For performing the vibration analysis, an appropriate mesh must be developed in the problem domain. Considering that the composite plate has a side-to-thickness ratio higher than 10, shell elements are an effective option for meshing the model. Here, 4-node, quadrilateral, stress/displacement shell elements with reduced integration and a large-strain formulation are used [37].

According to the ASTM D7264 standard, $w \times l \times t$, with $w, l > 10\,t$ rectangular composite plates are modelled, where w is the width, l is the length, and t is the thickness of the plate. The plates consist of 8 unidirectional layers. The thickness and the stacking sequence of all plates under consideration are 2 mm and $[0/+45/-45/90]s$, respectively. The model has been tested for its convergence in terms of mesh density and a mesh with a size of 0.5 mm for each element is finally chosen.

The hybrid composite plates of these characteristics, as well as different carbon nanostructure inclusion types and volume fractions, have been successfully validated [34] for bending loading conditions. Since the method is validated, it can be expanded for the solution of the free vibration problem. The elemental matrices and displacements are written in the global coordinate system by applying the appropriate transformation matrices. After generating the global stiffness matrix (K) and the global mass matrix (M), assembled from the elemental matrices using conventional finite element procedures and considering the undamped free vibrations of the composite plate, the equation of motion can be assembled as

$$M\ddot{d} + Kd = 0 \tag{17}$$

where d is the assembled displacement vector. By applying the boundary conditions on the composite plate, the eigenvalue problem can be solved using common finite element procedures, which reveal the natural frequencies and corresponding mode shapes of vibration.

3. Results and Discussion

In this work, polyester resin, graphene, and carbon fiber were selected as matrix material, nano inclusion and reinforcement, respectively. Physical properties of the components constituting the hybrid composite material are presented in Table 2. These properties are the basis for the estimation of the lamina physical properties presented in the next section.

Table 2. Physical properties of components of the composite material.

Carbon Fiber [38]		Polyester Resin [27]		Graphene (Equation (1))	
Ef_{11} (MPa)	230,000	E_m (MPa)	3000	E_{gr} (MPa)	735,000
Ef_{22} (MPa)	15,410	v_m	0.316	w_{gr} (nm)	10.02
vf_{23}	0.46	G_m (MPa)	1139.8	l_{gr} (nm)	9.97
vf_{12}	0.29	ρ_m (g/cm^3)	1.20	ρ_{gr} (g/cm^3)	2.26
Gf_{12} (MPa)	10,040				
Gf_{23} (MPa)	5287				
ρ_f (g/cm^3)	1.80				

3.1. Effect of Graphene on Composite Lamina Physical Properties

Given the physical properties of the components, the physical properties of graphene-added unidirectional carbon fiber-reinforced lamina are evaluated theoretically, based on the proposed method, and the outcome is provided in Table 3. Note that E_1, E_2, G_{12}, G_{23}, as well as ρ_c increase as the volume fraction of graphene increases. Specifically, there is an 34.6%, 542%, 533%, and 686% increase in E_1, E_2, G_{12}, and G_{23} with the addition of 50 vol.% graphene, respectively. Simultaneously, there is only 13.6% increase in the mass density of the lamina due to the presence of graphene since it is denser than polyester resin.

3.2. Vibrational Characteristics of the Composite Plate

A numerical calculation of the natural frequencies and mode shapes of vibration can be carried out for a composite plate, according to the procedure described in the previous section for different boundary conditions. Here, we assume two types of supports, (a) the longitudinal and transversal edges of the rectangular plate are clamped, CCCC (clamped—clamped—clamped—clamped), and (b) the longitudinal and transversal edges of the plate are simply supported SSSS (simple—simple—simple—simple). Figure 2 depicts several basic bending mode shapes of vibration of the rectangular plate. The general nomenclature for the mode shapes, here, is (*a,b*), where *a* and *b* correspond to the number of ridges or valleys existing in the mode at longitudinal and transverse direction of the plate. Note that Figure 2 depicts the transverse modes of the CCCC case; however, the general shape (*a,b*) exists in the SSSS case as well. The difference in the shape is mainly on the boundaries, where the rotation of the degrees of freedom are free and fixed in SSSS and CCCC boundary condition, respectively. The structure of the first three (basic) modes of vibration depends on the plate geometric and stiffness characteristics. In general, the presence of graphene into the composite seems not to influence the shape of the vibration modes; however, it affects their sequence.

Table 3. Physical properties of components of the composite material.

V_{gr}	E_1 (GPa)	E_2 (GPa)	v_{12}	v_{23}	G_{12} (GPa)	G_{23} (GPa)	ρ_c (g/cm^3)
0.00	139.20	7.799	0.300	0.397	4.924	3.493	1.560
0.05	142.21	13.280	0.300	0.397	8.451	5.732	1.581
0.10	145.47	16.673	0.300	0.397	10.589	7.466	1.602
0.15	149.04	19.828	0.300	0.397	12.552	9.233	1.624
0.20	152.95	23.073	0.300	0.397	14.559	11.124	1.645
0.25	157.25	26.537	0.300	0.397	16.693	13.182	1.666
0.30	162.00	30.305	0.300	0.397	19.010	15.445	1.687
0.35	167.29	34.453	0.300	0.397	21.558	17.953	1.708
0.40	173.20	39.064	0.300	0.397	24.388	20.752	1.730
0.45	179.86	44.237	0.300	0.397	27.562	23.901	1.751
0.50	187.42	50.094	0.300	0.397	31.154	27.473	1.772

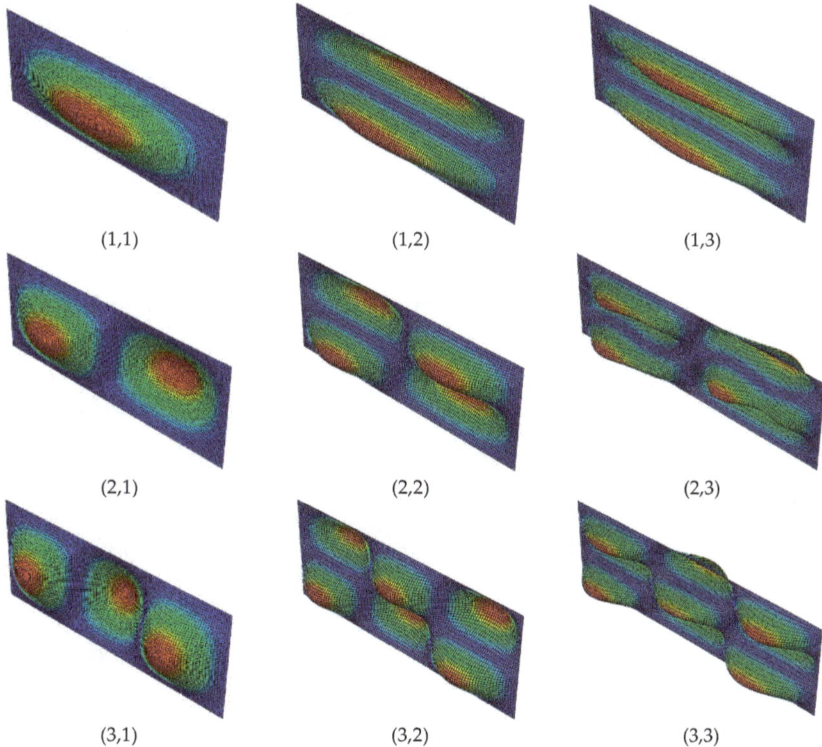

Figure 2. Mode shapes of vibration of a composite plate with $w = 25$ mm, $l = 64$ mm, and CCCC boundary conditions.

Figure 3 depicts the natural frequency, f_0, of the composite plate without the presence of graphene considering various configurations. Specifically, the width remains constant $w = 25$ mm, while the length is $l = 25, 30, 35, 45, 85, 145, 245$ mm. Figure 3a illustrates the natural frequency variation concerning CCCC boundary conditions. Figure 3b describes the natural frequency variation regarding SSSS boundary conditions. The natural frequencies of the CCCC case are greater than the corresponding ones of SSSS, as expected.

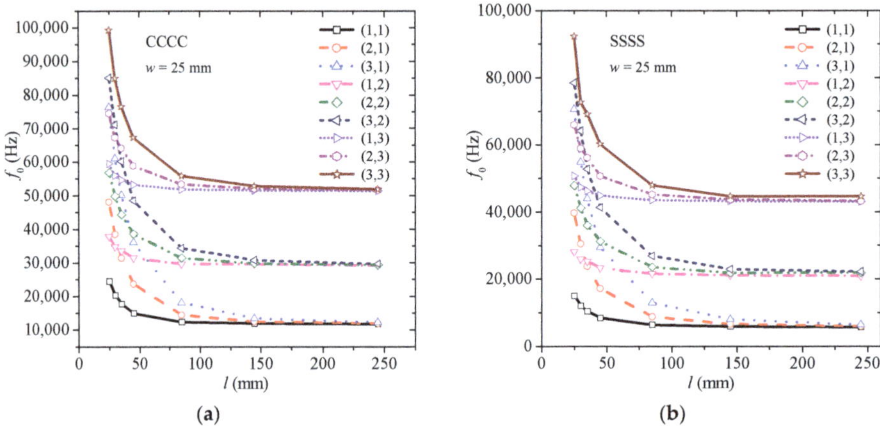

Figure 3. Natural frequency of the composite plate without the presence of graphene considering various configurations for (**a**) CCCC, and (**b**) SSSS boundary conditions.

3.3. Effect of Graphene on Natural Frequancies Assuming under CCCC Boundary Conditions

In order to study the effects of graphene inclusion on vibration behavior of laminated composite plates under CCCC boundary conditions, various values of volume fraction of graphene are studied. Specifically, eleven different cases are considered, i.e., $V_{gr} = 0, 0.05, 0.10, 0.15, 0.20, 0.25, 0.30, 0.35, 0.40, 0.45,$ and 0.50. Figure 4 illustrates the effect of graphene volume fraction on (1,1) (Figure 4a), (1,2) (Figure 4b), (2,1) (Figure 4c), and (2,2) (Figure 4d) transverse vibrations. As observed, the higher the volume fraction of graphene, the higher the frequency of graphene. This observation is true for all the different transverse modes. Specifically, there is an up to 73.2%, 82.7%, 72.6%, and 82.4% increase in (1,1), (1,2), (2,1), and (2,2) mode shape, respectively, with the addition of 50 vol.% graphene.

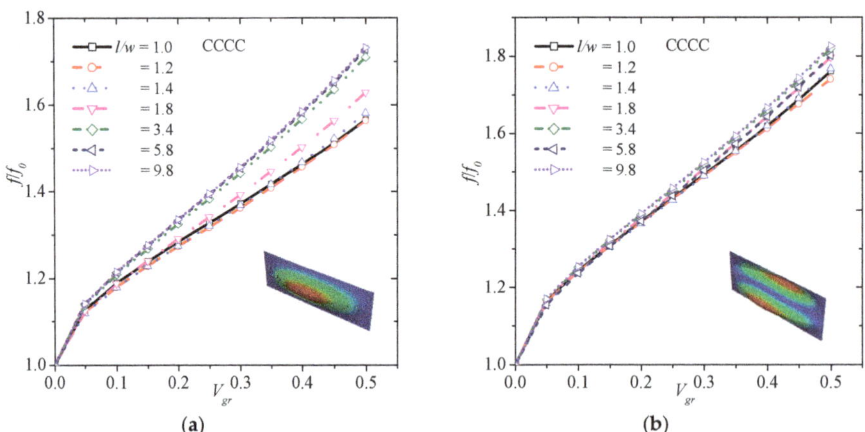

Figure 4. *Cont.*

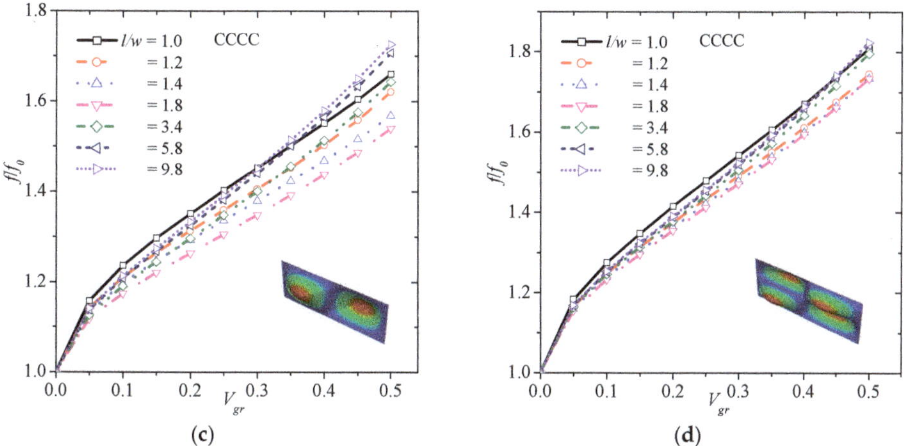

(c)

(d)

Figure 4. Normalized frequency versus volume fraction of graphene considering (**a**) (1,1), (**b**) (1,2), (**c**) (2,1), and (**d**) (2,2) vibration mode for the CCCC boundary condition.

Figure 5 shows the effect of graphene volume fraction on (1,3) (Figure 5a), (3,1) (Figure 5b), (2,3) (Figure 5c), (3,2) (Figure 5d), and (3,3) (Figure 5e) transverse vibrations. As described previously, the higher the volume fraction of graphene, the higher the natural frequency of graphene. This observation is true for all different transverse modes. Specifically, there is an up to 91.5%, 71.5%, 91.4%, 82% and 90.6% increase in (1,3), (3,1), (2,3), (3,2), and (3,3), respectively, with the addition of 50 vol.% graphene. It is noted that the graphene increases slightly more the normalized frequency of higher order modes than basic ones.

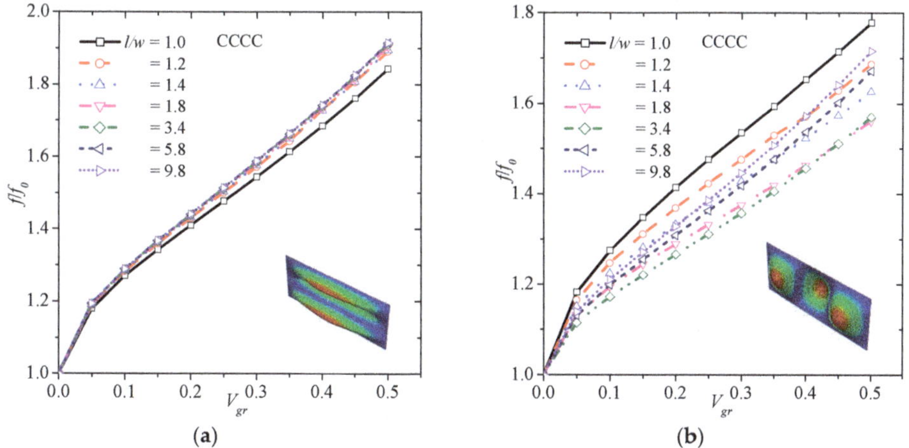

(a)

(b)

Figure 5. *Cont.*

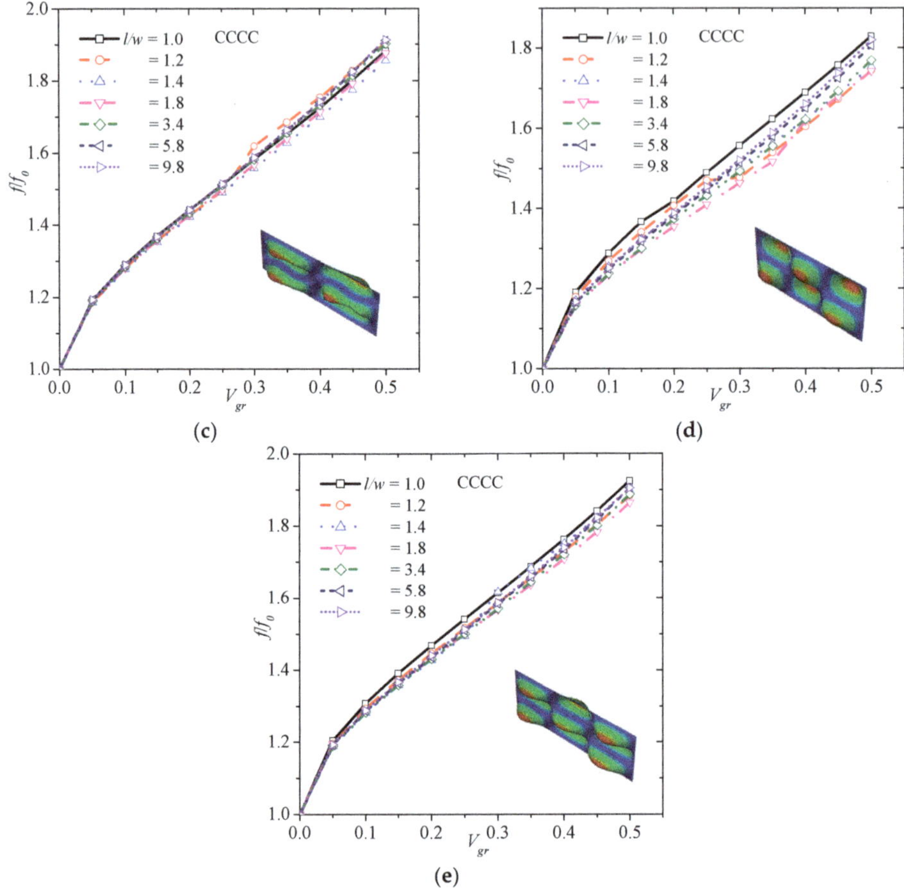

Figure 5. Normalized frequency versus volume fraction of graphene considering (**a**) (1,3), (**b**) (3,1), (**c**) (2,3), (**d**) (3,2), and (**e**) (3,3) vibration mode for CCCC boundary condition.

3.4. Effect of Graphene on Natural Frequancies Assuming SSSS Boundary Conditions

In order to study the effects of graphene inclusion on vibration behavior of laminated composite plates under SSSS boundary conditions, various values of volume fraction of graphene are considered.

Once again, 11 different cases are considered, i.e., V_{gr} = 0, 0.05, 0.10, 0.15, 0.20, 0.25, 0.30, 0.35, 0.40, 0.45, and 0.50. Figure 6 illustrates the effect of graphene volume fraction on (1,1) (Figure 6a), (1,2) (Figure 6b), (2,1) (Figure 6c), and (2,2) (Figure 6d) transverse vibrations. As observed, the higher the volume fraction of graphene, the higher the frequency of graphene for all different transverse modes. Specifically, there is an up to 64.9%, 60.9%, 63.4%, and 70.1% increase in (1,1), (1,2), (2,1), and (2,2) mode shape, respectively, with the addition of 50 vol.% graphene.

Figure 7 shows the effect of graphene volume fraction on (1,3) (Figure 7a), (3,1) (Figure 7b), (2,3) (Figure 7c), (3,2) (Figure 7d), and (3,3) (Figure 7e) transverse vibrations. As previously, the higher the volume fraction of graphene, the higher the frequency of graphene. This observation is true for all the different transverse modes. Specifically, there is an 79.3%, 60.9%, 79.2%, 69.1% and 80% increase in (1,3), (3,1), (2,3), (3,2), and (3,3), respectively, with the addition of 50 vol.% graphene. As in SSSS case, the graphene seems to increase slightly more the normalized frequency of higher order modes than basic ones.

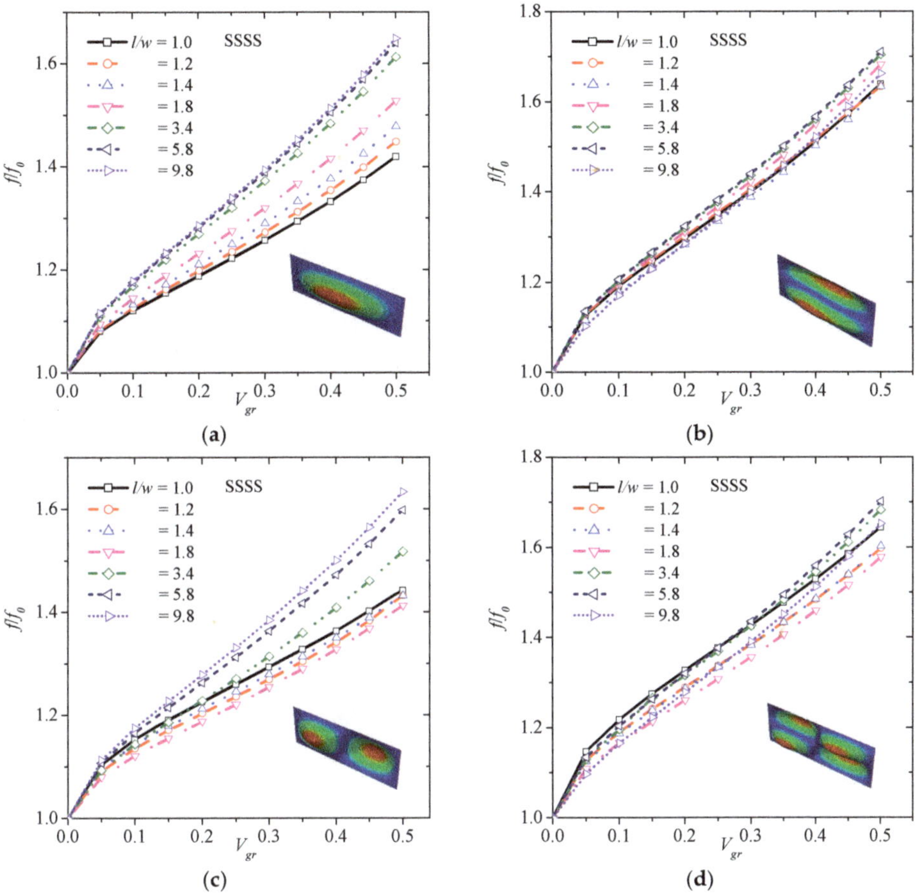

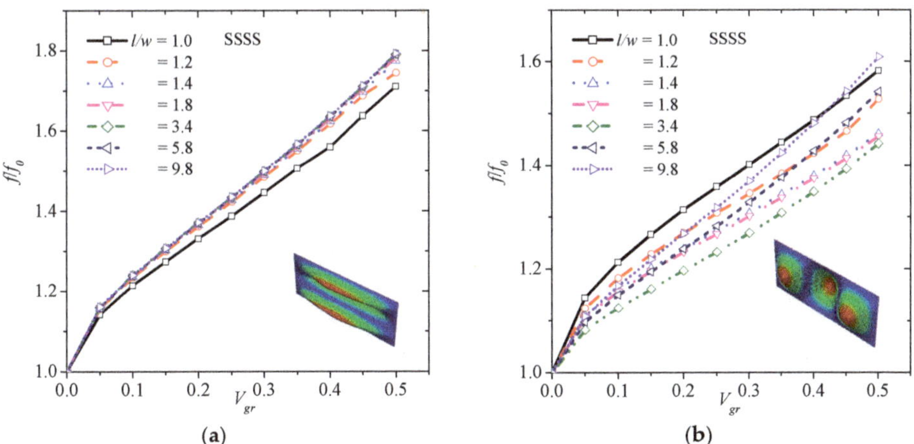

Figure 6. Normalized frequency versus volume fraction of graphene considering (**a**) (1,1), (**b**) (1,2), (**c**) (2,1), and (**d**) (2,2) vibration mode for SSSS boundary condition.

Figure 7. *Cont.*

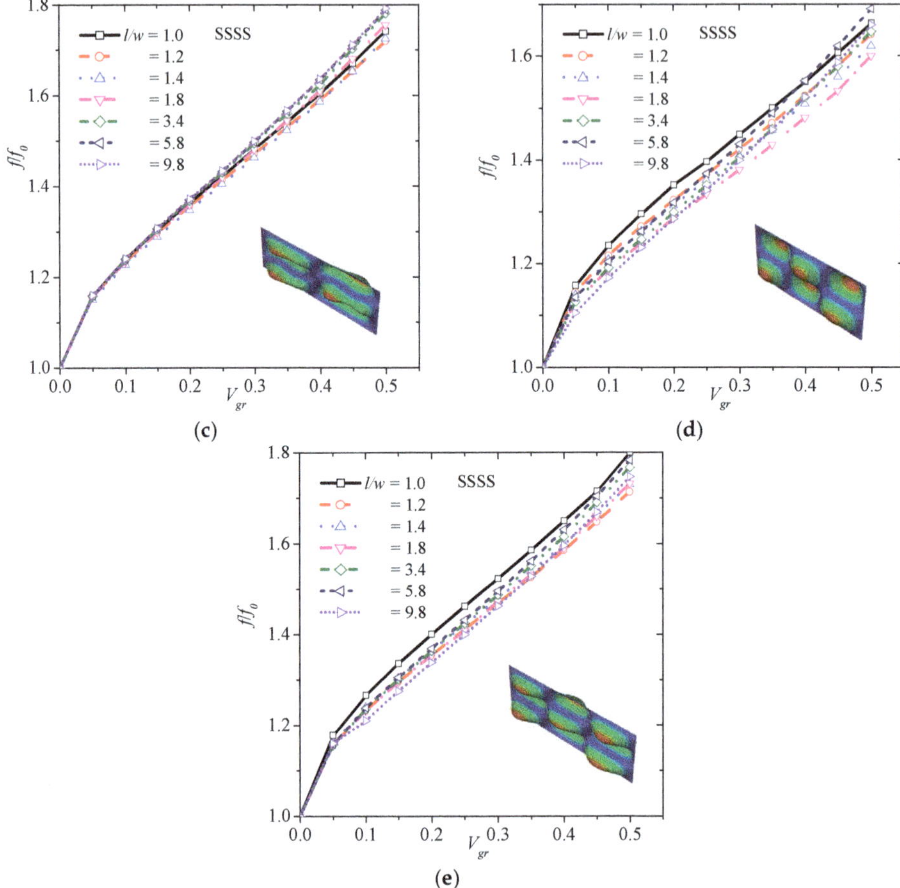

Figure 7. Normalized frequency versus volume fraction of graphene considering (**a**) (1,3), (**b**) (3,1), (**c**) (2,3), (**d**) (3,2), and (**e**) (3,3) vibration mode for the SSSS boundary condition.

In the previous cases (SSSS and CCCC), the behavior of the normalized frequency of various vibration mode versus the graphene volume fraction has been depicted concerning different L/w ratio. It is observed that the slope of this curve changes with different L/w ratio, and the trend is also different for the different vibration mode. The effect is more obvious in lower order than higher order modes. This phenomenon may be explained by the variation of the stiffness and mass (density) characteristics of the structure due to the presence of graphene that differs from each other (Table 3). Nevertheless, as is well known, mass and stiffness ratio directly affect the vibration frequency.

3.5. Average Effect of Graphene on Natural Frequencies

In order to estimate the average effect of graphene on the natural frequencies of the composite plate, the mean value of the normalized frequency is calculated for every graphene volume fraction considering all modes shape of vibration. Figure 8 depicts the average effect of graphene concerning both CCCC and SSSS boundary conditions. It is obvious that the effect of graphene is greater in CCCC than SSSS boundary conditions. For a rough estimation of the effect for V_{gr} greater than zero, a linear function can be used, i.e.,

$$\frac{f}{f_0} = cV_{gr} + d \tag{18}$$

where c is the slope, and d is the intercept. In the CCCC case, $c = 1.4164$ and $d = 1.0745$ with $R^2 = 0.9848$. In the SSSS case, $c = 1.1797$ and $d = 1.0533$ with $R^2 = 0.9885$.

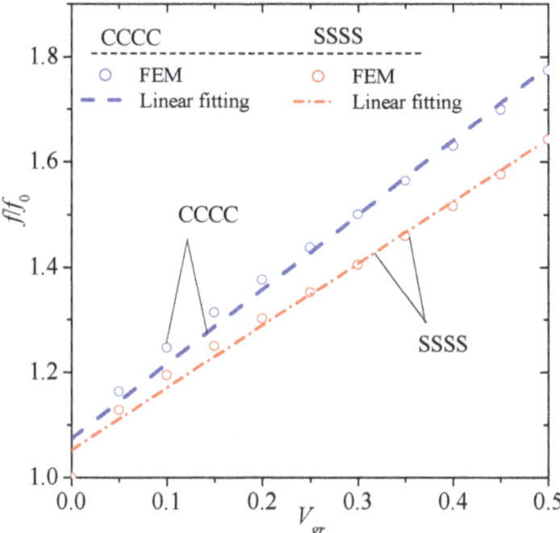

Figure 8. Average effect of graphene on natural frequency of the composite plate of graphene for CCCC, and SSSS boundary conditions.

4. Conclusions

A multi-scale procedure has been applied in the investigation of the vibration behavior of laminated composite plates reinforced by carbon fibers and graphene. According to analysis results, the following important general conclusions have been drawn:

- The presence of graphene enhanced considerably the orthotropic mechanical properties of the polymer composite lamina;
- The presence of graphene into the composite does not influence the shape of the vibration modes, however it affects their sequence;
- The presence of graphene into the composite increase significantly natural frequencies of the structure;
- The presence of graphene into the composite increase slightly more the normalized frequency of higher order than basic modes;
- The presence of graphene into the composite increase more natural frequencies of CCCC than SSSS boundary condition.

The dispersion of graphene particles in a polymer matrix is a big challenge, and there exists a practical restriction on the addition of graphene above a certain level (maximum of 10 wt.%), owing to the development of agglomerates [39]. Here, we considered a uniform distribution of the graphene in the polymer matrix concerning low- and high-volume fractions. Therefore, the present approach may be used as a design tool for low-volume fractions and be able to reveal the optimum vibrational performance for higher ones.

Future work concerns the attempt of introduction of a correction in the present computational approach considering the non-uniform distribution of nanoparticles in the polymer matrix.

Author Contributions: Conceptualization, S.G.; methodology, S.G.; software, S.G., G.G. and S.M.; validation, S.G., G.G. and S.M.; investigation, S.G.; writing—original draft preparation, S.G.; writing—review and editing, G.G. and S.M.; visualization, S.G., G.G. and S.M.; supervision, S.G.; All authors have read and agreed to the published version of the manuscript.

Funding: This research received no external funding

Conflicts of Interest: The authors declare no conflict of interest.

References

1. Rubino, F.; Nisticò, A.; Tucci, F.; Carlone, P. Marine Application of Fiber Reinforced Composites: A Review. *J. Mar. Sci. Eng.* **2020**, *8*, 26. [CrossRef]
2. Awad, Z.K.; Aravinthan, T.; Zhuge, Y.; Gonzalez, F. A Review of Optimization Techniques Used in the Design of Fibre Composite Structures for Civil Engineering Applications. *Mater. Des.* **2012**, *33*, 534–544. [CrossRef]
3. Ye, L.; Lu, Y.; Su, Z.; Meng, G. Functionalized Composite Structures for New Generation Airframes: A Review. *Compos. Sci. Technol.* **2005**, *65*, 1436–1446. [CrossRef]
4. Narayana, K.J.; Burela, R.G. A Review of Recent Research on Multifunctional Composite Materials and Structures with Their Applications. *Mater. Today-Proc.* **2018**, *5*, 5580–5590. [CrossRef]
5. Mohammed, L.; Ansari, M.N.; Pua, G.; Jawaid, M.; Islam, M.S. A Review on Natural Fiber Reinforced Polymer Composite and Its Applications. *Int. J. Polym. Sci.* **2015**, *2015*, 243947. [CrossRef]
6. Rajak, D.K.; Pagar, D.D.; Menezes, P.L.; Linul, E. Fiber-Reinforced Polymer Composites: Manufacturing, Properties, and Applications. *Polymers* **2019**, *11*, 1667. [CrossRef]
7. Georgantzinos, S.K. A New Finite Element for An Efficient Mechanical Analysis of Graphene Structures Using Computer Aided Design/Computer Aided Engineering Techniques. *J. Comput. Theor. Nanosci.* **2017**, *14*, 5347–5354. [CrossRef]
8. Cao, G. Atomistic Studies of Mechanical Properties of Graphene. *Polymers* **2014**, *6*, 2404–2432. [CrossRef]
9. Georgantzinos, S.K.; Katsareas, D.E.; Anifantis, N.K. Graphene Characterization: A Fully Non-Linear Spring-Based Finite Element Prediction. *Physica E Low Dimens. Syst. Nanostruct.* **2011**, *43*, 1833–1839. [CrossRef]
10. Lee, J.-H.; Park, S.-J.; Choi, J.-W. Electrical Property of Graphene and Its Application to Electrochemical Biosensing. *Nanomaterials* **2019**, *9*, 297. [CrossRef]
11. Georgantzinos, S.K.; Giannopoulos, G.I.; Fatsis, A.; Vlachakis, N.V. Analytical Expressions for Electrostatics of Graphene Structures. *Physica E Low Dimens. Syst. Nanostruct.* **2016**, *84*, 27–36. [CrossRef]
12. Prasher, R. Graphene Spreads the Heat. *Science* **2010**, *328*, 185–186. [CrossRef] [PubMed]
13. Falkovsky, L.A. Optical Properties of Graphene. *J. Phys. Conf. Ser.* **2008**, *129*, 012004. [CrossRef]
14. Liu, J.; Khan, U.; Coleman, J.; Fernandez, B.; Rodriguez, P.; Naher, S.; Brabazon, D. Graphene Oxide and Graphene Nanosheet Reinforced Aluminium Matrix Composites: Powder Synthesis and Prepared Composite Characteristics. *Mater. Des.* **2016**, *94*, 87–94. [CrossRef]
15. Li, G.; Xiong, B. Effects of Graphene Content on Microstructures and Tensile Property of Graphene-Nanosheets/Aluminum Composites. *J. Alloys Compd.* **2017**, *697*, 31–36. [CrossRef]
16. Abbasi, H.; Antunes, M.; Velasco, J.I. Effects of Carbon Nanotubes/Graphene Nanoplatelets Hybrid Systems on the Structure and Properties of Polyetherimide-Based Foams. *Polymers* **2018**, *10*, 348. [CrossRef] [PubMed]
17. Irez, A.B.; Bayraktar, E.; Miskioglu, I. Fracture Toughness Analysis of Epoxy-Recycled Rubber-Based Composite Reinforced with Graphene Nanoplatelets for Structural Applications in Automotive and Aeronautics. *Polymers* **2020**, *12*, 448. [CrossRef]
18. Spanos, K.N.; Georgantzinos, S.K.; Anifantis, N.K. Mechanical Properties of Graphene Nanocomposites: A Multiscale Finite Element Prediction. *Compos. Struct.* **2015**, *132*, 536–544. [CrossRef]
19. Lin, F.; Xiang, Y.; Shen, H.S. Temperature Dependent Mechanical Properties of Graphene Reinforced Polymer Nanocomposites–A Molecular Dynamics Simulation. *Compos. Part B Eng.* **2017**, *111*, 261–269. [CrossRef]
20. Montazeri, A.; Rafii-Tabar, H. Multiscale Modeling of Graphene-And Nanotube-Based Reinforced Polymer Nanocomposites. *Phys. Lett. A* **2011**, *375*, 4034–4040. [CrossRef]
21. Heidarhaei, M.; Shariati, M.; Eipakchi, H.R. Analytical Investigation of Interfacial Debonding In Graphene-Reinforced Polymer Nanocomposites with Cohesive Zone Interface. *Mech. Adv. Mater. Struct.* **2019**, *26*, 1008–1017. [CrossRef]
22. Roach, C.C.; Lu, Y.C. Analytical Modeling of Effect of Interlayer on Effective Moduli of Layered Graphene-Polymer Nanocomposites. *J. Mater. Sci. Technol.* **2017**, *33*, 827–833. [CrossRef]
23. Heidarhaei, M.; Shariati, M.; Eipakchi, H. Experimental and Analytical Investigations of the Tensile Behavior of Graphene-Reinforced Polymer Nanocomposites. *Mech. Adv. Mater. Struct.* **2018**. [CrossRef]
24. Kumar, A.; Sharma, K.; Dixit, A.R. A Review of the Mechanical and Thermal Properties of Graphene and Its Hybrid Polymer Nanocomposites for Structural Applications. *J. Mater. Sci.* **2019**, *54*, 5992–6026. [CrossRef]

25. Rafiee, M.; Nitzsche, F.; Laliberte, J.; Thibault, J.; Labrosse, M.R. Simultaneous Reinforcement of Matrix and Fibers for Enhancement of Mechanical Properties of Graphene-Modified Laminated Composites. *Polym. Compos.* **2019**, *40*, E1732–E1745. [CrossRef]

26. Kostagiannakopoulou, C.; Loutas, T.H.; Sotiriadis, G.; Markou, A.; Kostopoulos, V. On the Interlaminar Fracture Toughness of Carbon Fiber Composites Enhanced with Graphene Nano-Species. *Compos. Sci. Technol.* **2015**, *118*, 217–225. [CrossRef]

27. Taş, H.; Soykok, I.F. Effects of Carbon Nanotube Inclusion into the Carbon Fiber Reinforced Laminated Composites on Flexural Stiffness: A Numerical and Theoretical Study. *Compos. Part B Eng.* **2019**, *159*, 44–52. [CrossRef]

28. Guo, H.; Cao, S.; Yang, T.; Chen, Y. Vibration of laminated composite quadrilateral plates reinforced with graphene nanoplatelets using the element-free IMLS-Ritz method. *Int. J. Mech. Sci.* **2018**, *142*, 610–621. [CrossRef]

29. Shen, H.S.; Xiang, Y.; Lin, F. Nonlinear vibration of functionally graded graphene-reinforced composite laminated plates in thermal environments. *Comput. Methods Appl. Mech. Eng.* **2017**, *319*, 175–193. [CrossRef]

30. Shen, H.S.; Xiang, Y.; Fan, Y.; Hui, D. Nonlinear vibration of functionally graded graphene-reinforced composite laminated cylindrical panels resting on elastic foundations in thermal environments. *Compos. Part B Eng.* **2018**, *136*, 177–186. [CrossRef]

31. Shen, H.S.; Xiang, Y.; Fan, Y. Nonlinear vibration of functionally graded graphene-reinforced composite laminated cylindrical shells in thermal environments. *Compos. Struct.* **2017**, *182*, 447–456. [CrossRef]

32. Wang, M.; Xu, Y.G.; Qiao, P.; Li, Z.M. A two-dimensional elasticity model for bending and free vibration analysis of laminated graphene-reinforced composite beams. *Compos. Struct.* **2019**, *211*, 364–375. [CrossRef]

33. Giannopoulos, G.I.; Liosatos, I.A.; Moukanidis, A.K. Parametric Study of Elastic Mechanical Properties of Graphene Nanoribbons by a New Structural Mechanics Approach. *Physica E Low Dimens. Syst. Nanostruct.* **2011**, *44*, 124–134. [CrossRef]

34. Georgantzinos, S.K.; Markolefas, S.I.; Mavrommatis, S.A.; Stamoulis, K.P. Finite element modelling of carbon fiber-carbon nanostructure-polymer hybrid composite structures. *MATEC Web Conf.* **2020**, *314*, 02004. [CrossRef]

35. Gibson, R. *Mechanics of Composite Materials*, 4th ed.; CRC press: Boca Raton, FL, USA, 2016; pp. 107–112.

36. Affdl, J.H.; Kardos, J.L. The Halpin-Tsai equations: A review. *Polym. Eng. Sci.* **1976**, *16*, 344–352. [CrossRef]

37. Hibbitt, H.D.; Karlsson, B.I.; Sorensen, P. *ABAQUS Theory Manual, Version 6*; ABAQUS Inc.: Pawtucket, RI, USA, 2006.

38. Kara, Y.; Akbulut, H. Mechanical behavior of helical springs made of carbon nanotube additive epoxy composite reinforced with carbon fiber. *J. Fac. Eng. Archit. Gaz.* **2017**, *32*, 417–427.

39. Mohan, V.B.; Lau, K.T.; Hui, D.; Bhattacharyya, D. Graphene-based materials and their composites: A review on production, applications and product limitations. *Compos. Part B Eng.* **2018**, *142*, 200–220. [CrossRef]

Article

Thermomechanical Response of Fullerene-Reinforced Polymers by Coupling MD and FEM

Georgios I. Giannopoulos [1,*] , Stelios K. Georgantzinos [2] and Nick K. Anifantis [1]

[1] Department of Mechanical Engineering and Aeronautics, School of Engineering, University of Patras, GR-26500 Patras, Greece; nanifantis@upatras.gr
[2] Laboratory for Advanced Structures and Smart Systems, General Department, National and Kapodistrian University of Athens, 15772 Psachna, Greece; sgeor@uoa.gr
* Correspondence: giannopoulos@upatras.gr

Received: 30 August 2020; Accepted: 14 September 2020; Published: 17 September 2020

Abstract: The aim of the present study is to provide a computationally efficient and reliable hybrid numerical formulation capable of characterizing the thermomechanical behavior of nanocomposites, which is based on the combination of molecular dynamics (MD) and the finite element method (FEM). A polymeric material is selected as the matrix—specifically, the poly(methyl methacrylate) (PMMA) commonly known as Plexiglas due to its expanded applications. On the other hand, the fullerene C_{240} is adopted as a reinforcement because of its high symmetry and suitable size. The numerical approach is performed at two scales. First, an analysis is conducted at the nanoscale by utilizing an appropriate nanocomposite unit cell containing the C_{240} at a high mass fraction. A MD-only method is applied to accurately capture all the internal interfacial effects and accordingly its thermoelastic response. Then, a micromechanical, temperature-dependent finite element analysis takes place using a representative volume element (RVE), which incorporates the first-stage MD output, to study nanocomposites with small mass fractions, whose atomistic-only simulation would require a substantial computational effort. To demonstrate the effectiveness of the proposed scheme, numerous numerical results are presented while the investigation is performed in a temperature range that includes the PMMA glass transition temperature, T_g.

Keywords: nanocomposite; PMMA; fullerene; finite element method; molecular dynamics; multiscale

1. Introduction

Commonly, the nanocomposite materials applications are associated with the simultaneous action of more than one type of loading. Especially, the investigation of nanocomposites subjected to both thermal as well as mechanical loads is perhaps one of most interesting fields for research, since high-temperature applications are very frequent. Recently, polymers that are reinforced with carbon nanomaterials have greatly attracted the scientific interest because of their enhanced material properties such as high strength-to-weight ratio. Evidently, the characterization of the thermomechanical performance of such nanocomposites may offer versatile design solutions for a variety of novel applications. In an effort to highlight significant innovations and potential applications in this research area, Burgaz [1] has investigated the current status of thermomechanical properties of polymers containing nanofillers in the form of nanocylinders, nanospheres, and nanoplatelets.

Since the experimental procedures intended for an adequate characterization of nanostructured composites are complicated and require extensive resources and time, the development and introduction of new computational approaches for simulating nanocomposites may be considered as a valuable, if not necessary, alternative. Perhaps, the MD method is the most popular tool for analyzing nanomaterial-reinforced polymers due to its ability to capture, with high accuracy, all the interatomic phenomena with respect to the temperature and pressure.

Discussion on some of the most interesting MD studies associated with the thermomechanical response of nanoreinforced polymers is essential. In a relatively early study, Cho and Yang [2] have performed a parametric study to investigate the effects of composition variables on the thermal and mechanical properties of carbon nanotube (CNT) reinforced polymers using MD simulations. Given that the glass transition temperature T_g is a key property for polymers, Allaoui and Bounia [3] have reviewed and analyzed various literature results dealing with the effect of unmodified multiwall carbon nanotubes (MWCNT) on the cure kinetics and T_g of their epoxy composites. Aiming at a similar goal, Herasati et al. [4] have investigated the effects of polymer chain branches, crystallinity, and CNT additives on the glass transition temperature of polyethylene (PE). Targeting a different matrix material, Mohammadi et al. [5] have investigated the effect of alumina and modified alumina nanoparticles on the glass transition behavior of a PMMA/alumina nanocomposite by MD simulations. The effect of inorganic particles have been studied by Zhang et al. [6], who have established silica–epoxy nanocomposite models to investigate the influence of a silane-coupling agent on the structure and thermomechanical properties of the nanocomposites through MD simulation. An extended study has been performed by Pandey et al. [7] focusing on the computation of viscoelastic, thermal, electrical, and mechanical properties of graphite flake-reinforced high-density PE composites. Recently, Dikshit and Engle [8] have employed MD simulations to study the mechanical properties of epoxy bisphenol A diglycidyl ethe (DGEBA) with and without the reinforcement of CNT, while in a similar attempt, Dikshit et al. [9] have performed a MD study to investigate the mechanical properties of graphene-reinforced epoxy nanocomposite. Aiming on the study of functionalizing polymer carbon nanofillers, Xue [10] have performed a cooling process by MD simulation to predict the T_g of graphene/PMMA composites. On the other hand, for the first time, Park et al. [11] investigated the thermomechanical characteristics of silica-mineralized nitrogen-doped CNT-reinforced PMMA nanocomposites by MD simulations. An interesting investigation regarding the interfacial behavior of functionalized CNT/PE nanocomposites at different temperatures has been performed by Singh and Kumar [12] using MD simulations and the second-generation polymer consistent force field (PCFF). Experimenting in a differently nanostructured reinforcing agent, Zhang et al. [13] have investigated via MD simulations the thermomechanical properties of nanocomposites consisting of weaved PE and CNT junctions.

Although there have been numerous efforts to investigate the influence of dispersing CNTs and graphene nanoribbons in polymers, fewer studies are available on the relevant effects of spherical carbon nanoparticles. In a study distinguished because of the kind of carbon allotrope that is used as a nanoreinforcement, Jeyranpour et al. [14] have adopted MD to carry out a comparative study regarding the effects of fullerenes on the thermomechanical properties of a specialized resin epoxy. Izadi et al. considered a similar nanocomposite [15] when estimating the elastic properties of PMMA reinforced with C_{60} fullerene and $C_{60}@C_{240}$ carbon onion by using MD simulations; however, they did not consider the effect of temperature.

All the above investigations have been realized via MD, which is a method that demands extensive computational power. Due to the high pre-processing and main-processing computational times required for analyzing material components at the nanoscale, several multiscale techniques [16–18] have been proposed that combine the benefits of molecular and continuum modeling. Characteristically, Montazeria and Rafii-Tabar [16] presented a combination of MD, molecular mechanics (MM), and the finite element method (FEM) that is capable of computing the elastic constants of a polymeric nanocomposite embedded with graphene sheets and carbon nanotubes at various temperatures. Similarly, Tsiamaki and Anifantis [17] have utilized a multiscale model based on MM and FEM to analyze the thermomechanical behavior of graphene/PMMA nanocomposites. Recently, Giannopoulos [18] proposed a formulation combining MD and FEM to predict the mechanical behavior of fullerene-reinforced nylon-12; however, this was at room temperature only. An interesting review on the recent developments in multiscale modeling of the thermal and mechanical properties of advanced nanocomposite systems has been given by Reddy et al. [19].

Apart from the more common carbon nanomaterials such as CNTs and graphene, which have unquestionably attracted the most attention in recent years, many researchers have started to explore the effects of reinforcing polymers with fullerenes [14,15,18]. Especially giant fullerenes such as C_{240} present unique characteristics that have been widely studied in the recent years. Giant fullerenes have already been experimentally observed and successfully produced. Very early, Ruoff et al. [20] utilized mass-spectrometric techniques to demonstrate the presence of carbon clusters C_{2n} with n as high as 300, in carbon soot material produced using the arc-synthesis method. On the other hand, Shinohara et al. [21] successfully extracted a series of very large all-carbon molecules, including C_{240}, with quinoline from fullerene-rich carbon soot produced by the vaporization of graphite in a helium atmosphere using the contact arc method. In an effort to provide generalized geometrical relationships describing the structure of giant fullerenes, Wang and Chiu [22] have also shown that the C_{240} giant fullerene cage has the same I_h symmetry as C_{60} and that it has twelve pentagonal faces in icosahedra alignment. Having a similar aim, Schwerdtfeger et al. [23] recently presented a general overview of recent topological and graph theoretical developments in fullerene research over the past two decades, describing both solved and open problems. In the theoretical field, Kim and Tomnek [24] reported an MD simulation of melting and evaporation of the carbon fullerenes C_{20}, C_{60}, and C_{240}. Finally, focusing on the C_{240}, Cabrera-Trujillo [25] used density functional theory (DFT) to study the electronic structure and binding of Na clusters encapsulated inside the fullerene cage.

Considering the exceptional structural and physical properties of giant fullerenes, which have been extensively discussed in some of the aforementioned studies, the reinforcing capability of C_{240} when compounded with polymers is computationally investigated in the present study over a wide temperature range. The symmetric fullerene C_{240} is preferred as a reinforcing agent because of its high symmetry, which may allow the achievement of an almost isotropic nanocomposite behavior. In addition, the PMMA is selected as the matrix material due to its high stiffness and wide range of applications. Moreover, its sensitivity on the temperature around its glass transition point, which has been investigated in several experimental [26–28] and MD studies [29], may permit the drawing of more illustrative conclusions about the fullerene reinforcement impact under different loading and environmental conditions. The adopted numerical technique is performed at two scales. At the first scale, MD simulations [30] of a low computational cost are performed by using a periodic unit cell to extract the temperature-dependent properties of the pure PMMA as well as the C_{240}/PMMA nanocomposite at a high mass fraction of 20%. The Condensed Phase Optimized Molecular Potential (COMPASS) [31] is adopted in view of its superiority over other potential models describing polymers [32]. Then, at a second scale, an RVE is developed and simulated via FEM [33] by using the data outputs from the MD-only analysis, in order to investigate nanocomposites with small C_{240} mass fractions, whose analysis via MD would not be computationally feasible by utilizing typical computer resources. A variety of diagrams are presented that depict the variation of nanocomposite properties such as elastic modulus, Poisson's ratio, and linear coefficient of thermal expansion with temperature and C_{240} mass fraction. Comparisons with relevant predictions found elsewhere are attempted, where possible. To the author's best knowledge, it is the first time that the temperature-dependent mechanical properties of the C_{240}/PMMA nanocomposite are predicted via a multiscale technique based on MD and FEM.

2. Multiscale Analysis

It is well known that MD is a numerical simulation method that is capable of predicting the time evolution of a system of interacting atoms. It is based on the generation of atomic trajectories via the numerical integration of Newtown's equation of motion by utilizing a specific interatomic potential, initial conditions, and boundary conditions. Although the MD method may accurately represent all the interatomic phenomena, it entails a substantial computational cost, which is dramatically increased with the number of the interacting atoms [18], due to the numerical integrations over long time intervals that are usually required to reach equilibrium states. Thus, the analysis of large systems such as the one tested here, i.e., a C_{240}/PMMA nanocomposite, leads to the necessity of combining atomistic with continuum numerical approaches. The use of a multiscale technique becomes a must when dealing with composites reinforced with low mass fractions of nanoparticles, since their MD-only analysis would require extremely large periodic unit cells.

Let us assume that the investigated C_{240}/PMMA nanocomposite is characterized by a uniform and periodic reinforcement distribution. Given the spherical and symmetric shape of C_{240} nanoparticles, the system domain may be fully described by a cubic periodic volume of the system domain illustrated in Figure 1, which contains a centrally located fullerene surrounded by a number of PMMA chains, i.e., the matrix material.

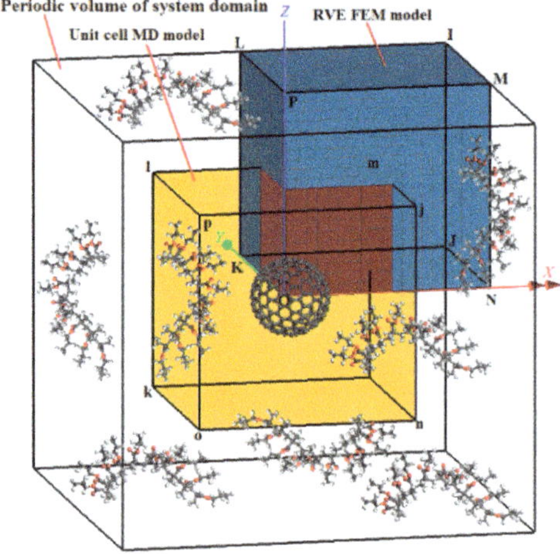

Figure 1. The nanocomposite representative volume element (RVE) (1/8 of the periodic volume of the system domain) with a small fullerene mass fraction and the nanocomposite unit cell with a fullerene mass fraction of 0.2, modeled via the finite element method (FEM) and molecular dynamics (MD), respectively.

The adopted numerical analysis is contacted at two scales. The MD-only method is utilized at the first scale while a CM method, realized via FEM and by using the previous MD output data, is performed at the second scale. The MD method offers the important precise representation of interfacial interactions and stress transfer mechanisms between the fullerene and the matrix while the CM method, based on FEM, provides modeling simplicity and a low computational cost.

At the first scale, MD simulations of the pure PMMA with respect to the temperature are initially performed to extract the necessary temperature-dependent property curves for the matrix material. Then, another periodic unit cell is developed and simulated that represents a nanocomposite with a high weight concentration of C_{240} equal to $w_{C240} = 0.2$. The MD-only simulation of the specific nanocomposite unit cell, whose topology is defined by the yellow colored ijklmnop cubic domain in Figure 1, leads to the computation of corresponding temperature-dependent property curves.

At the second scale, a CM-based RVE, denoted as ijklmnop in Figure 1, is developed for three small mass fractions of C_{240}, i.e., $w_{C240} = 0.01$, 0.03, and 0.05, and then, it is simulated through FEM using appropriate boundary conditions of symmetry and loading. Note that only the 1/8 of the periodic volume of the whole system domain is required to be represented due to the symmetry of the nanoparticle and its assumed uniform distribution within PMMA. Evidently, the red-colored subdomain of the RVE close to the vertex O (common volume between ijklmnop and IJKLMNOP cubes) is governed by the thermomechanical behavior of the nanocomposite with $w_{C240} = 0.2$, while the remaining blue-colored subdomain of the RVE represents the pure PMMA matrix. The temperature-dependent material properties, extracted from the MD-only simulations at the first scale analysis, are utilized in order to enable the finite element simulation of both RVE subdomains. The material properties in both subdomains are considered isotropic elastic, given that very small static strains are applied for the requirement of the present study.

3. First Simulation Scale: MD-Only Formulation

3.1. Unit Cells Construction

At the first scale of the analysis, as aforementioned, two different simulation stages take place. Initially, the pure PMMA is analyzed at various temperatures by utilizing a large enough unit cell to ensure convergence. Evidently, as the pure PMMA unit cell becomes larger, the number of polymer chains increases. Furthermore, when the simulation box contains a high number of polymer chains, its response becomes statistically independent of the relative chain nanostructural positioning and alignment. As a result, the MD-based numerical solution remains stable for large unit cells. Here, a series of polymeric chains of 10 monomers each are adopted in order to represent the PMMA, as Figure 2a depicts.

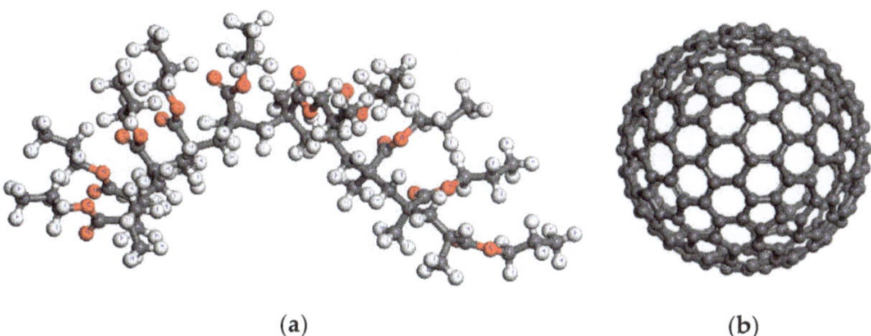

(a)   (b)

Figure 2. Molecular structure of the: (a) poly(methyl methacrylate) (PMMA) chain and (b) C_{240} fullerene.

The pure PMMA unit cell is analyzed according to a global Cartesian coordinate system (x, y, z) for each tested temperature level. When performing MD simulations, it is very convenient to initially adopt a small unit cell density in order to obtain a sparse molecular distribution. Then, common equilibrium algorithms are applied at each time point, to obtain the actual density of the unit cell as well as its equilibrated configuration. Here, it is assumed that the PMMA has an initial density at the room temperature $T = 300$ K equal to 0.6 g/cm^3. According to this PMMA density value, by utilizing 20 polymer chains and by taking into account the molecular weight of each chain, a cubic periodic unit cell may be defined [30]. It should be noted that the adoption of more than 20 chains inside the unit cell has been shown to have a negligible effect on the computed thermomechanical response of the pure PMMA. After conducting the full MD procedure described in the following section, the converged variations of volume, density, elastic modulus, and Poisson's ratio with the temperature are obtained. The equilibrated amorphous unit cell of the pure PMMA at 300 K is illustrated in Figure 3a.

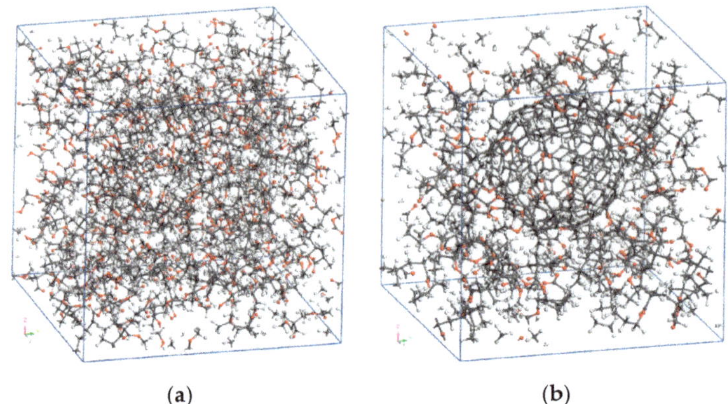

Figure 3. Equilibrated unit cell at the temperature T = 300 K of the (**a**) pure PMMA and (**b**) fullerene-reinforced PMMA at a mass fraction of w_{C240} = 0.2.

Secondly, the initial structure of the nanocomposite unit cell with w_{C240} = 0.2 is defined in a more complicated manner. First of all, the fullerene C_{240} of Figure 2b is maintained at the center of the unit cell at all times. The average radius of the specific fullerene is about r_{C240} = 7.07 Å [22,23]. In addition, its wall thickness is assumed to be equal to the usual distance between two successive carbon layers in graphite, i.e., t = 3.35 Å. Given, the specific wall thickness and the almost spherical shape of the tested fullerene, its density at the room temperature may be approximated by the following equation:

$$\rho_{C240} = \frac{4}{3} \frac{240 m_C}{\pi (r_{C240} + t/2)^3} \tag{1}$$

where m_C = 1.9927 × 10^{-23} g is the mass of a carbon atom.

In order to enable packing [30] of the PMMA chains into the unit cell, an initial nanocomposite density of 0.6 g/cm^3 is beforehand assumed for the room temperature. Then, the initial nanocomposite unit cell volume may be estimated by the following relationship:

$$V = \frac{m_{C240}}{w_{C240} \rho_{C240}} \tag{2}$$

Finally, before the initial packing of PMMA chains inside the unit cell and around the central positioned fullerene, the following nanocomposite unit cell length may be assumed:

$$L = \sqrt[3]{V} \tag{3}$$

After having defined the size of the three-dimensional (3d) nanocomposite unit cell for the room temperature, a number of PMMA chains are inserted into it, while the packing algorithm evenly increases their population until the initial assumed density is achieved. The equilibrated unit cell of the nanocomposite with w_{C240} = 0.2 at 300 K is shown in Figure 3b.

3.2. Geometry Optimization of Molecular Structures and Potential Model

Firstly, geometric optimization (GO) [30] is performed for each initially assumed molecular structure, i.e., the main PMMA chain as well as the C_{240} fullerene, which are depicted in Figure 2a,b, respectively. During the GO, energy minimization is achieved by using the steepest descent algorithm [30]. It is assumed that convergence is accomplished when the absolute difference of the computed system energy and force between two subsequent iterations becomes less than 0.001 Kcal/mol and 0.5 Kcal/mol/Å, respectively. The required numerical calculations are based on the COMPASS potential, which consists of the ten valence terms and two non-bonded interaction terms given below [31].

$$
\begin{aligned}
U =\ & \sum_{bond}[k_{b2}(b-b_0)^2 + k_{b3}(b-b_0)^3 + k_{b4}(b-b_0)^4] + \sum_{angle}[k_{a2}(\theta-\theta_0)^2 + k_{a3}(\theta-\theta_0)^3 + k_{a4}(\theta-\theta_0)^4] \\
& + \sum_{torsion}[k_{t1}(1-\cos\phi) + k_{t2}(1-\cos 2\phi) + k_{t3}(1-\cos 3\phi)] + \sum_{\substack{out\ of\ plane\ angle}} k_\chi(\chi-\chi_0)^2 + \sum_{bond/bond} k_{bb}(b-b_0)(b'-b'_0) \\
& + \sum_{bond/angle} k_{ba}(b-b_0)(\theta-\theta_0) + \sum_{angle/angle} k_{aa}(\theta-\theta_0)(\theta'-\theta'_0) + \\
& + \sum_{bond/torsion}(b-b_0)[k_{bt1}\cos\phi + k_{bt2}\cos 2\phi + k_{bt3}\cos 3\phi] + \sum_{angle/torsion}(\theta-\theta_0)[k_{at1}\cos\phi + k_{at2}\cos 2\phi + k_{at3}\cos 3\phi] \\
& + \sum_{angle/torsion/angle} k_{ata}(\theta-\theta_0)(\theta'-\theta'_0)\cos\phi + \sum_{nonbond}\varepsilon_{ij}\left[2\left(\frac{r_{ij0}}{r_{ij}}\right)^9 - 3\left(\frac{r_{ij0}}{r_{ij}}\right)^6\right] + \sum_{nonbond}\frac{q_i q_j}{4\pi\varepsilon_0 r_{ij}}
\end{aligned} \tag{4}
$$

In the last equation, the first four sums denote the energies required to stretch bonds (b), bend angles (θ), change torsion angles (ϕ) by twisting atoms about the bond axis, and distort atoms out of the plane (χ) formed by the atoms to which they are bonded. The next six sums denote the energies between the four types of internal coordinates described as functions of the Cartesian atomic coordinates [31]. The final two sums that contain functions of the atom pair distance q_{ij} denote the Lennard–Jones-based van der Waals (vdW) non-bond interactions and the Coulomb's electrostatic non-bond interactions due to the charges q_i and q_j, respectively. The subscript 0 found in some parameters denotes corresponding reference values. The constant ε_0 is the well-known vacuum permittivity. Depending on the atom-type combinations, the COMPASS force field predefines the stiffness-like parameters k_{b2}, k_{b3}, k_{b4}, k_{a2}, k_{a3}, k_{a4}, k_{t1}, k_{t2}, k_{t3}, k_χ, k_{bb}, k_{ba}, k_{aa}, k_{bt1}, k_{bt2}, k_{bt3}, k_{at1}, k_{at2}, k_{at3}, and k_{ata} as well as the functional form of each term q_i, q_j, ε_{ij}, and r_{ij0}. Here, the vdW contributions are computed according to the atom-based summation method using a cut-off radius of 12.5 Å and long-range corrections, while the electrostatic contributions are computed by adopting the Ewald summation method with an accuracy of 0.001 kcal/mol [34].

Evidently, the relevant positioning of the molecules is performed after computing the interactions between neighbor atoms via the COMPASS force field whereas the single chain conformations, ring spearing, and close contacts are constantly monitored. To achieve a minimized initial unit cell state, low-energy sites are preferred over high-energy sites for each molecular structure. A GO process, as the one described earlier, is executed to additionally reduce the overall potential energy of the 3D problem domain.

3.3. NPT Dynamic Analysis of the Unit Cells

All the MD simulations take place under the NPT ensemble and by using a time step of 1 fs. The external pressure of the unit cell is maintained at 1 atm throughout each dynamic analysis. After the finalization of the procedure at a specific temperature level, the relaxed equilibrium state, true final density, and side lengths of the unit cell are obtained. Performing a dynamic analysis by introducing additional time intervals under different ensembles such as NVT or using a time step lower than 1 fs has no observable effect on the final numerical solutions. Due to the dynamic nature of the simulation, in order to keep the system under a specific temperature and pressure level, the Andersen thermostat and Berendsen barostat are utilized, respectively [34].

3.4. Thermomechanical Properties Calculation

After achieving equilibrium at a given temperature T, the elastic properties are computed by applying to the 3D unit cells a set of three pairs of uniaxial tension/compression and three pairs of pure shear static strains of a maximum amplitude of ±0.001.

The stresses at each strain level may be estimated through the following average virial stress of a system of particles [34]:

$$
\sigma_{av} = \frac{1}{2V}\sum_{j(\neq i)} \mathbf{r}_{ij} \otimes \mathbf{f}_{ij} \tag{5}
$$

where V is the volume of the system, i and j denote two particles at positions $\mathbf{r}_i$ and $\mathbf{r}_j$, respectively, $\mathbf{r}_{ij}$ is equal to $\mathbf{r}_j - \mathbf{r}_i$, and $\mathbf{f}_{ij}$ is the inter-particle force applied on particle i by particle j.

By considering the symmetry of the stress, strain, and stiffness tensors, Hooke's law may be expressed as:

$$\sigma = \mathbf{C}\varepsilon \tag{6}$$

Since the nanocomposite is assumed to be isotropic, the Lamé coefficients λ and μ may be defined by diagonal stiffness coefficients of $\mathbf{C}$ as:

$$\lambda = \frac{1}{3}(C_{11} + C_{22} + C_{33}) - \frac{2}{3}(C_{44} + C_{55} + C_{66}) \tag{7}$$

$$\mu = \frac{1}{3}(C_{44} + C_{55} + C_{66}) = G \tag{8}$$

where G is the shear modulus.

Evidently, the elastic modulus E and the Poisson's ratio v may be calculated, respectively, by the following equations:

$$E = \frac{\mu(3\lambda + 2\mu)}{\lambda + \mu} \tag{9}$$

$$v = \frac{\lambda}{2(\lambda + \mu)} \tag{10}$$

The computation of the initial and final unit cell volume V_0 and V_1, respectively, at a reference temperature T_0 and an arbitrary temperature $T_1 > T_0$, respectively, enables the estimation of the volume coefficient of thermal expansion a_V via the equation:

$$a_V(T_0 \leq T \leq T_1) = \frac{V_1 - V_0}{T_1 - T_0} \frac{1}{V_0} \tag{11}$$

Finally, the linear coefficient of thermal expansion a_L for an isotropic medium may be approximated by:

$$a_L = a_V/3 \tag{12}$$

4. Second Simulation Scale: FEM Formulation

4.1. Geometry Definition and Finite Element Discretization

At the second scale, nanocomposites with small mass fractions of C_{240} are modeled and simulated through FEM. A representative FEM model of the RVE, which corresponds to the case $w_{C240} = 0.05$, is illustrated in Figure 4 and defined by the IJKLMNOP cubic domain. The problem is analyzed according to a global Cartesian coordinate system (x, y, z) positioned at the vertex O. As depicted in the figure, the RVE is consisted of the C_{240}/PMMA subdomain with $w_{C240} = 0.2$ and the outer pure PMMA subdomain.

The edge length of the nanocomposite subdomain with $w_{C240} = 0.2$ is taken equal to $L/2$, where L is the corresponding unit cell length computed via the MD-only simulation at the first scale analysis. On the other hand, Equations (1) to (2) may be combined in order to estimate the length of the RVE L_{RVE} as follows:

$$L_{RVE} = \frac{1}{2} \sqrt[3]{\frac{m_{C240}}{w_{C240}\rho_{C240}}} \tag{13}$$

Both subdomains are discretized with isoparametric, hexahedral, linear, eight-noded finite elements that have four degrees of freedom per node, i.e., the displacements u_x, u_y, u_z, and the temperature T [33]. The finite element meshes for the three case studies $w_{C240} = 0.01, 0.03$, and 0.05 are depicted in Figure 5a–c, respectively.

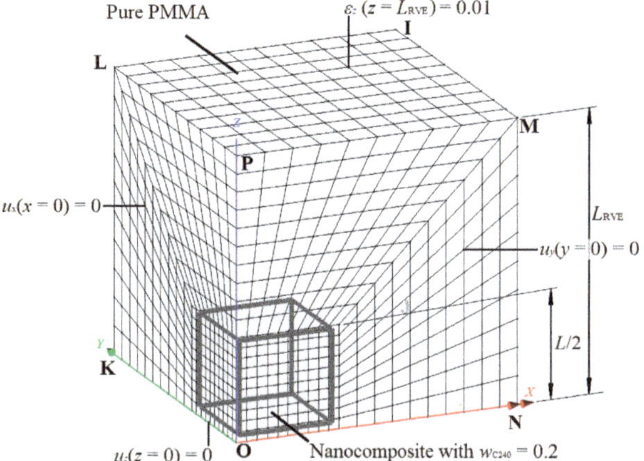

Figure 4. Geometry, boundary conditions, and finite element discretization of the nanocomposite RVE with a mass fraction of $w_{C240} = 0.05$ (1/8 of the periodic volume of the whole system domain).

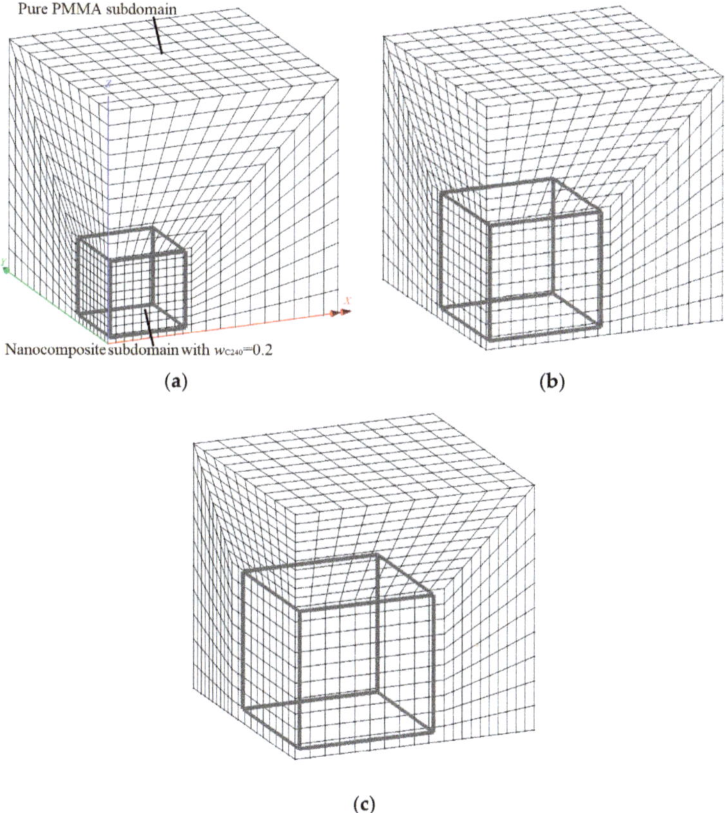

Figure 5. Finite element discretization of the nanocomposite RVE with a mass fraction of (**a**) $w_{C240} = 0.01$, (**b**) $w_{C240} = 0.03$, and (**c**) $w_{C240} = 0.05$ (1/8 of the periodic volume of the whole system domain).

4.2. Material Properties Input and Output

The properties of both subdomains of each RVE FEM model are considered as temperature-dependent elastic. The elastic modulus and Poisson's ratio of both the nanocomposite with $w_{C240} = 0.2$ and the pure PMMA are input as functions of temperature, i.e., $E(T)$ and $v(T)$. These functions are determined by fitting corresponding data points computed by the MD-only simulations at the first scale of the analysis. In order to compute the elastic properties of the nanocomposite RVE, appropriate boundary conditions are applied. For the calculation of the elastic modulus $E_{RVE}(T)$ and the Poisson's ratio $v_{RVE}(T)$, a uniform strain of $\varepsilon_z(T) = 0.001$ is applied on the edge $z = L_{RVE}$. Simultaneously, the constraints $u_x = 0$, $u_y = 0$, and $u_z = 0$ are applied on the edges $x = 0$, $y = 0$, and $z = 0$, respectively, while the edges $x = L_{RVE}$ and $y = L_{RVE}$ are kept parallel to their original shape by nodal coupling, since the symmetry implies that shear stresses on these edges should be zero. Then, the elastic modulus of the nanocomposite $E_{RVE}(T)$ is calculated from the ratio of average stress $\sigma_{zav}(T)$, which are obtained from the sum of reactions in the ground edge $z = 0$ to the applied strain $\varepsilon_z(T) = 0.001$. Finally, the Poisson's ratio $v_{RVE}(T)$ is estimated by the ratio of the arisen average transverse strain $\varepsilon_{xav}(T) = \varepsilon_{yav}(T)$ to the applied normal strain $\varepsilon_z(T) = 0.001$.

Considering the FEM simulation of the RVE thermal expansion behavior, a different load case is applied. Evidently, the constraints remain the same in accordance with the symmetry. The linear coefficient of thermal expansion $a_L(T)$ of both subdomains is defined according to the corresponding output from the MD-only simulations again. Then, a small temperature load increment $\Delta T = T_1 - T_0$ is applied in the whole RVE to compute its arisen edge length increment $\Delta L_{RVE} = L_{RVE1} - L_{RVE0}$. As a result, the linear coefficient of thermal expansion of the RVE may be estimated by the following relationship:

$$a_{L_{RVE}}(T_0 \leq T \leq T_1) = \frac{L_{RVE1} - L_{RVE0}}{T_1 - T_0} \frac{1}{L_{RVE0}} \tag{14}$$

5. Results and Discussion

5.1. First Simulation Scale

For both material cases under investigation, i.e., the pure PMMA and the PMMA reinforced with fullerene C_{240} at a mass fraction of 0.2, the MD simulations are conducted under the NPT ensemble with a target pressure of 1 atm and a time-dependent temperature T, which varies from 300 to 560 K as defined in Figure 6. For the figure, it may be seen that initially, the target temperature is considered stable at 300 K for a time interval of 1000 ps, so that both the pure PMMA unit cell and the nanocomposite unit cell with $w_{C240} = 0.2$ reach a minimized energy and an equilibrium state. Then, it is assumed that temperature exhibits a step increment with time. Specifically, at each step, the temperature remains stable for 300 ps and then increases by 10 K. Following such a technique, the equilibrium is achieved much faster at each temperature level, expect for the first investigated temperature level at 300 K for which a longer time interval is required for convergence.

The density variation of the pure PMMA and the C_{240}/PMMA nanocomposite unit cells during the dynamic analysis may be seen in Figure 6. Note that the initial density of 0.6 g/cm^3 assumed for both unit cells increases to reach its proper value at room temperature and then decreases as the temperature elevates. The variation of the potential and kinetic energies of the two cells with temperature is illustrated in Figure 7. All variations are almost linear ascending, while the kinetic energy of the pure PMMA is higher for the whole temperature range and presents a higher gradient in comparison with the nanocomposite unit cell. On the other hand, the pure PMMA presents a lower potential energy up to 500 K in contrast with its reinforced version with $w_{C240} = 0.2$.

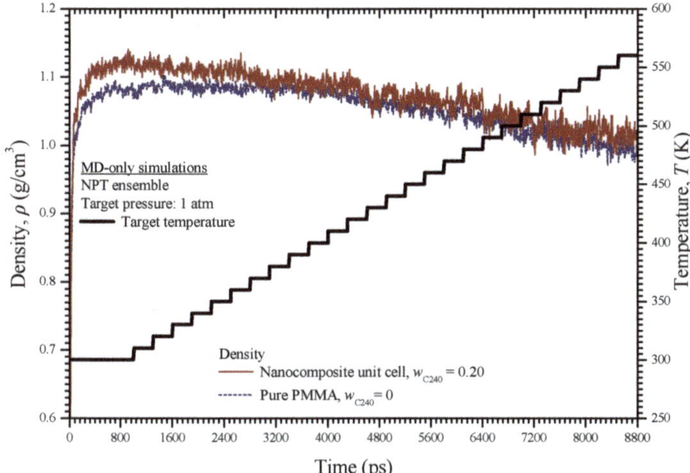

Figure 6. Density of the pure PMMA and nanocomposite with $w_{C240} = 0.2$ versus time in contrast with the time-dependent temperature during the MD simulations.

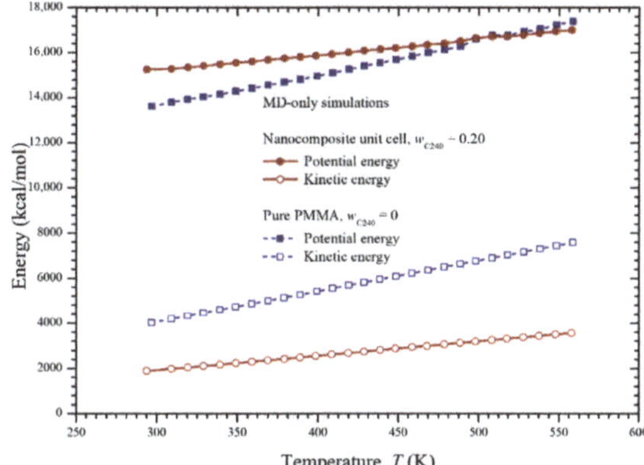

Figure 7. Potential and kinetic energy of the pure PMMA and nanocomposite with $w_{C240} = 0.2$ with respect to the temperature.

Figure 8 depicts the volume change of the two unit cells versus temperature. As it can be seen, the volume variation of each unit cell is characterized by two different regions, i.e., the glassy and the rubbery one. At each region, the MD data points imply a linear behavior of a different slope. A linear regression is applied on the set of data of the glassy and rubbery region of each medium. Then, the glass transition temperatures T_g are estimated from the intersection of the arisen lines at 421 and 462 K for the pure and reinforced PMMA, respectively. The glass transition temperature is considerably increased by reinforcing the PMME with fullerene C_{240} at a weight concentration of 20%. A similar phenomenon has been observed regarding the case of PMMA filled with functionalized graphene [10]. The computed T_g regarding the pure PMMA is in good agreement with the corresponding values 411.4 and 430 K, which have been predicted in the MD-based studies [10] and [29], respectively. The details about all the linear regressions shown in Figure 8 may be found in Table 1. Using the fitting parameters of the table, one may define the volume of the unit cells as a function of temperature.

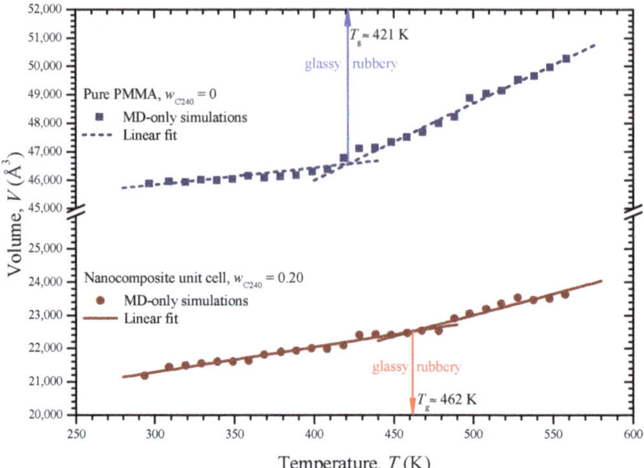

Figure 8. Volume of the pure PMMA and nanocomposite with $w_{C240} = 0.2$ with respect to temperature and the corresponding glass transition temperatures T_g arisen from the intersection of appropriate linear fittings.

Table 1. Definition of all the fitting functions adopted throughout the analysis.

		Fitting of MD Data with Functions of Temperature			
Property	**Fitting Equation ($T\rightarrow$K)**	**Nanocomposite with $w_{C240} = 0.02$**		**Pure PMMA**	
Volume, V (Å³)	Linear $V = f + gT$	*T* range	$300 \leq T \leq 462$	$462 < T \leq 560$	$300 \leq T \leq 421$	$421 \leq T \leq 560$
		f	19,038.73973	16,537.08047	44,046.02025	35,216.40195
		g	7.50775	13.10966	6.0055	26.96125
		Adjusted R^2	0.96339	0.89922	0.83912	0.98609
Density, ρ (g/cm³)	Linear $\rho = h + lT$	*T* range	$300 \leq T \leq 462$	$462 < T \leq 560$	$300 \leq T \leq 421$	$421 < T \leq 560$
		h	1.2401	1.33445	1.13239	1.3125
		l	-3.81086×10^{-4}	-5.84103×10^{-4}	-1.40928×10^{-4}	-5.71409×10^{-4}
		Adjusted R^2	0.96111	0.91912	0.85399	0.98549
Elastic modulus, E (GPa)	6th order polynomial $E = A + B_1T + B_2T^2 + B_3T^3 + B_4T^4 + B_5T^5 + B_6T^6$	*T* range	$300 \leq T \leq 500$		$300 \leq T \leq 500$	
		A	-2209.3885		409.32795	
		B_1	33.54518		-5.76718	
		B_2	-0.20948		0.0332	
		B_3	6.90037×10^{-4}		-9.88824×10^{-5}	
		B_4	-1.26453×10^{-6}		1.60037×10^{-7}	
		B_5	1.22205×10^{-9}		-1.33008×10^{-10}	
		B_6	-4.86522×10^{-13}		4.40984×10^{-14}	
		Adjusted R^2	0.99509		0.99509	
Poisson's ratio, ν	Boltzmann $\nu = (D_1 - D_2)/\{1 + \exp[(T - \tau_0)/\tau]\} + D_2$	*T* range	$300 \leq T \leq 500$		$300 \leq T \leq 500$	
		D_1	0.2731		0.28715	
		D_2	0.46094		0.49459	
		τ_0	497.04191		414.53197	
		τ	14.45546		12.68674	
		Adjusted R^2	0.9982		0.99698	

The linear coefficient of thermal expansion a_L of the pure PMMA and the nanocomposite with $w_{C240} = 0.2$ with respect to the temperature is illustrated in Figure 9. The relevant estimations are based on Equations (11) and (12). As observed for both materials, the linear coefficient of thermal expansion has a lower and constant value for the temperatures below T_g, while it exhibits a notable constant increase after this temperature point and up to 560 K. The presence of C_{240} within the PMMA matrix, as expected [10], appears to lead to a rise in thermal expansion, especially for temperatures below 421 K, i.e., the glass transition point of the matrix.

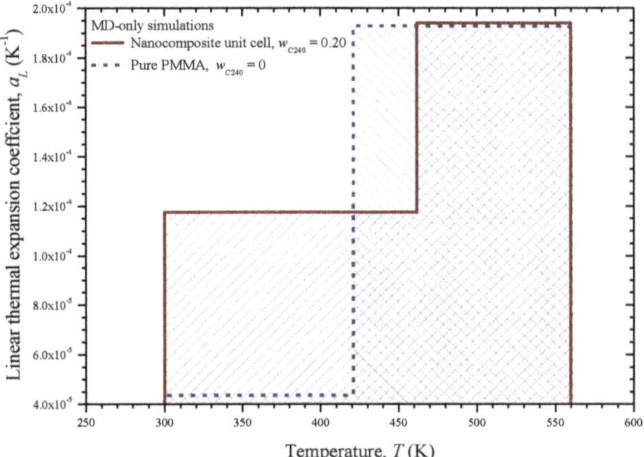

Figure 9. Linear thermal expansion coefficient of the pure PMMA and the nanocomposite with $w_{C240} = 0.2$ unit cell with respect to the temperature.

The calculated a_L of the pure PMMA is 4.4×10^{-5} and 1.9×10^{-4} K^{-1} in the glassy and rubbery state, respectively, which are in decent agreement with the corresponding values 7.3×10^{-5} K^{-1} and 2.6×10^{-4} K^{-1}, as reported in a different MD analysis [10]. Note that experimental evidence [35] suggests that the pure PMMA presents a coefficient of linear thermal expansion in the range from 5×10^{-5} to 9×10^{-5} K^{-1} at room temperature (glassy state).

The dependence of the density ρ on the temperature T is illustrated for the pure PMMA and nanocomposite unit cell in Figure 10. The density for both cases follows a linear decrease characterized by two slopes. The kink positions reveal the corresponding glass transition points, which are identical with those found from Figure 8. At all temperature levels, the density of the nanocomposite remains higher. Again, a two-slope linear regression of the density–temperature variations is performed, the results of which may be found in Table 1.

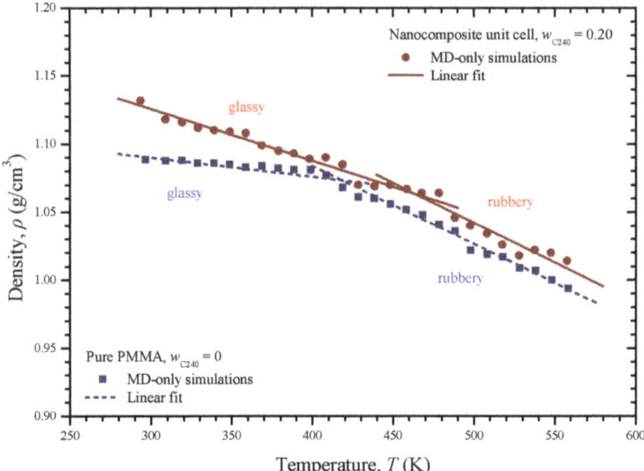

Figure 10. Density of the pure PMMA and nanocomposite with $w_{C240} = 0.2$ with respect to the temperature and corresponding linear fittings.

The elastic modulus of the two unit cells with respect to the temperature is shown in Figure 11a. It becomes obvious that the C_{240}/PMMA nanocomposite presents an advanced stiffness. This is due to the enhanced stiffness of the carbon nanoparticle. As expected, a stress relaxation is observed at the glass transition points for both materials, which is implied by the significant drop of elastic modulus. The elastic modulus–temperature nonlinear variations are fitted well with polynomial functions of 6th degree that are fully defined in Table 1 for temperatures up to 500 K.

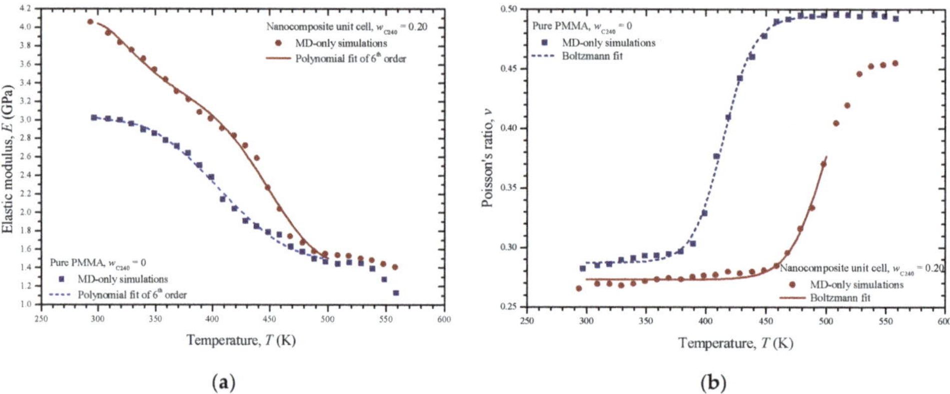

Figure 11. The (a) elastic modulus and (b) Poisson ratio of the pure PMMA and the nanocomposite with $w_{C240} = 0.2$ with respect to the temperature and corresponding nonlinear fittings up to 500 K.

The temperature dependence of Poisson's ratio of the PMMA and nanocomposite is illustrated in Figure 11b. The Poisson ratio tends to reach the value of 0.5 as the temperature increases for the pure PMMA case. On the other hand, the Poisson ratio of the PMMA reinforced with C_{240} at a mass fraction of 20% presents lower values due to the effects of the fullerene constituent. An obvious intense Poisson ratio increase occurs nearby the T_g point. Table 1 includes details about the fitting of the two Poisson ratio–temperature variations for the temperature range from 300 to 500 K with the Boltzmann sigmoid function, which are defined in the table as well.

In order to evaluate the performance of the MD-only simulations, Table 2 is presented. The table includes some comparisons between the present results regarding the pure PMMA mechanical properties at the room temperature with other corresponding predictions.

Table 2. Comparison of the elastic properties of the pure PMMA computed here via MD, with corresponding results from other studies.

Study	Materials Properties of Pure PMMA at $T = 300$ K	
	Elastic Modulus, E (GPa)	Poisson's Ratio, ν
Present MD-only formulation	3.065	0.286
Different MD formulation [15]	3.052	0.257
Experimental [26]	3.400	-

5.2. Second Simulation Scale

Three small C_{240} loadings of 1%, 3%, and 5% by weight are investigated by using the FEM models shown in Figure 5a–c. The linear coefficient of thermal expansion, elastic modulus, and Poisson's ratio of the pure PMMA subdomain and nanocomposite subdomain with a C_{240} concentration of 20 wt % are inserted into the model by utilizing the temperature-dependent functions provided from the first-scale MD analysis and illustrated in Figure 11a,b and Figure 9, respectively. The same number of finite elements is used in all cases. Denser meshes than the ones depicted in Figure 5 lead to negligibly different numerical solutions.

Figure 12a,b presents the elastic modulus E and the Poisson ratio v of the tested C_{240}/PMMA nanocomposites, respectively. The limit cases for $w_{C240} = 0$ and $w_{C240} = 0.2$ treated with the MD-only method are included in these figures for comparison reasons. A nonlinear reduction of the mechanical performance, as expressed by the elastic modulus decrease and Poisson ratio increase, is observed for all the materials as the temperature rises. Contrary, the higher the fullerene mass fraction, the higher the elastic modulus and the lower the Poisson ratio. For a given temperature, almost a linear change occurs in these properties as the reinforcement concentration increases.

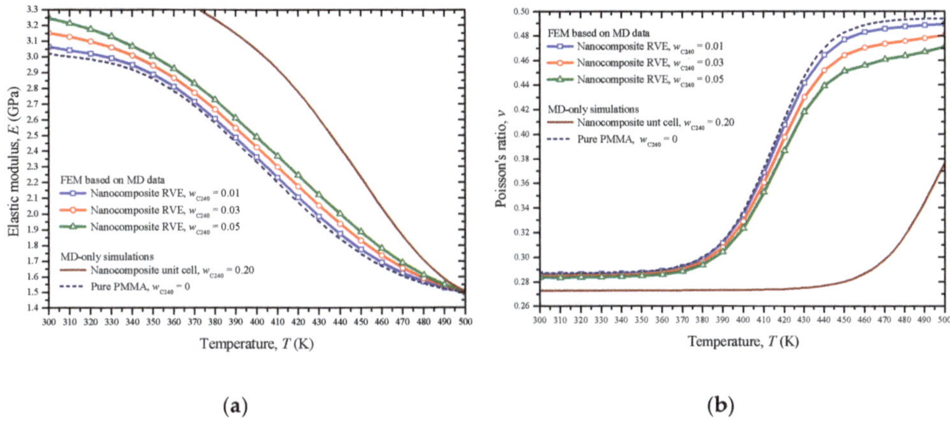

Figure 12. The (**a**) elastic modulus and (**b**) Poisson ratio of the nanocomposite with small fullerene mass fractions with respect to the temperature, predicted by combining FEM and MD.

Table 3 shows a qualitative comparison between some present estimations via the proposed multiscale analysis and others predicted via MD [15] in which a rather dissimilar fullerene structure such as the carbon onion C_{60}@C_{240} has been considered. Unfortunately, to the author's best knowledge, there is not any relevant experimental contribution regarding the stiffness of the C_{240}/PMMA nanocomposite, in order to provide a more comprehensive assessment. However, regarding the elastic modulus of the pure PMMA, a comparison with a corresponding experimental value is included in Table 2.

Table 3. Present elastic properties of the nanocomposite with $w_{C240} = 0.05$ in contrast with some comparable results from another theoretical study.

Theoretical Study	Material Properties at $T = 300$ K	
	Elastic Modulus, E (GPa)	Poisson's Ratio, v
Present FEM combined with MD of PMMA reinforced with C_{240} at 5.00 wt %	3.247	0.284
Different MD simulation [15] of PMMA reinforced with onion C_{60}@C_{240} at 5.01 wt %	3.590	0.271

The contours of the von Mises equivalent stress of the nanocomposite with $w_{C240} = 0.01, 0.03,$ 0.05 and for a temperature of 420 K are depicted in Figure 13a–c. As seen, the maximum equivalent stress, which is located in the nanocomposite subdomain with $w_{C240} = 0.2$, rises as the fullerene concentration is increased. This reveals an enhanced capability of the fullerene to carry loads.

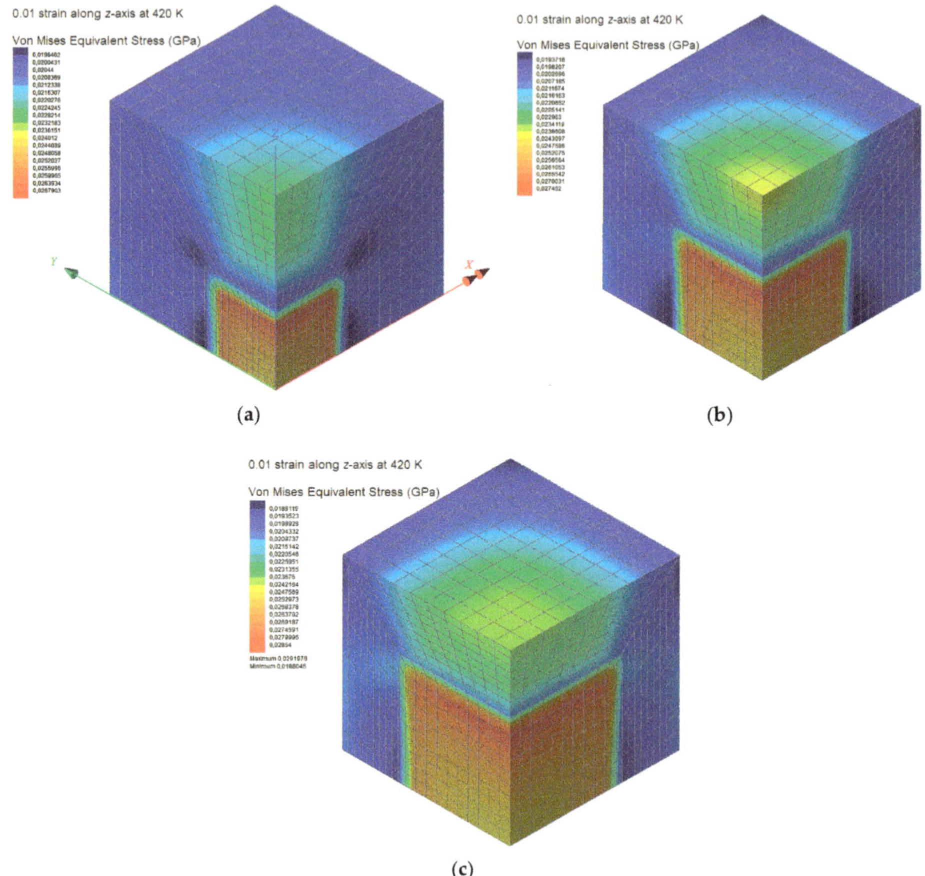

Figure 13. Contours of the resultant von Mises equivalent stress of the nanocomposite RVE at a temperature of 400 K, with mass fractions of (**a**) $w_{C240} = 0.01$, (**b**) $w_{C240} = 0.03$, and (**c**) $w_{C240} = 0.05$.

The proposed FEM analysis is not efficient enough to compute straightforwardly the linear coefficient of thermal expansion a_L for the whole temperature range from 300 to 500 K, since complex molecular phenomena occur in the interphase zone near the phase transition temperature, which may only be described via atomistic models. Thus, in order to assure that the FEM computations take place exclusively in the glassy or rubbery state of the nanocomposite with $w_{C240} = 0.01, 0.03,$ and 0.05, a targeted temperature change is applied from 300 to 301 K or from 499 to 500 K, respectively. Accordingly, the two boundary values of a_L in the temperature interval [300 K, 500 K] are obtained through Equation (14). These two FEM data points are inserted in Figure 14, which also includes the step functions $a_L(T)$ for the limit cases, investigated via MD only, where $w_{C240} = 0$ and $w_{C240} = 0.2$. To assure safe estimations for the cases $w_{C240} = 0.01, 0.03,$ and 0.05, a linear interpolation is required, which is graphically realized by interconnecting the two lower corner points $(T_g, a_L(T_g^-))$ of the two step functions $a_L(T)$ defined by MD. Then, some good approximations of the a_L of the nanocomposites with $w_{C240} = 0.01, 0.03,$ and 0.05 around their T_g may be graphically derived by defining specific intersection points and making linear interpolations as Figure 14 describes in detail.

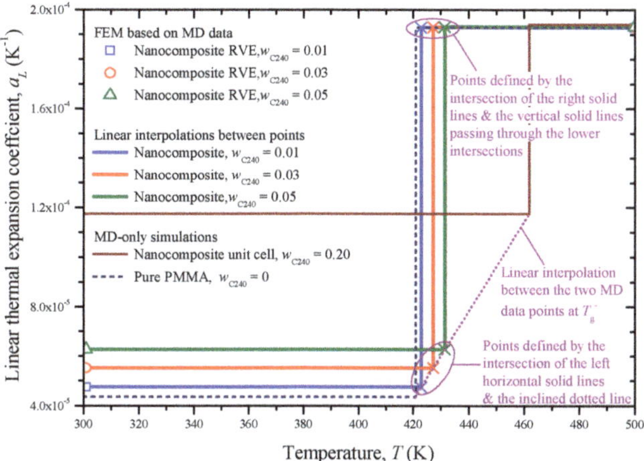

Figure 14. Linear thermal expansion coefficient of the nanocomposite with small fullerene mass fractions with respect to the temperature, predicted by combining FEM and MD as well as using linear interpolations.

Note that the T_g points for the small fullerene concentrations are also indirectly revealed by the arisen intersection points. The good performance of the proposed graphical procedure, which is grounded on the utilization of both FEM and MD data points, in predicting T_g is demonstrated in Figure 15, where a theoretical estimation via the well known Flory–Fox equation [36] is included. The specific equation is defined as follows:

$$\frac{1}{T_g(w_{C240})} = \frac{w_{C240}}{T_{gC240}} + \frac{1 - w_{C240}}{T_{gPMMA}} \tag{15}$$

where T_{gC240} are T_{gPMMA} is the glass transition temperature of the component C_{240} and pure PMMA, respectively, while the nanocomposite $T_g(w_{C240})$ function is assumed to pass through the two MD data points.

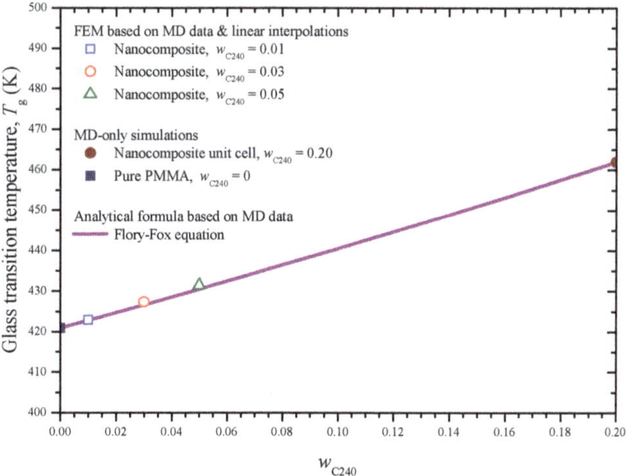

Figure 15. Glass transition temperature of the nanocomposite for small fullerene mass fractions, predicted by combining FEM and MD as well as using linear interpolations.

Figure 16a,b present the colored distributions of the resultant displacement due to a temperature change from 300 to 301 K (glassy state) and from 499 to 500 K (rubbery state), respectively, for the nanocomposite with $w_{C240} = 0.01$. It becomes obvious that the consequent expansions are more intense at a higher temperature level since the coefficients of thermal expansion for all subdomains are higher in their glassy state. Similar contours are presented in Figures 17 and 18 for the cases $w_{C240} = 0.03$ and 0.05, respectively, leading to analogous conclusions.

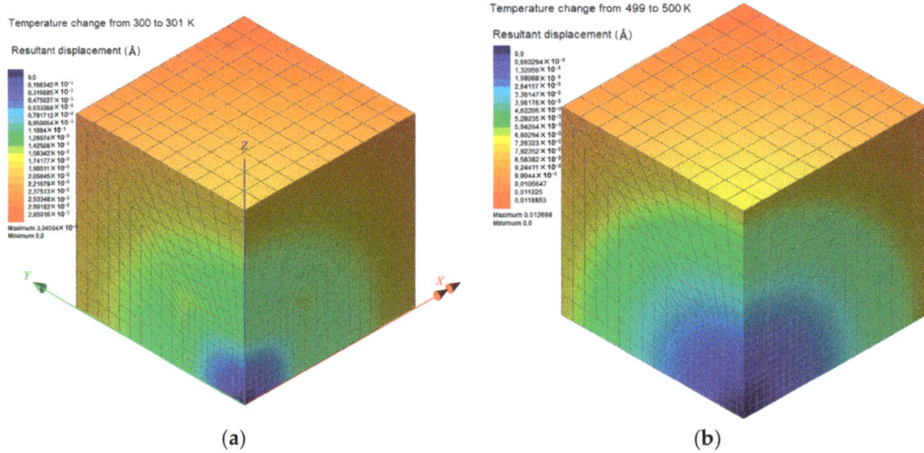

Figure 16. Contours of the resultant displacement of the nanocomposite RVE with a mass fraction with $w_{C240} = 0.01$, for a temperature change (**a**) from 300 to 301 K and (**b**) from 499 to 500 K.

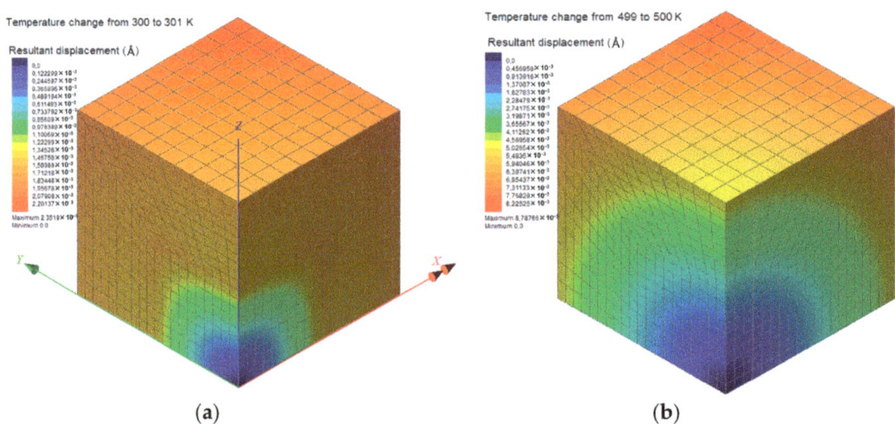

Figure 17. Contours of the resultant displacement of the nanocomposite RVE with a mass fraction with $w_{C240} = 0.03$, for a temperature change (**a**) from 300 to 301 K and (**b**) from 499 to 500 K.

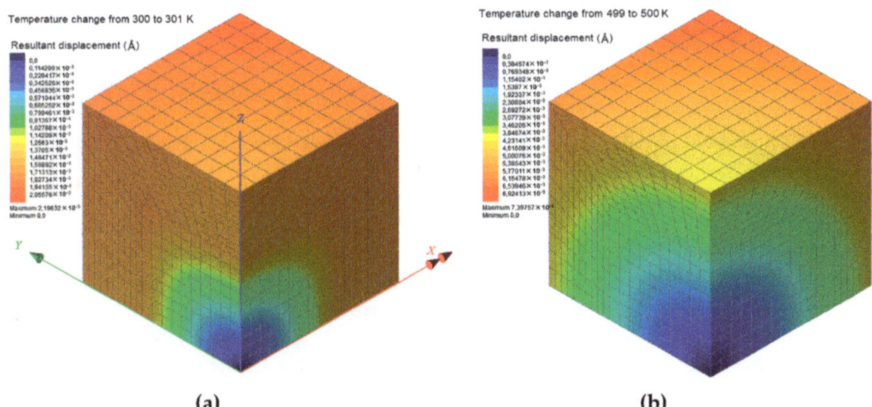

Figure 18. Contours of the resultant displacement of the nanocomposite RVE with a mass fraction with $w_{C240} = 0.05$, for a temperature change (**a**) from 300 to 301 K and (**b**) from 499 to 500 K.

6. Conclusions

A theoretical attempt was made to provide a multiscale numerical formulation for the efficient prediction of the thermoelastic response of nanomaterial/polymer composites. The aim was to provide an accurate numerical tool of a low computational cost that is capable of treating large problem domains, which would require substantial resources if treated via atomistic methods alone. To deal with such problems, the proposed method is applied into two phases. It starts from the molecular scale via MD and ends up to the continuum scale via FEM. Thus, the hybrid simulation is capable of capturing the complicated atomistic and interphase phenomena as well as reducing the computational effort simultaneously.

For the purpose of the study, the fullerene C_{240} and the PMMA were utilized as the reinforcement and the matrix material, respectively. A MD formulation was initially developed in order to predict the temperature-dependent elastic modulus, Poisson ratio, and linear coefficient of thermal expansion of the pure PMMA and the C_{240}/PMMA with a high fullerene mass fraction. The glass transition temperature of the specific media was also defined by the change in the slope of relevant thermal expansion curves. The extracted data points were fitted via appropriate functions and inserted into several FEM models to simulate nanocomposites with smaller fullerene mass fractions. The computations showed that the proposed multiscale formulation may perform well for low nanofiller contents up to 5 wt %.

The FEM computations led to the full definition of the same properties for the whole investigated temperature and fullerene mas fraction range. It was demonstrated that for a given temperature level, the rise of the fullerene mass fraction leads to an almost linear increase of the nanocomposite stiffness but also to an analogous decrease of its Poisson's ratio. The linear coefficient of thermal expansion of the nanocomposite was found to be constant before and after the glass transition temperature. Its value was significantly higher for the glassy state, while it showed a nearly linear increase with the increase of the nanofiller mass fraction. Finally, a drastic drop of the nanocomposite mechanical performance was observed near the glass transition point due to the stress relation.

A further investigation is planned to be made in a future work, where the effects of utilizing different fullerene sizes, several combinations of fullerene types, and non-uniform nanofiller distributions will be extensively studied.

Author Contributions: Conceptualization, G.I.G.; methodology, G.I.G.; software, G.I.G., S.K.G. and N.K.A.; validation, G.I.G., S.K.G. and N.K.A.; investigation, G.I.G.; writing—original draft preparation, G.I.G.; writing—review and editing, S.K.G. and N.K.A.; visualization, G.I.G., S.K.G. and N.K.A.; supervision, G.I.G.; All authors have read and agreed to the published version of the manuscript.

Funding: This research received no external funding.

Conflicts of Interest: The authors declare no conflict of interest.

References

1. Burgaz, E. Thermomechanical analysis of polymer nanocomposites. *Polym. Nanocompos. Electr. Therm. Prop.* **2016**, 191–242.
2. Cho, M.; Yang, S. Atomistic simulations for the thermal and mechanical properties of CNT-polymer nanocomposites. In Proceedings of the 48th AIAA/ASME/ASCE/AHS/ASC Structures, Structural Dynamics, and Materials Conference, Honolulu, HI, USA, 23–26 April 2007; Volume 5, pp. 4718–4726.
3. Allaoui, A.; El Bounia, N. How carbon nanotubes affect the cure kinetics and glass transition temperature of their epoxy composites?—A review. *Express Polym. Lett.* **2009**, *3*, 588–594. [CrossRef]
4. Herasati, S.; Ruan, H.H.; Zhang, L.C. Effect of chain morphology and carbon-nanotube additives on the glass transition temperature of polyethylene. *J. Nano Res.* **2013**, *23*, 16–23. [CrossRef]
5. Mohammadi, M.; Davoodi, J.; Javanbakht, M.; Rezaei, H. Glass transition temperature of PMMA/modified alumina nanocomposite: Molecular dynamic study. *Mater. Res. Express* **2019**, *6*, 035309. [CrossRef]
6. Zhang, X.; Wen, H.; Wu, Y. Computational thermomechanical properties of silica-epoxy nanocomposites by molecular dynamic simulation. *Polymers* **2017**, *9*, 430. [CrossRef] [PubMed]
7. Pandey, A.K.; Singh, K.; Kar, K.K. Thermo-mechanical properties of graphite-reinforced high-density polyethylene composites and its structure–property corelationship. *J. Compos. Mater.* **2017**, *51*, 1769–1782. [CrossRef]
8. Dikshit, M.K.; Engle, P.E. Investigation of mechanical properties of CNT reinforced epoxy nanocomposite: A molecular dynamic simulations. *Mater. Phys. Mech.* **2018**, *37*, 7–15.
9. Dikshit, M.K.; Nair, G.S.; Pathak, V.K.; Srivastava, A.K. Investigation of mechanical properties of graphene reinforced epoxy nanocomposite using molecular dynamics. *Mater. Phys. Mech.* **2019**, *42*, 224–233.
10. Xue, Q.; Lv, C.; Shan, M.; Zhang, H.; Ling, C.; Zhou, X.; Jiao, Z. Glass transition temperature of functionalized graphene-polymer composites. *Comput. Mater. Sci.* **2013**, *71*, 66–71. [CrossRef]
11. Park, C.; Jung, J.; Yun, G.J. Thermomechanical properties of mineralized nitrogen-doped carbon nanotube/polymer nanocomposites by molecular dynamics simulations. *Compos. Part B Eng.* **2019**, *161*, 639–650. [CrossRef]
12. Singh, A.; Kumar, D. Temperature effects on the interfacial behavior of functionalized carbon nanotube–polyethylene nanocomposite using molecular dynamics simulation. *Proc. Inst. Mech. Eng. Part N J. Nanomater. Nanoeng. Nanosyst.* **2019**, *233*, 3–15. [CrossRef]
13. Zhang, B.; Li, J.; Gao, S.; Liu, W.; Liu, Z. Comparison of Thermomechanical Properties for Weaved Polyethylene and Its Nanocomposite Based on the CNT Junction by Molecular Dynamics Simulation. *J. Phys. Chem. C* **2019**, *123*, 19412–19420. [CrossRef]
14. Jeyranpour, F.; Alahyarizadeh, G.H.; Minuchehr, A. The thermo-mechanical properties estimation of fullerene-reinforced resin epoxy composites by molecular dynamics simulation-A comparative study. *Polymer* **2016**, *88*, 9–18. [CrossRef]
15. Izadi, R.; Ghavanloo, E.; Nayebi, A. Elastic properties of polymer composites reinforced with C_{60} fullerene and carbon onion: Molecular dynamics simulation. *Phys. B Condens. Matter* **2019**, *574*, 311636. [CrossRef]
16. Montazeri, A.; Rafii-Tabar, H. Multiscale modeling of graphene- and nanotube-based reinforced polymer nanocomposites. *Phys. Lett. Sect. A Gen. At. Solid State Phys.* **2011**, *375*, 4034–4040. [CrossRef]
17. Tsiamaki, A.; Anifantis, N. Simulation of the thermomechanical behavior of graphene/PMMA nanocomposites via continuum mechanics. *Int. J. Struct. Integr.* **2019**, *11*, 655–669. [CrossRef]
18. Giannopoulos, G.I. Linking MD and FEM to predict the mechanical behaviour of fullerene reinforced nylon-12. *Compos. Part B Eng.* **2019**, *161*, 455–463. [CrossRef]
19. Reddy, J.N.; Unnikrishnan, V.U.; Unnikrishnan, G.U. Recent developments in multiscale thermomechanical analysis of nanocomposites. In *Advances in Nanocomposites*; Springer: Cham, Switzerland, 2016; pp. 177–188.
20. Ruoff, R.S.; Thornton, T.; Smith, D. Density of fullerene containing soot as determined by helium pycnometry. *Chem. Phys. Lett.* **1991**, *186*, 456–458. [CrossRef]

21. Shinohara, H.; Sato, H.; Saito, Y.; Izuoka, A.; Sugawara, T.; Ito, H.; Sakurai, T.; Matsuo, T. Extraction and mass spectroscopic characterization of giant fullerences up to C_{500}. *Rapid Commun. Mass Spectrom.* **1992**, *6*, 413–416. [CrossRef]

22. Wang, B.-C.; Chiu, Y.-N. Geometry and orbital symmetry relationships of giant and hyperfullerenes: C_{240}, C_{540}, C_{960} and C_{1500}. *Synth. Met.* **1993**, *56*, 2949–2954. [CrossRef]

23. Schwerdtfeger, P.; Wirz, L.N.; Avery, J. The topology of fullerenes. *Wiley Interdiscip. Rev. Comput. Mol. Sci.* **2015**, *5*, 96–145. [CrossRef] [PubMed]

24. Kim, S.G.; Tomnek, D. Melting the fullerenes: A molecular dynamics study. *Phys. Rev. Lett.* **1994**, *72*, 2418–2421. [CrossRef] [PubMed]

25. Cabrera-Trujillo, J.; Alonso, J.; Iñiguez, M.; López, M.; Rubio, A. Theoretical study of the binding of Na clusters encapsulated in the fullerene. *Phys. Rev. B-Condens. Matter Mater. Phys.* **1996**, *53*, 16059–16066. [CrossRef]

26. Li, Z.; Lambros, J. Strain rate effects on the thermomechanical behavior of polymers. *Int. J. Solids Struct.* **2001**, *38*, 3549–3562.

27. Van Loock, F.; Fleck, N.A. Deformation and failure maps for PMMA in uniaxial tension. *Polymer* **2018**, *148*, 259–268. [CrossRef]

28. Swallowe, G.M.; Lee, S.F. Quasi-static and dynamic compressive behaviour of poly(methyl methacrylate) and polystyrene at temperatures from 293 K to 363 K. *J. Mater. Sci.* **2006**, *41*, 6280–6289. [CrossRef]

29. Mohammadi, M.; fazli, H.; karevan, M.; Davoodi, J. The glass transition temperature of PMMA: A molecular dynamics study and comparison of various determination methods. *Eur. Polym. J.* **2017**, *91*, 121–133. [CrossRef]

30. Accelrys, I. *Materials Studio*; Accelrys Software Inc.: San Diego, CA, USA, 2010.

31. Sun, H. Compass: An ab initio force-field optimized for condensed-phase applications-Overview with details on alkane and benzene compounds. *J. Phys. Chem. B* **1998**, *102*, 7338–7364.

32. Chen, X.P.; Yuan, C.A.; Wong, C.K.Y.; Koh, S.W.; Zhang, G.Q. Validation of forcefields in predicting the physical and thermophysical properties of emeraldine base polyaniline. *Mol. Simul.* **2011**, *37*, 990–996.

33. Horrigan, D.P.W.; Williams, J.F. *LUSAS Finite Element System User Examples Manual*; Department of Mechanical and Manufacturing Engineering, University of Melbourne: Melbourne, Australian, 1997.

34. Marx, D.; Hutter, J. *Ab Initio Molecular Dynamics: Basic Theory and Advanced Methods*; Cambridge University Press: New York, NY, USA, 2009.

35. Murakoshi, Y.; Shan, X.C.; Shimizu, T.; Maeda, R. Hot embossing characteristics of PEEK compared to PC and PMMA. In Proceedings of the Design, Test, Integration and Packaging of MEMS/MOEMS, Cannes, France, 7 May 2003; pp. 304–311.

36. Flory, P.J. *Principles of Polymer Chemistry*; Cornell University Press: New York, NY, USA, 1953.

 *materials*

Article

Parametric Investigation of Particle Swarm Optimization to Improve the Performance of the Adaptive Neuro-Fuzzy Inference System in Determining the Buckling Capacity of Circular Opening Steel Beams

Quang Hung Nguyen [1,*], **Hai-Bang Ly** [2,*], **Tien-Thinh Le** [3,*], **Thuy-Anh Nguyen** [2], **Viet-Hung Phan** [4], **Van Quan Tran** [2] and **Binh Thai Pham** [2]

1   Thuyloi University, Hanoi 100000, Vietnam
2   University of Transport Technology, Hanoi 100000, Vietnam; anhnt@utt.edu.vn (T.-A.N.); quantv@utt.edu.vn (V.Q.T.); binhpt@utt.edu.vn (B.T.P.)
3   Institute of Research and Development, Duy Tan University, Da Nang 550000, Vietnam
4   University of Transport and Communications, Ha Noi 100000, Vietnam; phanviethung@utc.edu.vn
*   Correspondence: hungwuhan@tlu.edu.vn (Q.H.N.); banglh@utt.edu.vn (H.-B.L.); letienthinh@duytan.edu.vn (T.-T.L.)

Received: 20 April 2020; Accepted: 6 May 2020; Published: 12 May 2020

Abstract: In this paper, the main objectives are to investigate and select the most suitable parameters used in particle swarm optimization (PSO), namely the number of rules (n_{rule}), population size (n_{pop}), initial weight (w_{ini}), personal learning coefficient (c_1), global learning coefficient (c_2), and velocity limits (f_v), in order to improve the performance of the adaptive neuro-fuzzy inference system in determining the buckling capacity of circular opening steel beams. This is an important mechanical property in terms of the safety of structures under subjected loads. An available database of 3645 data samples was used for generation of training (70%) and testing (30%) datasets. Monte Carlo simulations, which are natural variability generators, were used in the training phase of the algorithm. Various statistical measurements, such as root mean square error (RMSE), mean absolute error (MAE), Willmott's index of agreement (IA), and Pearson's coefficient of correlation (R), were used to evaluate the performance of the models. The results of the study show that the performance of ANFIS optimized by PSO (ANFIS-PSO) is suitable for determining the buckling capacity of circular opening steel beams, but is very sensitive under different PSO investigation and selection parameters. The findings of this study show that $n_{rule} = 10$, $n_{pop} = 50$, $w_{ini} = 0.1$ to 0.4, $c_1 = [1, 1.4]$, $c_2 = [1.8, 2]$, $f_v = 0.1$, which are the most suitable selection values to ensure the best performance for ANFIS-PSO. In short, this study might help in selection of suitable PSO parameters for optimization of the ANFIS model.

Keywords: particle swarm parameters; adaptive neuro-fuzzy inference system; circular opening steel beams; buckling capacity

---

1. Introduction

Circular opening steel beams have been increasingly acknowledged in structural engineering because of their many remarkable advantages [1], including their ability to bridge the span of a large aperture or their lighter weight compared with conventional steel beams. In general, the industrial approach to producing such a structural member is the rolled method, involving a single steel piece. This is then cut so that the two halves can be assembled, making an I-section, which is also called an H-section steel beam. Hoffman et al. [2] showed that the flexural stiffness and specific gravity per unit length was improved significantly in circular opening steel beams structures. In addition, economic and aesthetics factors are also beneficial points that deserve significant attention [3,4]. A typical structural member has a regular circular openings along its length [1–8], and is about 40–60% deeper and 40–60% stronger than a regular I-section [5,6]. Because of these advantages, circular beams are not only used in lightweight or large-span structures, but are also used for other complex civil engineering structures, such as bridges [9]. Due to the possibility of using circular opening steel beams in various engineering applications, investigation of the failure behavior is crucial to ensure the safety of structures. Several previously published studies on the failure modes of circular beams, for instance the work by Sonck et al. [3], have shown that the web openings are the leading causes of the complex failure behavior of cellular beams, including web post-buckling (WPB), the Vierendeel mechanism (VM), rupture of the web post-weld [1], local web buckling (LWB), and web distortional buckling (WDB) [5,6].

Miscellaneous analysis-related research studies have been conducted to study the behavior of circular opening steel beams [10–12], which have mainly focused on the web openings using various numerical approaches [7,9]. As an example, Chung et al. [11] used finite element models with material and geometrical nonlinearity to calculate the behavior of circular beams, resulting in approximately 15.8% of error. Numerical methods help create various case studies in order to gain more knowledge about the working principles of the structures. Taking the work of Panedpojaman and Thepchatri [4] as an example, the authors created a total of 408 nonlinear finite element models using ANSYS software to investigate the behavior of circular steel beams. The results indicated that there is always a small difference between the finite element model and the theoretical formulation. In another study, Sonck et al. [3] generated 597 numerical models, which were calibrated with laboratory tests for 14 geometrically different full-scale steel cellular beams and verified with 1948 numerical analyzes. The results showed that the experimental and numerical curves were identical, with a maximum load gap range of 5.1% to 6.5%. Typically, the numerical models are useful for evaluating the behavior of circular beams [1,3,6,9,13]. However, these model require much effort and the use of modern software and equipment.

Machine learning (ML) algorithms, a branch of artificial intelligence (AI) techniques, have been constantly developed during the past few decades due to the significant increase in computer science [14–21]. Various ML models have been effectively implemented to solve countless specific engineering problems, including in material sciences [22–24], geotechnical engineering [25–29], and especially structural engineering [18,30–32]. As an example, Vahid et al. [33] selected an artificial neural network (ANN) algorithm, the most popular ML model, to predict the shear capacity of a web opening steel I-beam. The proposed ANN model had better accuracy compared with other existing formulas or theoretical predictions derived from the ACI 318-08 standard. Abambres et al. [34] also used the ANN method to investigate the buckling load capacity of cellular beams under uniformly distributed vertical loads, using eight geometrical parameters. Good results were achieved by the ANN, giving 3.7% for the total error and 0.4% for the average relative error. Blachowski and Pnevmatikos [35] proposed an ANN model for the design and control of the vibration of structural elements under earthquake loading. In the same context of seismic excitation, Pnevmatikos and Thomos [36] employed a stochastic control approach to determine the influence of random characters on the dynamic behavior of engineering structures. The neuro-fuzzy system is another efficient ML algorithm, which has been employed in many structural and material engineering applications, including for steel structures. Seitllari and Naser [37] investigated the performance of an adaptive neuro-fuzzy inference system (ANFIS) in predicting a fire-induced spalling phenomenon in steel-reinforced concrete structures. Naser [38] derived a material model for steel structures, taking into account the dependency of temperature based on machine learning techniques. Basarir et al. [39] compared the performance between conventional regression techniques and ANFIS in predicting the ultimate pure bending of concrete-filled steel tubular members. Naderpour and Mirrashid [40] used ANFIS to predict the shear strength of beams that had been reinforced with steel stirrups. Mermerdaş et al. [41] applied ANFIS to evaluate the flexural behavior of steel circular hollow section (CHS) beams. It was stated that the ANFIS was a promising tool for quick and accurate evaluation of the mechanical behavior of steel-based engineering structures.

In general, the ML algorithms are excellent and effective for evaluating the behavior of structural members, including circular beams. However, their performance depends significantly on the selection of parameters used to learn the models [42]. Therefore, the process of determining such parameters is crucial to obtain highly reliable and accurate prediction results. Concerning the ANN, many parameters could be involved, such as the initial weights, biases to start the training phase, the learning rate, the stopping criterion, the choice of features in the training phase, the choice of the splitting dataset ratio, the number of hidden layers and the corresponding activation functions, the training algorithm, and the number of neurons in each hidden layer [43–45]. Considering the ANFIS, two groups of parameters can be considered, namely the nonlinear parameters of the antecedent membership function (MF) and linear parameters of the consequent MF, which depends on the partitioning of the fuzzy space, as well as the type of Sugeno model [46,47]. Besides, many optimization techniques, such as particle swarm optimization (PSO), differential evolution (DE), evolutionary algorithm (EA), genetic algorithm (GA), artificial bee colony (ABC). or cuckoo search (CS) techniques, have been proposed to optimize the parameters of the ML models [48,49]. Each optimization technique also possesses many different parameters that need to be tuned to obtain good prediction performances, inducing the time required to adjust the combination of these parameters [48,49]. Among the well-known optimization techniques, PSO is considered as one of the most popular and effective techniques [50]. Many hybrid ML algorithms have used PSO for the parameter tuning process, including ANN, ANFIS, and Support Vector Machine (SVM) algorithms [51–53]. In the literature, limited studies have used ANFIS optimized by PSO (ANFIS-PSO) to predict the mechanical properties of structural members. Moreover, a systematic investigation of ANFIS-PSO parameters under random sampling has not been performed, as the sampling method has been proven to greatly affect the accuracy of the ML algorithms [54].

In this study, the main purpose was to carry out a parametric investigation of PSO parameters to improve the performance of ANFIS in predicting the buckling capacity of circular opening steel beams, which is an important mechanical property that is crucial for the safety of structures under subjected loads. The database used in this work consisted of 3645 data samples, which were derived from numerical results using ANSYS and available in the literature. The parametric studies were carried out with the help of Monte Carlo simulations, which are natural variability generators, in the training phase of the algorithm. Various statistical measurements, such as the root mean square error (RMSE), mean absolute error (MAE), Willmott's index of agreement (IA), and Pearson's coefficient of correlation (R), were used to evaluate the performance of the model.

2. Novelty and Significance of This Study

As reported in the introduction, the estimation of the buckling capacity of circular opening steel beams is important for the safety of structures under subjected loads. As instability is a complex (nonlinear) problem that is affected by various parameters, the determination of the critical buckling load remains challenge for researchers (engineers) in the fields of mechanics and civil engineering. Despite various experimental works having investigated this problem, it is not easy to derive a generalized expression that considers all the parameters that govern the instability of circular opening steel beams. To overcome this difficulty, the use of ML techniques, such as ANFIS optimized by the PSO algorithm proposed in this study, could be a good choice as a surrogate model. This soft computing method could help to explore the nonlinear relationships between the buckling capacity and the input variables, especially the geometrical parameters of the beams. In addition, the investigation of PSO parameters based on the Monte Carlo random sampling technique could contribute to better knowledge on selection of suitable parameters to achieve better performance with the PSO algorithm, which could be further recommended for other problems. Finally, the proposed ML-based model could be a potential tool for researchers or structural engineers in accurately estimating the buckling capacity of circular opening steel beams, which could (i) work within the ranges of values used in this study for the input variables and (ii) save time and costs in development of other numerical schemes (i.e., finite element models).

3. Database Construction

The database in this study was obtained by analyzing 3645 different configurations of circular opening steel beams (Figure 1). It should be noted that the database was extracted from a validated finite element model, which was previously proposed in the literature by Abambres et al. [34]. It consisted of 8 input parameters, namely the length of the beam (denoted as L), the end opening distance (denoted as d_0), circular opening diameter (denoted as D), the inter-opening distance (denoted as d), the height of the section (denoted as H), the thickness of the web (denoted as t_{web}), the width of the flange (denoted as w_{flange}), the thickness of the flange (denoted as t_{flange}), and the buckling capacity, which was considered as the target variable (denoted as P_u). It should be pointed out that the database was generated for one material type (with a typical Young's modulus of 210 GPa and Poisson's ratio of 0.3). The results of the statistical analysis of the Pu and the corresponding influential parameters are presented in Table 1.

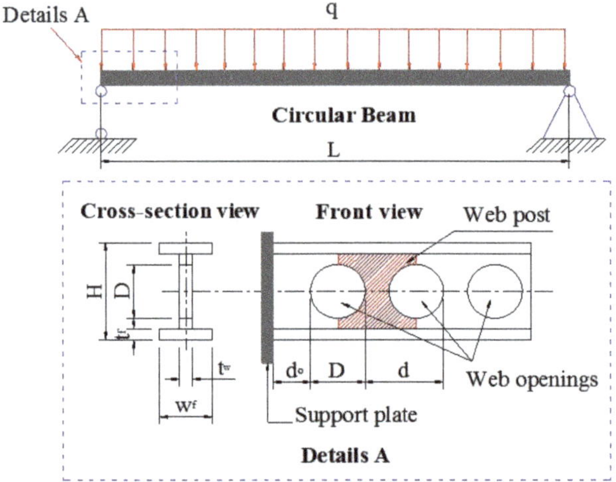

Figure 1. Diagram of circular opening steel beam under uniform loading and its geometrical parameters.

Table 1. Initial statistical analysis of the dataset.

Variable	Length of Beam	End Opening Distance	Opening Diameter	Inter-Opening Distance	Height of Section	Thickness of Web	Width of Flange	Thickness of Flange	Buckling Capacity
Symbol	L	d_0	D	d	H	t_{web}	w_{flange}	t_{flange}	P_u
Unit	m	mm	mm	mm	mm	mm	mm	mm	N/m
Role	Input	Input	Input	Input	Input	Input	Input	Input	Output
Min	4.0	12.0	247.0	24.70	420.00	9.0	162.0	15.0	26.4
MD [a]	6.0	256.5	373.0	108.17	560.00	12.0	216.0	20.0	169.3
Max	8.0	718.0	560.0	274.40	700.00	15.0	270.0	25.0	1361.7
Mean	6.0	265.4	383.6	112.51	560.00	12.0	216.0	20.0	225.7
SD [b]	1.4	157.5	93.0	68.51	114.33	2.5	44.1	4.1	182.5
CV [c]	23.6	59.3	24.2	60.90	20.42	20.4	20.4	20.4	80.9

[a] Median. [b] Standard deviation. [c] Coefficient of variation (%).

The input and target variables in this work were scaled in the range of [0, 1] to minimize the numerical bias of the dataset. After performing the simulation part, a transformation into the normal range was conducted to better interpret the obtained results. Concerning the development phase, the dataset was split into two parts, namely the training part (70% of the total data) and the testing part (the remaining 30% of the data), which served as the learning and validation phases of the proposed ANFIS-PSO model, respectively.

4. Machine Learning Methods

4.1. Adaptive Neuro-Fuzzy Inference System

Jang et al. [55] introduced the fuzzy adaptive system of adaptive neurology, called ANFIS, as an improved ML method and a data-driven modeling approach to evaluate the behavior of complex dynamic systems [56,57]. ANFIS aims to systematically generate unknown fuzzy rules from a given set of input and output data. ANFIS creates a functional map that approximates the internal system parameter estimation method [58–60]. Fuzzy systems are rule-based systems developed from a set of language rules. These systems can represent any system with good accuracy and are, therefore, considered to be universal approximators. Thus, ANFIS is the most popular neuro-fuzzy hybrid network used for the modeling of complex systems. The ANFIS model's main strength is that it is a universal approximator with the ability to request interpretable "if–then" rules [61]. In ANFIS, a Sugeno-type fuzzy system was used to construct the five-layer network.

4.2. Particle Swarm Optimization (PSO)

Eberhart and Kennedy developed the PSO algorithm in 1995. It is an evolutionary computing technique with a particular enhancement method, population collaboration, and competition based on the simulation of simplified social models, such as bird flocking, fish schooling, and swarming theory [62–65]. It is a biological-based algorithm that shapes bird flocking social dynamics large number of birds flock synchronously, suddenly change direction, iteratively scatter and group, and eventually perch on a target. The PSO algorithm supports simple rules for bird flocking and acts as an optimizer for nonlinear continuous functions [66]. PSO has gained much attention and has been successfully applied in various fields, especially for unconstrained continuous optimization problems [67]. Indeed, in PSO, a swarm member, also called a particle, is a potential solution, which is used as a search space point. The global equilibrium is known as the food position. The particle has a fitness value and a speed with which to change its flight path for the best swarm experiences to find the global optimum in the D-dimensional solution space. The PSO algorithm is easy to implement and many optimization problems have been empirically shown to perform well [68]. However, its performance depends significantly on the algorithm parameters described below.

4.2.1. Initial Weight (w_{ini})

The particle in the PSO is represented as a real-valued vector containing an instance of all parameters that characterize the problem of optimization. By flying a number of particles, called a swarm, the PSO explores the solution space. The initial swarm is generated at random, and generally consecutive iterations maintain a consistent swarm size. The swarm of particles looks for the optimum target solution in each iteration by referring to past experiences.

4.2.2. Cognition Learning Rate (Personal Learning Coefficient—c_1)

PSO enriches swarm intelligence by storing the best positions that each particle has visited so far. Particles I recall the best position among those it met, called pbest, and the best positions of its neighbors. There are two variants, namely lbest and gbest, used to hold the neighbors in the best position. The particle in the local version keeps track of the best lbest location obtained by its neighboring local particles. For the global version, any particles in the whole swarm will determine the best location for gbest. Therefore, the gbest model is the lbest model's special case.

4.2.3. Social Learning Rate (Global Learning Coefficient—c_2)

PSO starts with the random initialization in the search space of a population (swarm) of individuals (particles) and operates on the particles' social behavior in the swarm. Consequently, it finds the best global solution by simply adjusting each individual's trajectory to their own best location and to the best swarm particle in each phase (generation). Nevertheless, the trajectory of each particle in the search space is modified according to their own flying experience and the flying experience of the other particles in the search space by dynamically altering the velocity of each particle.

4.2.4. Number of Particles (Population Size—n_{pop})

The location and speed of the ith particle can be expressed in the dimensional search space. Every particle has its own best (pbest) location, according to the best personal objective value at the time t. The world's best particle (gbest) is the best particle found at time t in the entire swarm.

4.2.5. Velocity Limits (f_v)

Each particle's new speed is determined as follows:

$$y_{i,j}(t+1) = wy_{i,j}(t) + c_1 r_1 (p_{i,j} - x_{i,j}(t)) + c_2 r_2 (p_{g,j} - x_{i,j}(t)); \quad j = 1, 2, \ldots, d \tag{1}$$

where c_1 and c_2 are constants referred to as acceleration coefficients, w is referred to as the inertia factor, and r_1 and r_2 are two independent random numbers distributed evenly within the spectrum. The location of each particle is, thus, modified according to the following equation in each generation:

$$a_{i,j}(t+1) = a_{i,j}(t) + y_{i,j}(t+1), \quad j = 1, 2, 3, \ldots, d \tag{2}$$

In the standard PSO, Equation (1) is used to calculate the new velocity according to its previous velocity and to the distance of its current position from both its own best historical position and its neighbors' best positions. The value of each factor in Y_i can be clamped within the range to monitor excessive particles roaming outside the search area, then the particle flies toward a new location.

4.3. Monte Carlo Simulation

The Monte Carlo technique has been commonly used as a variability generator in the training phase of the algorithm, taking into account the randomness of the input space [69–72]. Hun et al. [73] studied the problem of crack propagation in heterogeneous media within a probabilistic context using Monte Carlo simulations. Additionally, Capillon et al. [74] investigated an uncertainty problem in structural dynamics for composite structures using Monte Carlo simulations. Overall, the Monte Carlo method has been successfully applied to take into account the randomness in the field of mechanics [75–80]. The key point of the Monte Carlo method is to repeat the simulations many times to calculate the output responses by randomly choosing values of the input variables in the corresponding space [81,82]. In this manner, all information about the fluctuations in the input space can be transferred to the output response. In this work, a massive numerical parallelization scheme was programmed to conduct the randomness propagation process. The statistical convergence of the Monte Carlo method reflects whether the number of simulations is sufficient, which can be defined as follows [83–85]:

$$f_{conv} = \frac{100}{m\underline{S}} \sum_{j=1}^{m} S_j \tag{3}$$

where m is the number of Monte Carlo iterations, S is the random variable considered, and $\underline{S}$ is the average value of S.

4.4. Quality Assessment Criteria

In the present work, three quality assessment criteria—the correlation coefficient (R), root mean squared error (RMSE), and mean absolute error (MAE)—have been used in order to validate and test the developed AI models. R^2 allows us to identify the statistical relationship between two data points and can be calculated using the following equation [86–92]:

$$R = \sqrt{\frac{\sum_{j=1}^{N} \left(y_{0,j} - \overline{y}\right)\left(y_{p,j} - \overline{y}\right)}{\sqrt{\sum_{j=1}^{N} \left(y_{0,j} - \overline{y}\right)^2 \sum_{j=1}^{N} \left(y_{p,j} - \overline{y}\right)^2}}} \tag{4}$$

where N is the number of observations, y_p and $\overline{y}$ are the predicted and mean predicted values, while y_0 and $\overline{y}$ are the measured and mean measured values of Young's modulus of the nanocomposite, respective $j = 1{:}N$. In the case of RMSE and MAE, which have the same units as the values being estimated, low value for RMSE and MAE basically indicate good accuracy of the models' prediction output [93,94]. In an ideal prediction, RMSE and MAE should be zero. RMSE and MAE are given by the following formulae [95–99]:

$$RMSE = \sqrt{\sum_{i=1}^{N} (y_0 - y_p)^2 / N} \tag{5}$$

$$MAE = \frac{1}{N} \sum_{i=1}^{N} |y_0 - y_p| \tag{6}$$

In addition, the Willmott's index of agreement (IA) has also been employed in this study. The formulation of IA is given by [100,101]:

$$IA = 1 - \frac{\sum_{i=1}^{N} \left(y_0 - y_p\right)^2}{\sum_{i=1}^{N} \left(|y_0 - \overline{y}| + |y_p - \overline{y}|\right)^2} \tag{7}$$

5. Results and Discussion

5.1. Description of Parametric Studies

In order to investigate the influence of PSO parameters on the performance of ANFIS, parametric studies were carried out by varying n_{rule}, n_{pop}, w_{ini}, c_1, c_2, and f_v, as indicated in Table 2. It is noteworthy that the proposed range was selected by considering both problem dimensionality (i.e., complexity) and computation time. As recommended by He et al. [102] and Chen et al. [48], the PSO initial weight should be carefully investigated. Therefore, a broad range of w_{ini} was proposed, ranging from 0.1 to 1.2. The number of populations varied from 20 to 300 with a nonconstant step, whereas the coefficients c_1 and c_2 ranged from 0.2 to 2 with a resolution of 0.2. The number of fuzzy rules varied from 5 to 40. Finally, the f_v ranged from 0.05 to 0.2.

The relationship between the number of fuzzy rules and the number of total ANFIS weight parameters is depicted in Figure 2. As can be seen, the relationship is linear, showing that as the number of fuzzy rules increases, the number of ANFIS weight parameters increases. For illustration purposes, the number of weight parameters increases from 50 to 370, while the number of fuzzy rules increases from 5 to 40. Additionally, the characteristics of the ANFIS structure are described in Table 3, showing that the Gaussian membership function was used to generate fuzzy rules.

Table 2. Values used for parameters in parametric studies.

Parameters	Values Used											
n_{rule}	5	10	15	20	30	40						
n_{pop}	20	40	60	80	100	150	200	250	300			
w_{ini}	0.1	0.2	0.3	0.4	0.5	0.6	0.7	0.8	0.9	1	1.1	1.2
c_1	0.2	0.4	0.6	0.8	1	1.2	1.4	1.6	1.8	2		
c_2	0.2	0.4	0.6	0.8	1	1.2	1.4	1.6	1.8	2		
f_v	0.05	0.1	0.15	0.2								

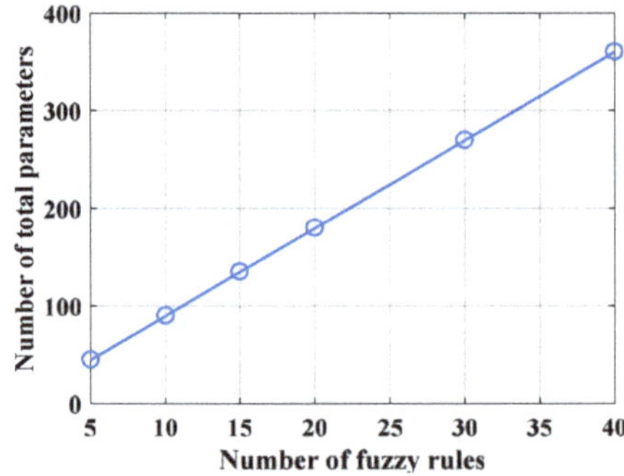

Figure 2. Influence of the number of fuzzy rules on the number of total ANFIS weight parameters to be optimized by PSO.

Table 3. Characteristics of ANFIS structure.

Parameter	Description
Number of inputs	8
Number of outputs	1
Input membership function type	Gaussian
Number of parameter per membership function	2
Number of fuzzy rules	n_{rule}
Output membership function type	Linear
Number of nonlinear parameters	$8 \times 2 \times n_{rule}$
Number of linear parameters	$9 \times n_{rule}$
Number of total parameters	$25 \times n_{rule}$

5.2. Preliminary Analyses

5.2.1. Computation Time

Figure 3 presents the influence of n_{rule} and swarm parameters on the computation time. It is worth noting that the running time was scaled with respect to the minimum value of the corresponding parameter. For instance, the computation time using $n_{rule} = 10$ is two times larger than the case using $n_{rule} = 5$. Additionally, in Figure 3, it is seen that n_{rule} and n_{pop} exhibited the highest slope (about 0.75), confirming that these two parameters required considerable computation time. For all other parameters, the computation time remained constant when increasing the value of the parameter.

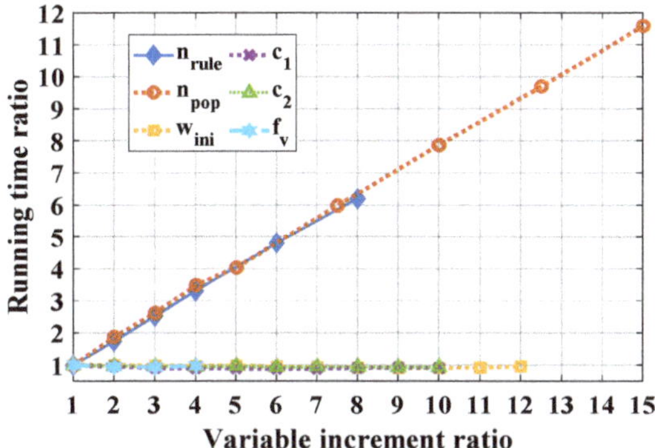

Figure 3. Influence of variable increment ratio on running time, noting that both n_{rule} and n_{pop} exhibit a slope coefficient of 0.75.

5.2.2. PSO Stopping Criterion

In this study, 1000 iterations were applied as a stopping criterion in the optimization problem for the weight parameters of ANFIS. Figure 4 shows the convergence of statistical criteria in the function of n_{rule}, whereas Figure 5 presents the convergence of these criteria regarding n_{pop}. For the evaluation of RMSE, MAE, and R over 1000 iterations in 6 cases for different n_{rule}, the training parts are given in Figure 4a–c, whereas the testing parts are displayed in Figure 4d–f. It was observed that at least 800 iterations were required to obtain convergence results for RMSE, MAE, and R for all the cases. However, no specific trend could be deduced in order to obtain the best n_{rule} parameter. Finally, it is worth noting that for all the cases of n_{rule}, the values of RMSE, MAE, and R for the testing part were very close. Indeed, the values of RMSE for the testing part ranged from 0.038 to 0.043, the values of MAE for the testing part varied from 0.015 to 0.022, and those of R ranged from 0.95 to 0.97. The evaluation of RMSE, MAE, R over 1000 iterations in 9 cases of n_{pope} is shown (Figure 5). Similar results were obtained as for n_{rule}. At least 800 iterations were needed to obtain the convergence results.

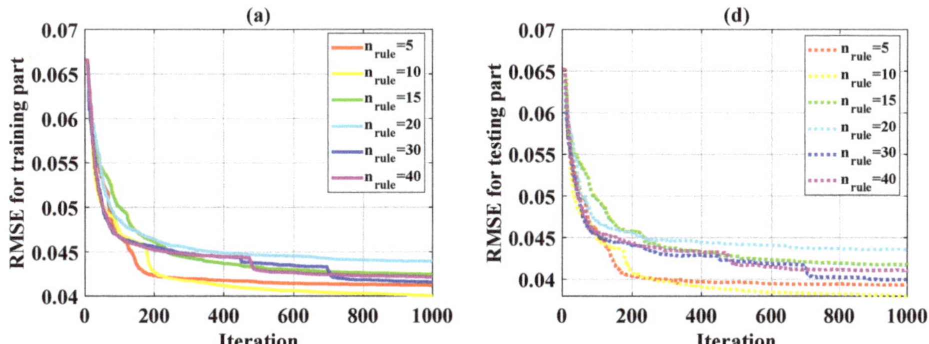

Figure 4. *Cont.*

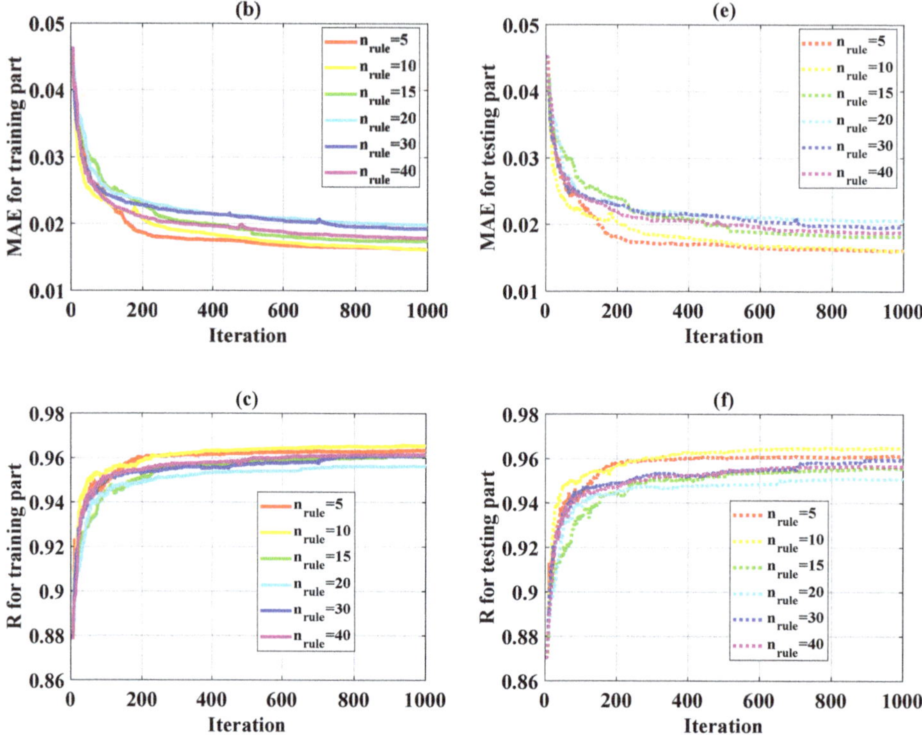

Figure 4. Convergence of several statistical criteria over 1000 iterations in terms of n_{rule} for the training part: (**a**) RMSE, (**b**) MAE, (**c**) R. Convergence of several statistical criteria over 1000 iterations in terms of n_{rule} for the testing part: (**d**) RMSE, (**e**) MAE, (**f**) R.

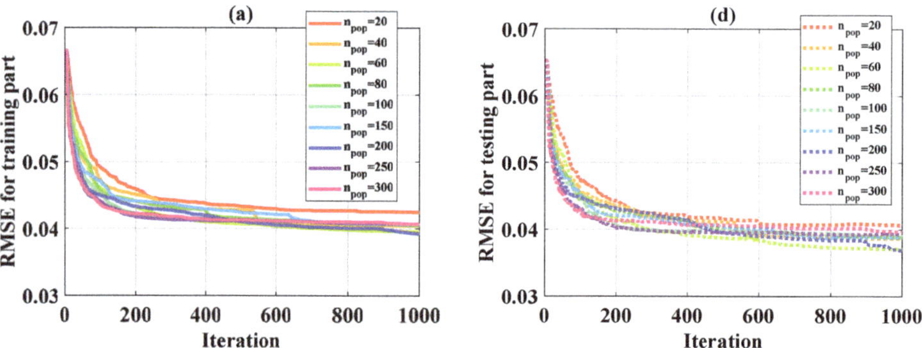

Figure 5. *Cont.*

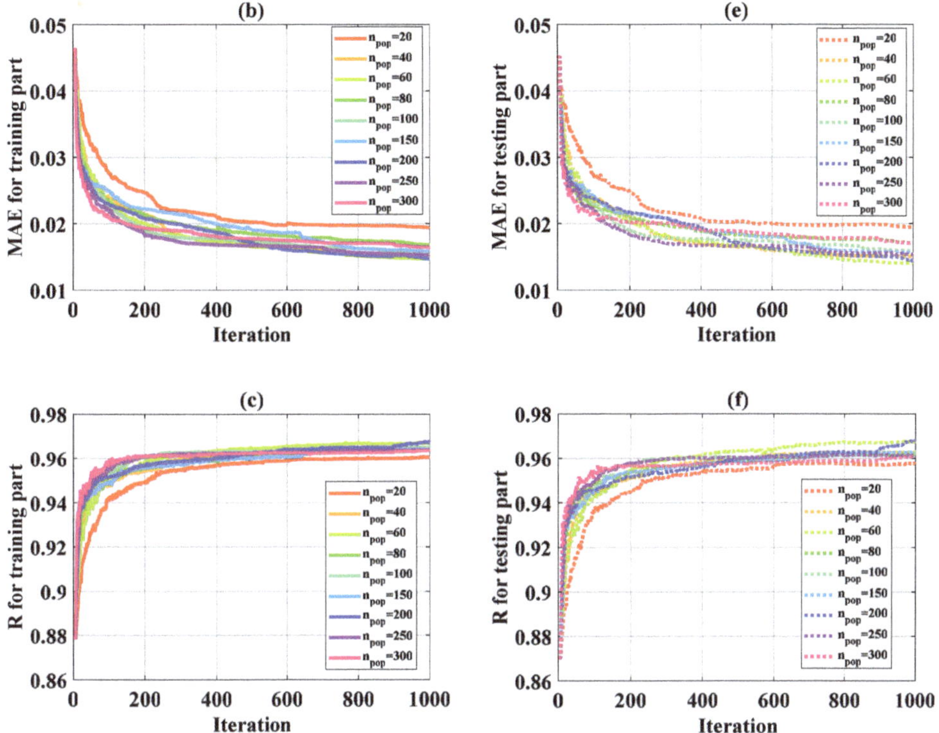

Figure 5. Convergence of several statistical criteria over 1000 iterations in terms of n_{pop} for the training part: (**a**) RMSE, (**b**) MAE, (**c**) R. Convergence of several statistical criteria over 1000 iterations in terms of n_{pop} for the testing part: (**d**) RMSE, (**e**) MAE, (**f**) R.

5.2.3. Statistical Convergence

In order to take into account variability in the input space, 200 random realizations were performed for each configuration. These realizations increased the influence of the probability density function of inputs on the optimization results. In terms of n_{rule}, Figure 6a–c indicate the statistical convergence of RMSE, MAE, and R for the training part, whereas Figure 6d–f present the statistical convergence of the same parameters for the testing part, respectively. It can be seen that after about 100 random realizations, statistical convergence was reached, which was correct for all the tested cases. Similarly, Figure 7 shows the statistical convergence in terms of n_{pop} for both training and testing parts. Similarly, 200 random realizations were observed to be sufficient to achieve reliable results.

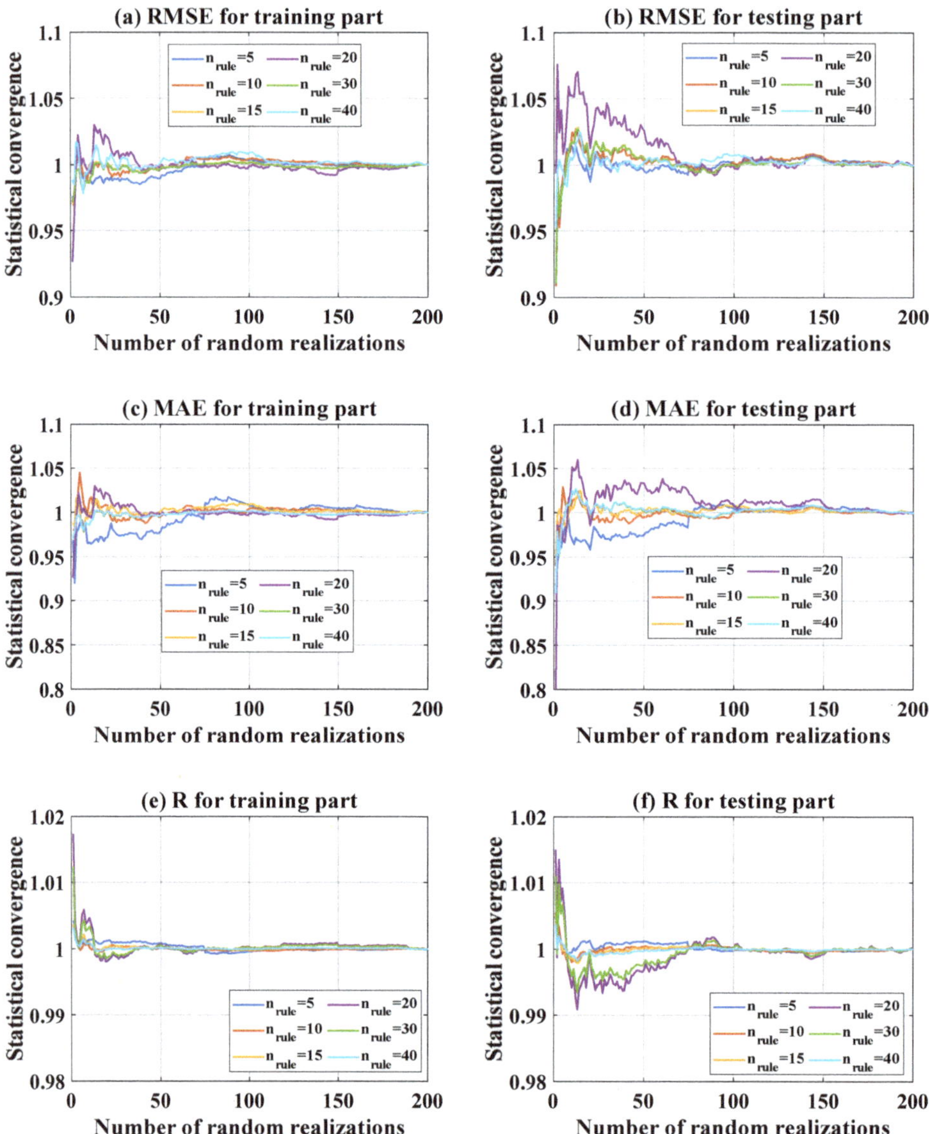

Figure 6. Statistical convergence over 200 random realizations in terms of n_{rule} for the training part: (**a**) RMSE, (**b**) MAE, (**c**) R. Statistical convergence over 200 random realizations in terms of n_{rule} for the testing part: (**d**) RMSE, (**e**) MAE, (**f**) R.

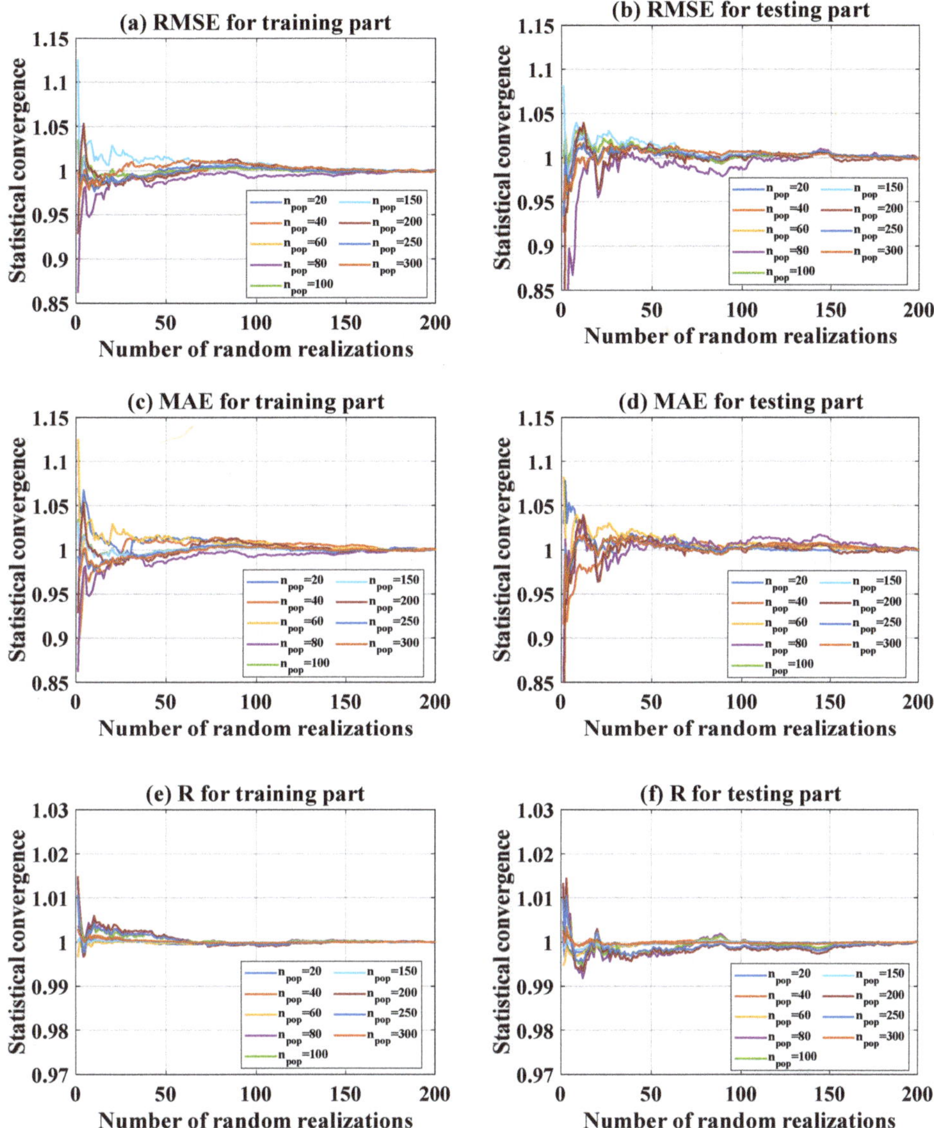

Figure 7. Statistical convergence over 200 random realizations in terms of n_{pop} for the training part: (**a**) RMSE, (**b**) MAE, (**c**) R. Statistical convergence over 200 random realizations in terms of n_{pop} for the testing part: (**d**) RMSE, (**e**) MAE, (**f**) R.

5.3. Parametric Performance

5.3.1. Influence of Number of Rules (n_{rule})

The evaluation of RMSE, MAE, R, and IA in the function of n_{rule} is presented in Figure 8a–d, respectively, for both training and testing parts. It can be seen that the accuracy of the ANFIS-PSO reduced when the number of n_{rule} increased (i.e., RMSE and MAE increased, while R and IA decreased). It is worth noting that the higher the number of rules, the larger dimensionality of the problem

(Figure 2). Therefore, regarding the total number of ANFIS weight parameters, the computation time, and the average value of the statistical criteria (RMSE, MAE, R, and IA), $n_{rule} = 10$ was considered as the most appropriate value.

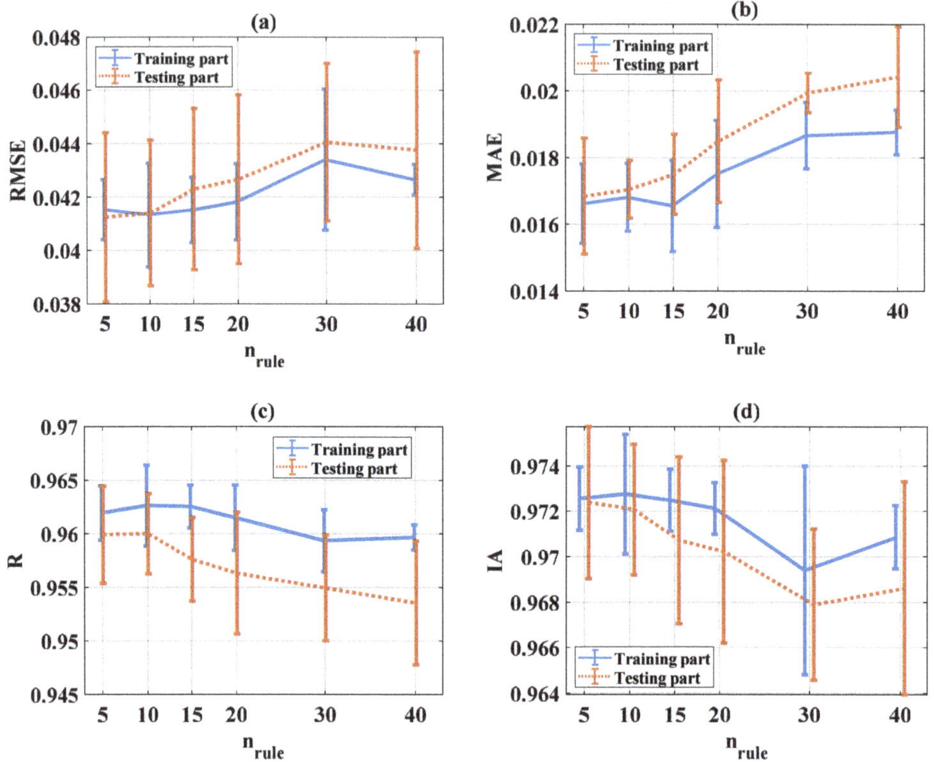

Figure 8. Evaluation of statistical criteria in the function of n_{rule}: (**a**) RMSE, (**b**) MAE, (**c**) R, and (**d**) IA.

5.3.2. Influence of Population Number (n_{pop})

The evaluation of statistical criteria in the function of n_{pop} for RMSE, MAE, R, and IA is shown in Figure 9a–d, respectively, for both training and testing parts. It can be seen that except for the low value for population size (i.e., $n_{pop} = 20$), all other n_{pop} values show good prediction results, especially for $n_{pop} = 200$. However, as introduced in the preliminary analyses for computation time, the higher the number of n_{pop}, the more time is consumed. Finally, $n_{pop} = 50$ was chosen as the most appropriate average value for statistical criteria and computation time.

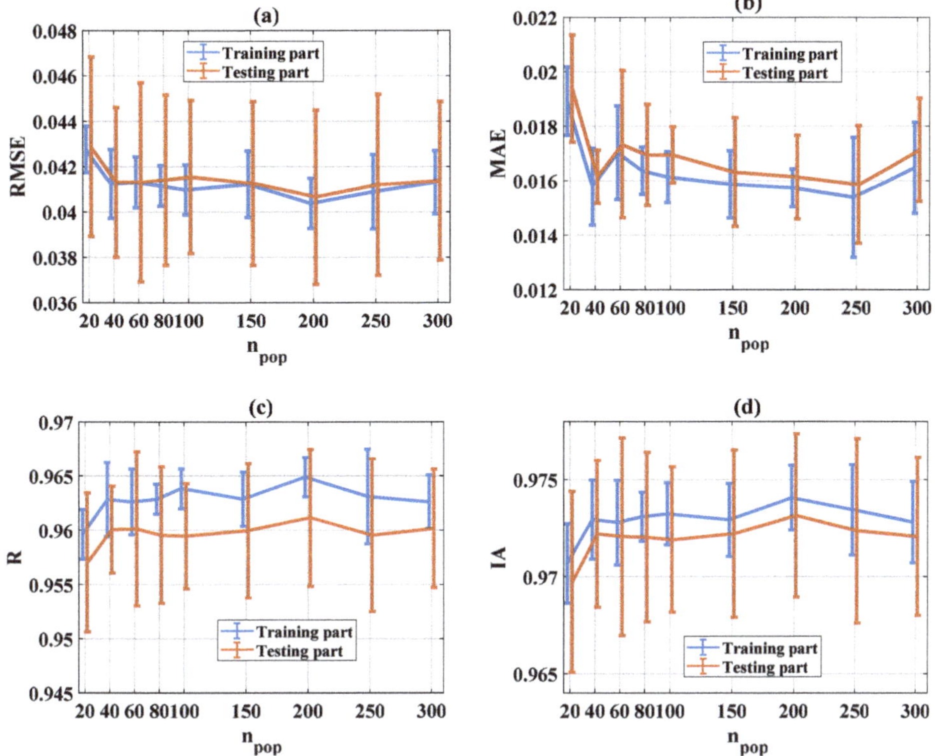

Figure 9. Evaluation of statistical criteria in the function of n_{pop}: (**a**) RMSE, (**b**) MAE, (**c**) R, and (**d**) IA.

5.3.3. Influence of Initial Weight (w_{ini})

The evaluation of statistical criteria in the function of w_{ini} for RMSE, MAE, R, and IA is shown in Figure 10a–d, respectively, for both training and testing parts. It can be seen that poor prediction performance was obtained when w_{ini} was larger than 0.5 (i.e., an increase of RMSE and MAE values and a decrease of R and IA values). Regarding the statistical criteria (RMSE, MAE, R, and IA), a w_{ini} value range of between 0.1 and 0.4 was the most appropriate.

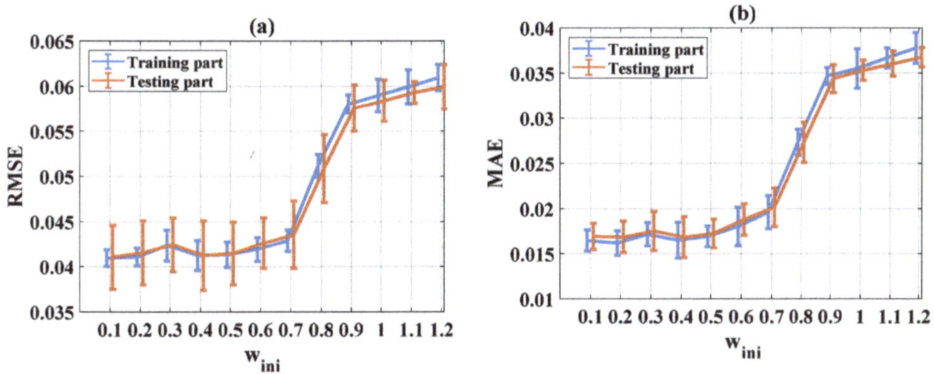

Figure 10. *Cont.*

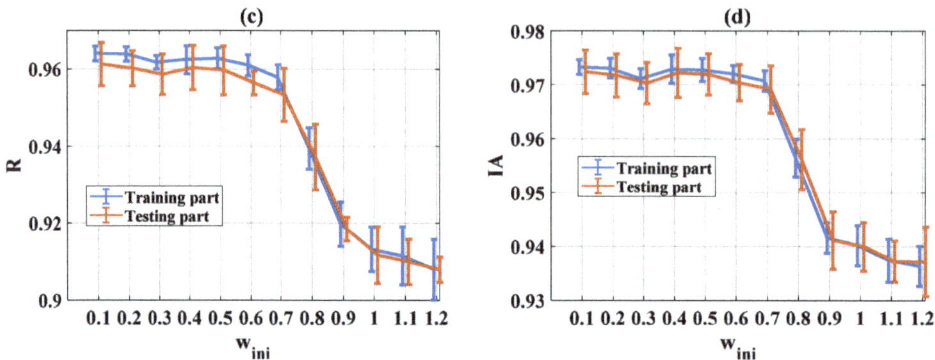

Figure 10. Evaluation of statistical criteria in the function of w_{ini}: (**a**) RMSE, (**b**) MAE, (**c**) R, and (**d**) IA.

5.3.4. Influence of Personal Learning Coefficient (c_1)

The evaluation of statistical criteria in the function of c_1 for RMSE, MAE, R, and IA is shown in Figure 11a–d, respectively, for both training and testing parts. It can be seen that good prediction performance was obtained when c_1 was in the range of [1, 1.4] for all statistical criteria (RMSE, MAE, R, and IA). Therefore, c_1 = [1, 1.4] was the most appropriate value.

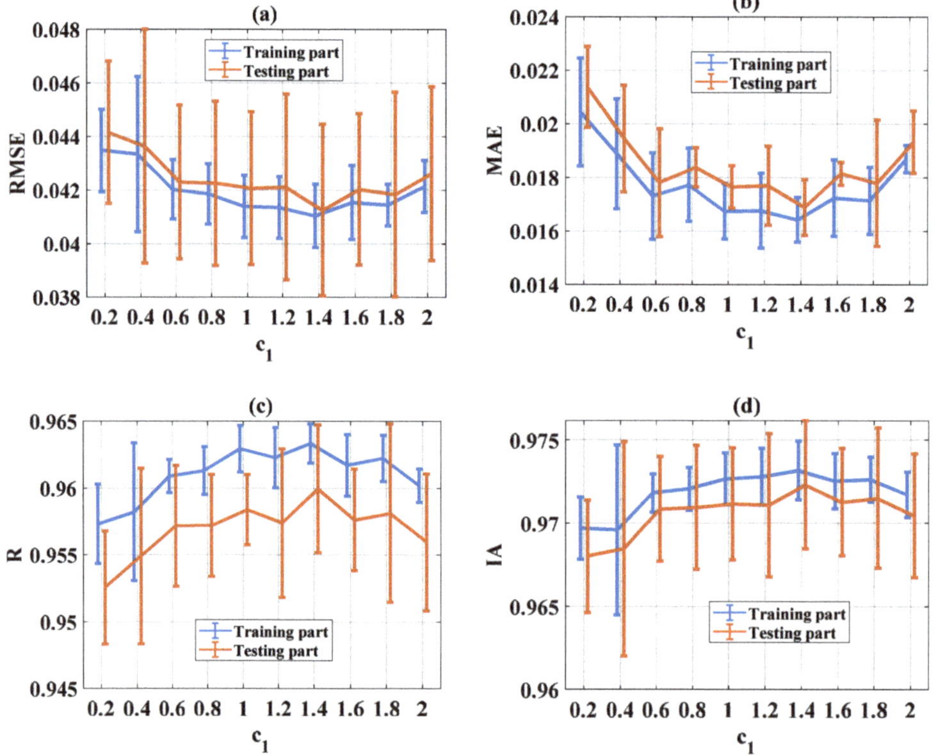

Figure 11. Evaluation of statistical criteria in the function of c_1: (**a**) RMSE, (**b**) MAE, (**c**) R, and (**d**) IA.

5.3.5. Influence of Global Learning Coefficient

The evaluation of statistical criteria in the function of c_2 for RMSE, MAE, R, and IA is shown in Figure 12a–d, respectively, for both training and testing parts. It can be seen that good prediction performance was obtained when c_2 was higher than 1.8 for all statistical criteria (RMSE, MAE, R, and IA). Therefore, $c_2 = [1.8, 2]$ was the most appropriate value.

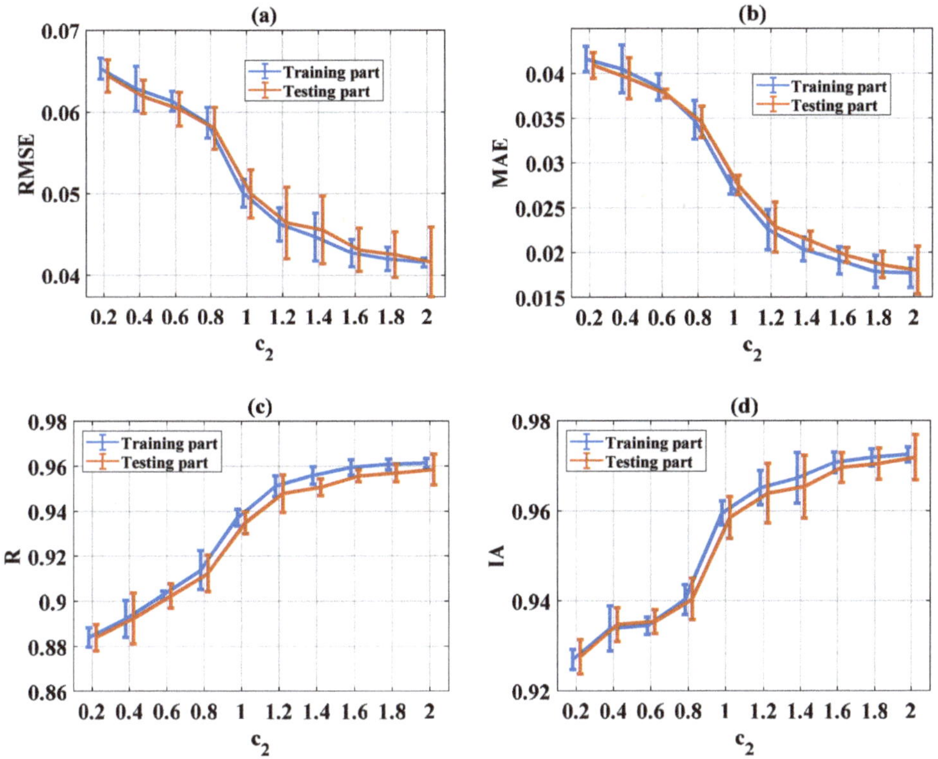

Figure 12. Evaluation of statistical criteria in the function of c_2: (**a**) RMSE, (**b**) MAE, (**c**) R, and (**d**) IA.

5.3.6. Influence of Velocity Limits

The evaluation of statistical criteria in the function of f_v for RMSE, MAE, R, and IA is shown in Figure 13a–d, respectively, for both training and testing parts. It can be seen that no influence could be established regarding all statistical criteria (RMSE, MAE, R, and IA). Therefore, $f_v = 0.1$ was finally chosen.

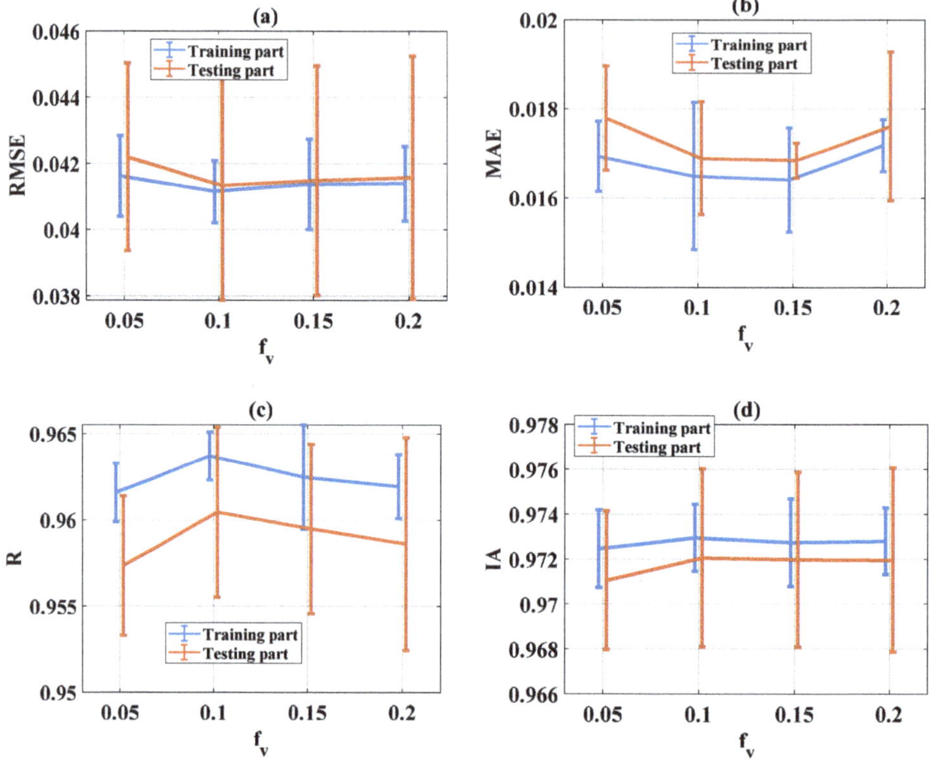

Figure 13. Evaluation of statistical criteria in the function of f_v: (**a**) RMSE, (**b**) MAE, (**c**) R and (**d**) IA.

5.4. Prediction Capability of the ANFIS-PSO Model using Optimal Configuration

Table 4 summarizes all of the optimal values, as identified previously. By using the optimal coefficient in Table 4, a regression graph between the real and predicted P_u (kN) is shown in Figure 14. The slope of the ideal fit was then used to measure the angle between the x-axis and the ideal fit, with angles closer than $45°$ showing better performance. Figure 14a shows the predictability when using the training set, whereas Figure 14b shows the same information applied to the testing set. In both cases, the angles generated by the predicted output had slopes close to that of the ideal fit. This showed that the performance of the proposed model was consistent. Figure 15 shows the error distribution graph using the training part, testing part, and all data. In short, using the selected number of fuzzy rules and PSO parameters, the prediction model gave excellent results (Table 5).

Table 4. Parameters used as optimum.

Parameter	Optimal Value	Final Selection
n_{rule}	10	10
n_{pop}	50	50
w_{ini}	0.1–0.4	0.4
c_1	1–1.4	1
c_2	1.8–2	2
f_v	0.1	0.1

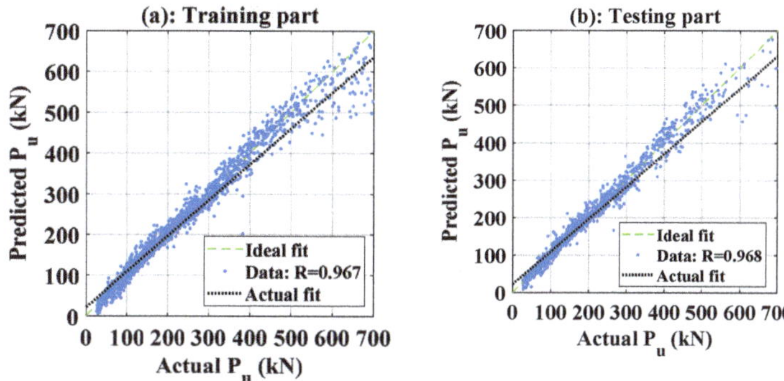

Figure 14. Graphs of regression plots between actual and predicted P_u (kN) for the (**a**) training part and (**b**) testing part.

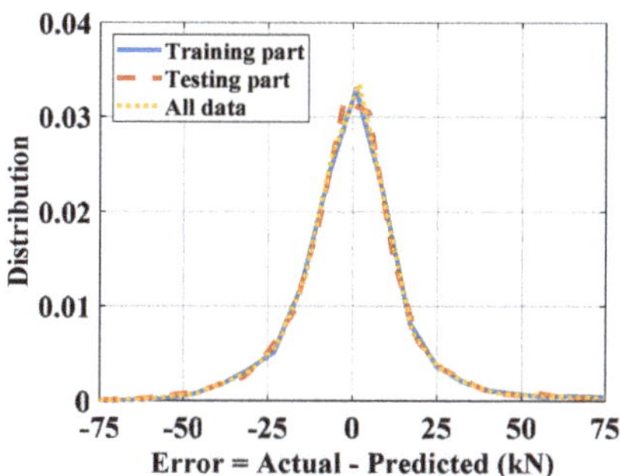

Figure 15. Distribution of errors.

Table 5. Prediction capability.

Part	RMSE	MAE	R	IA	Error Mean	Error Std	Running Time
Training	0.040	0.015	0.967	0.976	0.006	0.039	8 min
Testing	0.037	0.014	0.968	0.977	0.005	0.037	

5.5. Sensitivity Analysis

The sensitivity analysis was performed in order to explore the degree of importance of each input variable using the ANFIS-PSO model. For this, quantile values at 21 points (from 0% to 100%, with a step of 5%) of each input variable were collected from the database and served as a new dataset for the calculation of critical buckling load. More precisely, for a given input, its value varied from 0% to 100%, while all other inputs remained at their median (50%). This variation of values following the probability distribution allows the influence of each input variable to be explored based on their statistical behavior. The results of the sensitivity analysis are indicated in Figure 16 in a bar graph (scaled into the range of 0% to 100%). It can be seen that all variables influenced the prediction of critical buckling load through the ANFIS-PSO model. The most important input variables were L, w_{flange}, t_{web}, and t_{flange}, which gave degree of importance values of 33.9%, 21.7%, 18.6%, and 10.6%, respectively. This information is strongly relevant and in good agreement with the literature, in which the length of the beam and geometrical parameter of the cross-section are the most important parameters [3–5]. However, it can be seen in Figure 16 that the height of the beam does not seriously affect the buckling capacity of the structural members. It should be noted that only three independent values of the section's height were used to generate the database; for example, 420, 560, and 700 mm. Consequently, the linear correlation coefficient between the section's height and the buckling capacity was only −0.092.

On the contrary, the minimum value of the beam's length was 4000 mm (approximately 5.7 times larger than the maximum section's height) and five independent values were used to generate the database, ranging from 4000 to 8000 mm, with a step of 1000 mm. Thus, the linear correlation coefficient between the beam's length and the buckling capacity was −0.667 (approximately 7.25 times bigger than the linear correlation coefficient between the section's height and the buckling capacity). Consequently, a larger database should be considered in future studies to estimate the degree of importance of the section's height.

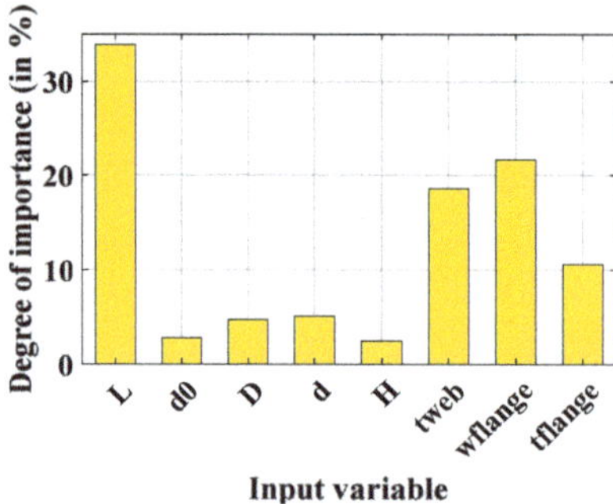

Figure 16. Bar graph showing the estimations of degree of importance values.

The sensitivity analysis presented above demonstrates that the ML technique could assist in the design phase for circular opening steel beams. In addition to reliable prediction of the critical buckling load, the ANFIS-PSO model can also assist in the creation of input–output maps, as illustrated in Figure 17. In particular, as L, t_{web}, w_{flange}, and t_{flange} were the most important variables, they are used for map illustrations in this section. The values of the remaining variables were kept constant. In Figure 17, four maps of critical buckling load are presented (with the same color range), involving the relationship between P_u and L-w_{flange}, L-t_{flange}, L-t_{web}, and w_{flange}-t_{flange}, respectively. As can be seen from the surface plots, the input–output relationship exhibited nonlinear behavior, which cannot be easily identified from the database. Figure 17a shows that a maximum value for the critical buckling load can be obtained if L reaches its minimum and w_{flange} reaches its maximum value. On the other hand, the critical buckling load reaches its minimum if L reaches its highest value and w_{flange} reaches its lowest value. This map confirms the negative effect of L, as pointed out in the literature [4]. In Figure 17b,c, the same results are obtained as in Figure 17a. This observation again confirms that the geometrical parameters of the cross-section are highly important [1,5]. Such quantitative information allows the design and analysis recommendations to be explored, as well as for new beam configurations to be generated (within the range of variables considered in this present study).

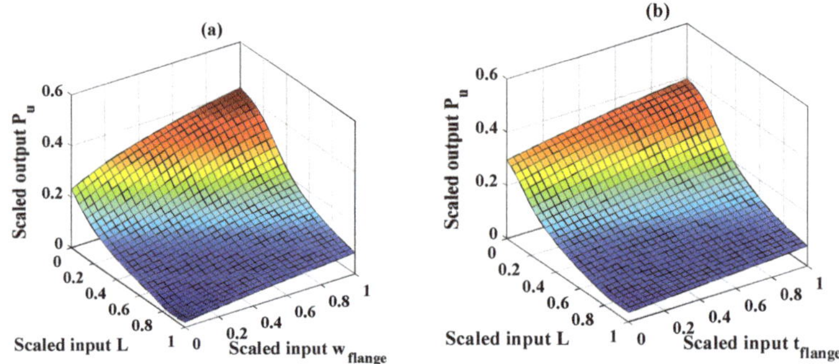

Figure 17. *Cont.*

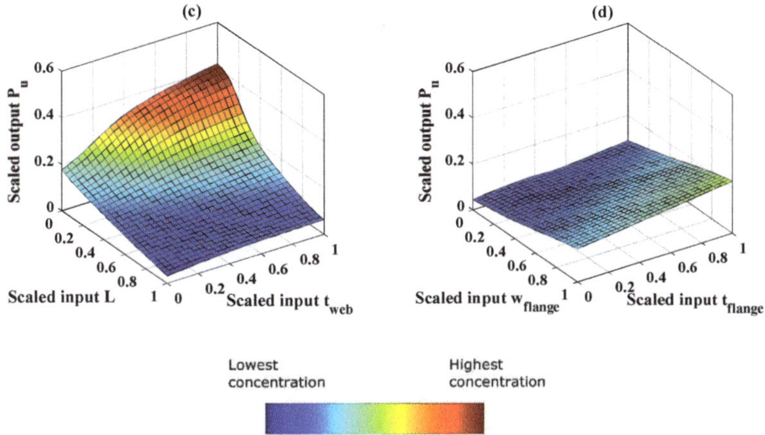

Figure 17. Three-dimensional scaled input–output maps: (**a**) L-w_{flange}, (**b**) L-t_{flange}, (**c**) L-t_{web}, and (**d**) w_{flange}-t_{flange}.

6. Conclusions

PSO is one of the most popular optimization techniques used to optimize and improve the performance of machine learning models in terms of classification and regression. However, its effectiveness depends significantly on the selection of parameters used to train this technique. In this paper, investigation and selection of PSO parameters was carried out to improve and optimize the performance of the ANFIS model, which is one of the most popular and effective ML models, for prediction of the buckling capacity of circular opening steel beams. Different parameters (n_{rule}, n_{pop}, w_{ini}, c_1, c_2, and f_v) of PSO were tuned on 3645 available data samples to determine the best values for optimization of the performance of ANFIS.

The results show that the performance of ANFIS optimized by PSO (ANFIS-PSO) is suitable for determining the buckling capacity of circular opening steel beams, but is very sensitive under different PSO investigation and selection parameters. The results also show that $n_{rule} = 10$, $n_{pop} = 50$, $w_{ini} = 0.1$ to 0.4, $c_1 = [1, 1.4]$, $c_2 = [1.8, 2]$, and $f_v = 0.1$ are the most suitable selection settings in order to get the best performance from ANFIS-PSO. The sensitivity analysis shows that L, w_{flange}, t_{web}, and t_{flange} are the most important input variables used for prediction of the buckling capacity of circular opening steel beams.

In short, this study might help in selection of the suitable PSO parameters for optimization of ANFIS in determining the buckling capacity of circular opening steel beams. It also helps in suitable selection of input variables for better prediction of the buckling capacity of circular opening steel beams. However, it is noted that the optimal values of PSO parameters found in this study are suitable for the ANFIS model in determining the buckling capacity of circular opening steel beams. Thus, it is suggested that these parameters should be validated with other ML models applied in other problems. Finally, variation in the mechanical properties of material used should be investigated in further research, as this is important from a physics perspective.

Author Contributions: Conceptualization, Q.H.N., H.-B.L., T.-T.L, and B.T.P; methodology, H.-B.L, T.-T.L, and B.T.P; validation, H.-B.L. and T.-T.L; formal analysis, Q.H.N., V.Q.T., T.-A.N., V.-H.P., and H.-B.L; data curation, V.Q.T., T.-A.N., and V.-H.P.; writing—original draft preparation, all authors; writing—review and editing, H.-B.L, T.-T.L, and B.T.P; visualization, H.-B.L., T.-A.N., V.-H.P. and T.-T.L; supervision, Q.H.N., H.-B.L, T.-T.L, and B.T.P; project administration, H.-B.L., T.-T.L, and B.T.P; funding acquisition, Q.H.N. All authors have read and agreed to the published version of the manuscript.

Funding: This research received no external funding

Conflicts of Interest: The authors declare no conflict of interest.

Abbreviations

Symbol	Explanation	SI Unit
ANFIS	Adaptive neuro-fuzzy inference system	
PSO	Particle swarm optimization	
R	Correlation coefficient	
RMSE	Root mean squared error	
MAE	Mean absolute error	
IA	Index of agreement	
CV	Coefficient of variation	%
n_{rule}	Number of fuzzy rules	
n_{pop}	PSO population size	
w_{ini}	PSO initial weight	
c_1	PSO cognition learning rate	
c_2	PSO social learning rate	
f_v	PSO velocity limits	

References

1. Grilo, L.F.; Fakury, R.H.; de Souza Veríssimo, G. Design procedure for the web-post buckling of steel cellular beams. *J. Constr. Steel Res.* **2018**, *148*, 525–541. [CrossRef]
2. Hoffman, R.; Dinehart, D.; Gross, S.; Yost, J. Analysis of stress distribution and failure behavior of cellular beams. In Proceedings of the International Ansys Conference, Pittsburgh, PA, USA, 2–4 May 2006.
3. Sonck, D.; Belis, J. Lateral–torsional buckling resistance of cellular beams. *J. Constr. Steel Res.* **2015**, *105*, 119–128. [CrossRef]
4. Panedpojaman, P.; Thepchatri, T. Finite element investigation on deflection of cellular beams with various configurations. *Int. J. Steel Struct.* **2013**, *13*, 487–494. [CrossRef]
5. Sweedan, A.M. Elastic lateral stability of I-shaped cellular steel beams. *J. Constr. Steel Res.* **2011**, *67*, 151–163. [CrossRef]
6. Ellobody, E. Nonlinear analysis of cellular steel beams under combined buckling modes. *Thin-Walled Struct.* **2012**, *52*, 66–79. [CrossRef]
7. Abidin, A.Z.; Izzuddin, B.A. Meshless local buckling analysis of steel beams with irregular web openings. *Eng. Struct.* **2013**, *50*, 197–206. [CrossRef]
8. Gandomi, A.H.; Tabatabaei, S.M.; Moradian, M.H.; Radfar, A.; Alavi, A.H. A new prediction model for the load capacity of castellated steel beams. *J. Constr. Steel Res.* **2011**, *67*, 1096–1105. [CrossRef]
9. Tsavdaridis, K.D.; D'Mello, C. Web buckling study of the behaviour and strength of perforated steel beams with different novel web opening shapes. *J. Constr. Steel Res.* **2011**, *67*, 1605–1620. [CrossRef]
10. Redwood, R.; Cho, S.H. Design of steel and composite beams with web openings. *J. Constr. Steel Res.* **1993**, *25*, 23–41. [CrossRef]
11. Chung, K.F.; Liu, T.C.H.; Ko, A.C.H. Investigation on Vierendeel mechanism in steel beams with circular web openings. *J. Constr. Steel Res.* **2001**, *57*, 467–490. [CrossRef]
12. Chung, K.F.; Liu, C.H.; Ko, A.C.H. Steel beams with large web openings of various shapes and sizes: An empirical design method using a generalised moment-shear interaction curve. *J. Constr. Steel Res.* **2003**, *59*, 1177–1200. [CrossRef]
13. Nseir, J.; Lo, M.; Sonck, D.; Somja, H.; Vassart, O.; Boissonnade, N. Lateral torsional buckling of cellular steel beams. In Proceedings of the Annual Stability Conference Structural Stability Research Council, Grapevine, TX, USA, 18–21 April 2012; pp. 18–21.
14. Ghasemain, B.; Asl, D.T.; Pham, B.T.; Avand, M.; Nguyen, H.D.; Janizadeh, S. Shallow landslide susceptibility mapping: A comparison between classification and regression tree and reduced error pruning tree algorithms. *Vietnam J. Earth Sci.* **2020**. [CrossRef]
15. Ly, H.-B.; Pham, B.T. Prediction of Shear Strength of Soil Using Direct Shear Test and Support Vector Machine Model. *Open Constr. Build. Technol. J.* **2020**, *14*.

16. Pham, B.T.; Phong, T.V.; Nguyen-Thoi, T.; Parial, K.; K. Singh, S.; Ly, H.-B.; Nguyen, K.T.; Ho, L.S.; Le, H.V.; Prakash, I. Ensemble modeling of landslide susceptibility using random subspace learner and different decision tree classifiers. *Geocarto Int.* **2020**, 1–23. [CrossRef]

17. Nguyen, P.T.; Ha, D.H.; Avand, M.; Jaafari, A.; Nguyen, H.D.; Al-Ansari, N.; Phong, T.V.; Sharma, R.; Kumar, R.; Le, H.V. Soft Computing Ensemble Models Based on Logistic Regression for Groundwater Potential Mapping. *Appl. Sci.* **2020**, *10*, 2469. [CrossRef]

18. Ly, H.-B.; Le, T.-T.; Vu, H.-L.T.; Tran, V.Q.; Le, L.M.; Pham, B.T. Computational Hybrid Machine Learning Based Prediction of Shear Capacity for Steel Fiber Reinforced Concrete Beams. *Sustainability* **2020**, *12*, 2709. [CrossRef]

19. Nguyen, P.T.; Ha, D.H.; Nguyen, H.D.; Van Phong, T.; Trinh, P.T.; Al-Ansari, N.; Le, H.V.; Pham, B.T.; Ho, L.S.; Prakash, I. Improvement of Credal Decision Trees Using Ensemble Frameworks for Groundwater Potential Modeling. *Sustainability* **2020**, *12*, 2622. [CrossRef]

20. Pham, B.T.; Bui, D.T.; Prakash, I.; Nguyen, L.H.; Dholakia, M. A comparative study of sequential minimal optimization-based support vector machines, vote feature intervals, and logistic regression in landslide susceptibility assessment using GIS. *Environ. Earth Sci.* **2017**, *76*, 371. [CrossRef]

21. Pham, B.T.; Tien Bui, D.; Pham, H.V.; Le, H.Q.; Prakash, I.; Dholakia, M.B. Landslide Hazard Assessment Using Random SubSpace Fuzzy Rules Based Classifier Ensemble and Probability Analysis of Rainfall Data: A Case Study at Mu Cang Chai District, Yen Bai Province (Viet Nam). *J Indian Soc. Remote Sens.* **2017**, *45*, 673–683. [CrossRef]

22. Feng, S.; Zhou, H.; Dong, H. Using deep neural network with small dataset to predict material defects. *Mater. Des.* **2019**, *162*, 300–310. [CrossRef]

23. Guo, S.; Yu, J.; Liu, X.; Wang, C.; Jiang, Q. A predicting model for properties of steel using the industrial big data based on machine learning. *Comput. Mater. Sci.* **2019**, *160*, 95–104. [CrossRef]

24. Nguyen, H.-L.; Le, T.-H.; Pham, C.-T.; Le, T.-T.; Ho, L.S.; Le, V.M.; Pham, B.T.; Ly, H.-B. Development of Hybrid Artificial Intelligence Approaches and a Support Vector Machine Algorithm for Predicting the Marshall Parameters of Stone Matrix Asphalt. *Appl. Sci.* **2019**, *9*, 3172. [CrossRef]

25. Dao, D.V.; Jaafari, A.; Bayat, M.; Mafi-Gholami, D.; Qi, C.; Moayedi, H.; Phong, T.V.; Ly, H.-B.; Le, T.-T.; Trinh, P.T.; et al. A spatially explicit deep learning neural network model for the prediction of landslide susceptibility. *Catena* **2020**, *188*, 104451. [CrossRef]

26. Qi, C.; Ly, H.-B.; Chen, Q.; Le, T.-T.; Le, V.M.; Pham, B.T. Flocculation-dewatering prediction of fine mineral tailings using a hybrid machine learning approach. *Chemosphere* **2020**, *244*, 125450. [CrossRef]

27. Dou, J.; Yunus, A.P.; Merghadi, A.; Shirzadi, A.; Nguyen, H.; Hussain, Y.; Avtar, R.; Chen, Y.; Pham, B.T.; Yamagishi, H. Different sampling strategies for predicting landslide susceptibilities are deemed less consequential with deep learning. *Sci. Total Environ.* **2020**, *720*, 137320. [CrossRef]

28. Qi, C.; Zhou, W.; Lu, X.; Luo, H.; Pham, B.T.; Yaseen, Z.M. Particulate matter concentration from open-cut coal mines: A hybrid machine learning estimation. *Environ. Pollut.* **2020**, *263*, 114517. [CrossRef]

29. Tian, J.; Qi, C.; Sun, Y.; Yaseen, Z.M.; Pham, B.T. Permeability prediction of porous media using a combination of computational fluid dynamics and hybrid machine learning methods. *Eng. Comput.* **2020**, 1–17. [CrossRef]

30. Nguyen, H.Q.; Ly, H.-B.; Tran, V.Q.; Nguyen, T.-A.; Le, T.-T.; Pham, B.T. Optimization of Artificial Intelligence System by Evolutionary Algorithm for Prediction of Axial Capacity of Rectangular Concrete Filled Steel Tubes under Compression. *Materials* **2020**, *13*, 1205. [CrossRef]

31. Ly, H.-B.; Le, L.M.; Duong, H.T.; Nguyen, T.C.; Pham, T.A.; Le, T.-T.; Le, V.M.; Nguyen-Ngoc, L.; Pham, B.T. Hybrid Artificial Intelligence Approaches for Predicting Critical Buckling Load of Structural Members under Compression Considering the Influence of Initial Geometric Imperfections. *Appl. Sci.* **2019**, *9*, 2258. [CrossRef]

32. Ly, H.-B.; Le, T.-T.; Le, L.M.; Tran, V.Q.; Le, V.M.; Vu, H.-L.T.; Nguyen, Q.H.; Pham, B.T. Development of Hybrid Machine Learning Models for Predicting the Critical Buckling Load of I-Shaped Cellular Beams. *Appl. Sci.* **2019**, *9*, 5458. [CrossRef]

33. KhalilzadeVahidi, E.; Rahimi, F. Investigation of Ultimate Shear Capacity of RC Deep Beams with Opening using Artificial Neural Networks. *Adv. Comput. Sci. Int. J.* **2016**, *5*, 57–65.

34. Abambres, M.; Rajana, K.; Tsavdaridis, K.; Ribeiro, T. Neural Network-based formula for the buckling load prediction of I-section cellular steel beams. *Computers* **2019**, *8*, 2. [CrossRef]

35. Blachowski, B.; Pnevmatikos, N. Neural Network Based Vibration Control of Seismically Excited Civil Structures. *Period. Polytech. Civ. Eng.* **2018**, *62*, 620–628. [CrossRef]

36. Pnevmatikos, N.G.; Thomos, G.C. Stochastic structural control under earthquake excitations. *Struct. Control Health Monit.* **2014**, *21*, 620–633. [CrossRef]

37. Seitllari, A.; Naser, M.Z. Leveraging artificial intelligence to assess explosive spalling in fire-exposed RC columns. *Comput. Concr.* **2019**, *24*, 271–282. [CrossRef]

38. Naser, M.Z. Deriving temperature-dependent material models for structural steel through artificial intelligence. *Constr. Build. Mater.* **2018**, *191*, 56–68. [CrossRef]

39. Basarir, H.; Elchalakani, M.; Karrech, A. The prediction of ultimate pure bending moment of concrete-filled steel tubes by adaptive neuro-fuzzy inference system (ANFIS). *Neural Comput. Appl.* **2019**, *31*, 1239–1252. [CrossRef]

40. Naderpour, H.; Mirrashid, M. Shear Strength Prediction of RC Beams Using Adaptive Neuro-Fuzzy Inference System. *Sci. Iran.* **2018**, *27*, 657–670. [CrossRef]

41. Güneyisi, E.M. A study on estimation of flexural overstrength factor for thin- walled CHS beams using ANFIS. In Proceedings of the 12th International Congress on Advances in Civil Engineering (ACE 2016), Istanbul, Turkey, 21–23 September 2016; Volume 8, pp. 1–7.

42. Oliveira, A.L.; Braga, P.L.; Lima, R.M.; Cornélio, M.L. GA-based method for feature selection and parameters optimization for machine learning regression applied to software effort estimation. *Inf. Softw. Technol.* **2010**, *52*, 1155–1166. [CrossRef]

43. Saini, L.M. Peak load forecasting using Bayesian regularization, Resilient and adaptive backpropagation learning based artificial neural networks. *Electr. Power Syst. Res.* **2008**, *78*, 1302–1310. [CrossRef]

44. Rahman, M.A.; Hoque, M.A. Online self-tuning ANN-based speed control of a PM DC motor. *IEEE/ASME Trans. Mechatron.* **1997**, *2*, 169–178. [CrossRef]

45. Liu, Y.; Yao, X. A population-based learning algorithm which learns both architectures and weights of neural networks. *Chin. J. Adv. Softw. Res.* **1996**, *3*, 54–65.

46. Salem, F.; Awadallah, M.A. Parameters estimation of photovoltaic modules: Comparison of ANN and ANFIS. *Int. J. Ind. Electron. Drives* **2014**, *1*, 121–129. [CrossRef]

47. Wei, M.; Bai, B.; Sung, A.H.; Liu, Q.; Wang, J.; Cather, M.E. Predicting injection profiles using ANFIS. *Inf. Sci.* **2007**, *177*, 4445–4461. [CrossRef]

48. Chen, W.; Panahi, M.; Pourghasemi, H.R. Performance evaluation of GIS-based new ensemble data mining techniques of adaptive neuro-fuzzy inference system (ANFIS) with genetic algorithm (GA), differential evolution (DE), and particle swarm optimization (PSO) for landslide spatial modelling. *Catena* **2017**, *157*, 310–324. [CrossRef]

49. Chen, M.-Y. A hybrid ANFIS model for business failure prediction utilizing particle swarm optimization and subtractive clustering. *Inf. Sci.* **2013**, *220*, 180–195. [CrossRef]

50. Kiranyaz, S.; Ince, T.; Gabbouj, M. *Multidimensional Particle Swarm Optimization for Machine Learning and Pattern Recognition*; Springer: Berlin/Heidelberg, Germany, 2014; ISBN 3-642-37846-3.

51. Ahmadi, M.A.; Soleimani, R.; Lee, M.; Kashiwao, T.; Bahadori, A. Determination of oil well production performance using artificial neural network (ANN) linked to the particle swarm optimization (PSO) tool. *Petroleum* **2015**, *1*, 118–132. [CrossRef]

52. Bagheri, A.; Peyhani, H.M.; Akbari, M. Financial forecasting using ANFIS networks with quantum-behaved particle swarm optimization. *Expert Syst. Appl.* **2014**, *41*, 6235–6250. [CrossRef]

53. Melgani, F.; Bazi, Y. Classification of electrocardiogram signals with support vector machines and particle swarm optimization. *IEEE Trans. Inf. Technol. Biomed.* **2008**, *12*, 667–677. [CrossRef] [PubMed]

54. Rini, D.P.; Shamsuddin, S.M.; Yuhaniz, S.S. Particle swarm optimization for ANFIS interpretability and accuracy. *Soft Comput.* **2016**, *20*, 251–262. [CrossRef]

55. Jang, J.-S. ANFIS: Adaptive-network-based fuzzy inference system. *IEEE Trans. Syst. Man Cybern.* **1993**, *23*, 665–685. [CrossRef]

56. Haykin, S.; Network, N. A comprehensive foundation. *Neural Netw.* **2004**, *2*, 41.

57. Bui, D.T.; Tsangaratos, P.; Ngo, P.-T.T.; Pham, T.D.; Pham, B.T. Flash flood susceptibility modeling using an optimized fuzzy rule based feature selection technique and tree based ensemble methods. *Sci. Total Environ.* **2019**, *668*, 1038–1054. [CrossRef] [PubMed]

58. Kisi, O. Suspended sediment estimation using neuro-fuzzy and neural network approaches/Estimation des matières en suspension par des approches neurofloues et à base de réseau de neurones. *Hydrol. Sci. J.* **2005**, *50*. [CrossRef]
59. Nguyen, H.-L.; Pham, B.T.; Son, L.H.; Thang, N.T.; Ly, H.-B.; Le, T.-T.; Ho, L.S.; Le, T.-H.; Tien Bui, D. Adaptive Network Based Fuzzy Inference System with Meta-Heuristic Optimizations for International Roughness Index Prediction. *Appl. Sci.* **2019**, *9*, 4715. [CrossRef]
60. Dao, D.V.; Ly, H.-B.; Trinh, S.H.; Le, T.-T.; Pham, B.T. Artificial Intelligence Approaches for Prediction of Compressive Strength of Geopolymer Concrete. *Materials* **2019**, *12*, 983. [CrossRef]
61. Mansouri, I.; Ozbakkaloglu, T.; Kisi, O.; Xie, T. Predicting behavior of FRP-confined concrete using neuro fuzzy, neural network, multivariate adaptive regression splines and M5 model tree techniques. *Mater. Struct.* **2016**, *49*, 4319–4334. [CrossRef]
62. Eberhart, R.; Kennedy, J. A new optimizer using particle swarm theory. In Proceedings of the MHS'95, Proceedings of the Sixth International Symposium on Micro Machine and Human Science, Nagoya, Japan, 4–6 October 1995; pp. 39–43.
63. Termeh, S.V.R.; Khosravi, K.; Sartaj, M.; Keesstra, S.D.; Tsai, F.T.-C.; Dijksma, R.; Pham, B.T. Optimization of an adaptive neuro-fuzzy inference system for groundwater potential mapping. *Hydrogeol. J.* **2019**, *27*, 2511–2534. [CrossRef]
64. Jaafari, A.; Panahi, M.; Pham, B.T.; Shahabi, H.; Bui, D.T.; Rezaie, F.; Lee, S. Meta optimization of an adaptive neuro-fuzzy inference system with grey wolf optimizer and biogeography-based optimization algorithms for spatial prediction of landslide susceptibility. *Catena* **2019**, *175*, 430–445. [CrossRef]
65. Zhou, J.; Nekouie, A.; Arslan, C.A.; Pham, B.T.; Hasanipanah, M. Novel approach for forecasting the blast-induced AOp using a hybrid fuzzy system and firefly algorithm. *Eng. Comput.* **2019**. [CrossRef]
66. Yildiz, A.R. A new hybrid particle swarm optimization approach for structural design optimization in the automotive industry. *Proc. Inst. Mech. Eng. Part D J. Automob. Eng.* **2012**, *226*, 1340–1351. [CrossRef]
67. He, Q.; Wang, L. An effective co-evolutionary particle swarm optimization for constrained engineering design problems. *Eng. Appl. Artif. Intell.* **2007**, *20*, 89–99. [CrossRef]
68. Pathak, N.N.; Mahanti, G.K.; Singh, S.K.; Mishra, J.K.; Chakraborty, A. Synthesis of thinned planar circular array antennas using modified particle swarm optimization. *Prog. Electromagn. Res.* **2009**, *12*, 87–97. [CrossRef]
69. Guilleminot, J.; Le, T.T.; Soize, C. Stochastic framework for modeling the linear apparent behavior of complex materials: Application to random porous materials with interphases. *Acta Mech. Sin.* **2013**, *29*, 773–782. [CrossRef]
70. Pham, B.T.; Nguyen, M.D.; Dao, D.V.; Prakash, I.; Ly, H.-B.; Le, T.-T.; Ho, L.S.; Nguyen, K.T.; Ngo, T.Q.; Hoang, V.; et al. Development of artificial intelligence models for the prediction of Compression Coefficient of soil: An application of Monte Carlo sensitivity analysis. *Sci. Total Environ.* **2019**, *679*, 172–184. [CrossRef]
71. Le, T.T.; Guilleminot, J.; Soize, C. Stochastic continuum modeling of random interphases from atomistic simulations. Application to a polymer nanocomposite. *Comput. Methods Appl. Mech. Eng.* **2016**, *303*, 430–449. [CrossRef]
72. Pham, B.T.; Nguyen-Thoi, T.; Ly, H.-B.; Nguyen, M.D.; Al-Ansari, N.; Tran, V.-Q.; Le, T.-T. Extreme Learning Machine Based Prediction of Soil Shear Strength: A Sensitivity Analysis Using Monte Carlo Simulations and Feature Backward Elimination. *Sustainability* **2020**, *12*, 2339. [CrossRef]
73. Hun, D.-A.; Guilleminot, J.; Yvonnet, J.; Bornert, M. Stochastic multiscale modeling of crack propagation in random heterogeneous media. *Int. J. Numer. Methods Eng.* **2019**, *119*, 1325–1344. [CrossRef]
74. Capillon, R.; Desceliers, C.; Soize, C. Uncertainty quantification in computational linear structural dynamics for viscoelastic composite structures. *Comput. Methods Appl. Mech. Eng.* **2016**, *305*, 154–172. [CrossRef]
75. Le, M.V.; Yvonnet, J.; Feld, N.; Detrez, F. The Coarse Mesh Condensation Multiscale Method for parallel computation of heterogeneous linear structures without scale separation. *Comput. Methods Appl. Mech. Eng.* **2020**, *363*, 112877. [CrossRef]
76. Tran, V.P.; Brisard, S.; Guilleminot, J.; Sab, K. Mori–Tanaka estimates of the effective elastic properties of stress-gradient composites. *Int. J. Solids Struct.* **2018**, *146*, 55–68. [CrossRef]
77. Staber, B.; Guilleminot, J.; Soize, C.; Michopoulos, J.; Iliopoulos, A. Stochastic modeling and identification of a hyperelastic constitutive model for laminated composites. *Comput. Methods Appl. Mech. Eng.* **2019**, *347*, 425–444. [CrossRef]

78. Tran, V.-P.; Guilleminot, J.; Brisard, S.; Sab, K. Stochastic modeling of mesoscopic elasticity random field. *Mech. Mater.* **2016**, *93*, 1–12. [CrossRef]

79. Soize, C.; Desceliers, C.; Guilleminot, J.; Le, T.-T.; Nguyen, M.-T.; Perrin, G.; Allain, J.-M.; Gharbi, H.; Duhamel, D.; Funfschilling, C. Stochastic representations and statistical inverse identification for uncertainty quantification in computational mechanics. In Proceedings of the UNCECOMP 2015—1st ECCOMAS Thematic Conference on Uncertainty Quantification in Computational Sciences and Engineering, Crete Island, Greece, 25–27 May 2015; pp. 1–26.

80. Le, T.-T.; Guilleminot, J.; Soize, C. Stochastic continuum modeling of random interphases from atomistic simulations. In Proceedings of the Euromech 559, Multi-Scale Computational Methods for Bridging Scales in Materials and Structures, Eindhoven, The Netherlands, 23–25 February 2015; pp. 1–2.

81. Christian, P.d.S. (Ed.) *Stochastic Models of Uncertainties in Computational Mechanics*; American Society of Civil Engineers: Reston, VA, USA, 2012; ISBN 978-0-7844-1223-7.

82. Dao, D.V.; Ly, H.-B.; Vu, H.-L.T.; Le, T.-T.; Pham, B.T. Investigation and Optimization of the C-ANN Structure in Predicting the Compressive Strength of Foamed Concrete. *Materials* **2020**, *13*, 1072. [CrossRef] [PubMed]

83. Le, T.-T. Modélisation Stochastique, en Mécanique des Milieux Continus, de l'Interphase Inclusion-Matrice à Partir de Simulations en Dynamique Moléculaire. Ph.D. Thesis, University of Paris-Est Marne-la-Vallée, Paris, France, 2015.

84. Le, T.-T.; Pham, B.T.; Le, V.M.; Ly, H.-B.; Le, L.M. A Robustness Analysis of Different Nonlinear Autoregressive Networks Using Monte Carlo Simulations for Predicting High Fluctuation Rainfall. In *Proceedings of the Micro-Electronics and Telecommunication Engineering*; Sharma, D.K., Balas, V.E., Son, L.H., Sharma, R., Cengiz, K., Eds.; Springer: Singapore, 2020; pp. 205–212.

85. Dao, D.V.; Adeli, H.; Ly, H.-B.; Le, L.M.; Le, V.M.; Le, T.-T.; Pham, B.T. A Sensitivity and Robustness Analysis of GPR and ANN for High-Performance Concrete Compressive Strength Prediction Using a Monte Carlo Simulation. *Sustainability* **2020**, *12*, 830. [CrossRef]

86. Pham, B.T.; Jaafari, A.; Prakash, I.; Bui, D.T. A novel hybrid intelligent model of support vector machines and the MultiBoost ensemble for landslide susceptibility modeling. *Bull. Eng. Geol. Environ.* **2019**, *78*, 2865–2886. [CrossRef]

87. Ly, H.-B.; Desceliers, C.; Le, L.M.; Le, T.-T.; Pham, B.T.; Nguyen-Ngoc, L.; Doan, V.T.; Le, M. Quantification of Uncertainties on the Critical Buckling Load of Columns under Axial Compression with Uncertain Random Materials. *Materials* **2019**, *12*, 1828. [CrossRef]

88. Le, L.M.; Ly, H.-B.; Pham, B.T.; Le, V.M.; Pham, T.A.; Nguyen, D.-H.; Tran, X.-T.; Le, T.-T. Hybrid Artificial Intelligence Approaches for Predicting Buckling Damage of Steel Columns Under Axial Compression. *Materials* **2019**, *12*, 1670. [CrossRef]

89. Pham, B.T.; Le, L.M.; Le, T.-T.; Bui, K.-T.T.; Le, V.M.; Ly, H.-B.; Prakash, I. Development of advanced artificial intelligence models for daily rainfall prediction. *Atmos. Res.* **2020**, *237*, 104845. [CrossRef]

90. Le, V.M.; Pham, B.T.; Le, T.-T.; Ly, H.-B.; Le, L.M. Daily Rainfall Prediction Using Nonlinear Autoregressive Neural Network. In *Proceedings of the Micro-Electronics and Telecommunication Engineering*; Sharma, D.K., Balas, V.E., Son, L.H., Sharma, R., Cengiz, K., Eds.; Springer: Singapore, 2020; pp. 213–221.

91. Pham, B.T.; Nguyen, M.D.; Ly, H.-B.; Pham, T.A.; Hoang, V.; Van Le, H.; Le, T.-T.; Nguyen, H.Q.; Bui, G.L. Development of Artificial Neural Networks for Prediction of Compression Coefficient of Soft Soil. In *Proceedings of the CIGOS 2019, Innovation for Sustainable Infrastructure*; Ha-Minh, C., Dao, D.V., Benboudjema, F., Derrible, S., Huynh, D.V.K., Tang, A.M., Eds.; Springer: Singapore, 2020; pp. 1167–1172.

92. Tran, V.Q.; Nguyen, H.L.; Dao, V.D.; Hilloulin, B.; Nguyen, L.K.; Nguyen, Q.H.; Le, T.-T.; Ly, H.-B. Temperature effects on chloride binding capacity of cementitious materials. *Mag. Concr. Res.* **2019**, 1–39. [CrossRef]

93. Pham, B.T.; Singh, S.K.; Ly, H.-B. Using Artificial Neural Network (ANN) for prediction of soil coefficient of consolidation. *Vietnam J. Earth Sci.* **2020**. [CrossRef]

94. Ly, H.-B.; Asteris, P.G.; Pham, B.T. Accuracy assessment of extreme learning machine in predicting soil compression soefficient. *Vietnam J. Earth Sci.* **2020**. [CrossRef]

95. Ly, H.-B.; Monteiro, E.; Le, T.-T.; Le, V.M.; Dal, M.; Regnier, G.; Pham, B.T. Prediction and Sensitivity Analysis of Bubble Dissolution Time in 3D Selective Laser Sintering Using Ensemble Decision Trees. *Materials* **2019**, *12*, 1544. [CrossRef] [PubMed]

96. Montavon, G.; Rupp, M.; Gobre, V.; Vazquez-Mayagoitia, A.; Hansen, K.; Tkatchenko, A.; Müller, K.-R.; Lilienfeld, O.A. von Machine learning of molecular electronic properties in chemical compound space. *New J. Phys.* **2013**, *15*, 095003. [CrossRef]

97. Ly, H.-B.; Le, L.M.; Phi, L.V.; Phan, V.-H.; Tran, V.Q.; Pham, B.T.; Le, T.-T.; Derrible, S. Development of an AI Model to Measure Traffic Air Pollution from Multisensor and Weather Data. *Sensors* **2019**, *19*, 4941. [CrossRef]

98. Ly, H.-B.; Pham, B.T.; Dao, D.V.; Le, V.M.; Le, L.M.; Le, T.-T. Improvement of ANFIS Model for Prediction of Compressive Strength of Manufactured Sand Concrete. *Appl. Sci.* **2019**, *9*, 3841. [CrossRef]

99. Le, T.-T.; Pham, B.T.; Ly, H.-B.; Shirzadi, A.; Le, L.M. Development of 48-hour Precipitation Forecasting Model using Nonlinear Autoregressive Neural Network. In *Proceedings of the CIGOS 2019, Innovation for Sustainable Infrastructure*; Ha-Minh, C., Dao, D.V., Benboudjema, F., Derrible, S., Huynh, D.V.K., Tang, A.M., Eds.; Springer: Singapore, 2020; pp. 1191–1196.

100. Qi, C.; Tang, X.; Dong, X.; Chen, Q.; Fourie, A.; Liu, E. Towards Intelligent Mining for Backfill: A genetic programming-based method for strength forecasting of cemented paste backfill. *Miner. Eng.* **2019**, *133*, 69–79. [CrossRef]

101. Kim, S.; Kim, H. A new metric of absolute percentage error for intermittent demand forecasts. *Int. J. Forecast.* **2016**, *32*, 669–679. [CrossRef]

102. He, X.; Guan, H.; Qin, J. A hybrid wavelet neural network model with mutual information and particle swarm optimization for forecasting monthly rainfall. *J. Hydrol.* **2015**, *527*, 88–100. [CrossRef]

Article

Optimization of Artificial Intelligence System by Evolutionary Algorithm for Prediction of Axial Capacity of Rectangular Concrete Filled Steel Tubes under Compression

Hung Quang Nguyen [1,*], **Hai-Bang Ly** [2,*], **Van Quan Tran** [2], **Thuy-Anh Nguyen** [2], **Tien-Thinh Le** [3,*] **and Binh Thai Pham** [2]

1 Thuyloi University, Hanoi 100000, Vietnam
2 University of Transport Technology, Hanoi 100000, Vietnam; quantv@utt.edu.vn (V.Q.T.); anhnt@utt.edu.vn (T.-A.N.); binhpt@utt.edu.vn (B.T.P.)
3 Institute of Research and Development, Duy Tan University, Da Nang 550000, Vietnam
* Correspondence: hungwuhan@tlu.edu.vn (H.Q.N); banglh@utt.edu.vn (H.-B.L.); letienthinh@duytan.edu.vn (T.-T.L.)

Received: 8 February 2020; Accepted: 4 March 2020; Published: 7 March 2020

Abstract: Concrete filled steel tubes (CFSTs) show advantageous applications in the field of construction, especially for a high axial load capacity. The challenge in using such structure lies in the selection of many parameters constituting CFST, which necessitates defining complex relationships between the components and the corresponding properties. The axial capacity (P_u) of CFST is among the most important mechanical properties. In this study, the possibility of using a feedforward neural network (FNN) to predict P_u was investigated. Furthermore, an evolutionary optimization algorithm, namely invasive weed optimization (IWO), was used for tuning and optimizing the FNN weights and biases to construct a hybrid FNN–IWO model and improve its prediction performance. The results showed that the FNN–IWO algorithm is an excellent predictor of P_u, with a value of R^2 of up to 0.979. The advantage of FNN–IWO was also pointed out with the gains in accuracy of 47.9%, 49.2%, and 6.5% for root mean square error (RMSE), mean absolute error (MAE), and R^2, respectively, compared with simulation using the single FNN. Finally, the performance in predicting the P_u in the function of structural parameters such as depth/width ratio, thickness of steel tube, yield stress of steel, concrete compressive strength, and slenderness ratio was investigated and discussed.

Keywords: axial capacity prediction; rectangular CFST columns; feedforward neural network; invasive weed optimization; hybrid machine learning

---

1. Introduction

Concrete and steel are the two most commonly used construction materials today. However, each material has different advantages and disadvantages [1–3]. Therefore, to be able to take advantages and minimize disadvantages, an optimal solution is to use a combination of both materials, such as a "combined steel concrete structure" or using a combination of concrete elements and steel elements in "composite structures". One of the combined steel concrete structures is a steel pipe composite structure filled with medium or high strength concrete. This type of structure is called a steel-concrete pipe.

In recent decades, concrete filled steel tubes (CFSTs) have been widely used in the construction of modern buildings and bridges [4], even in high seismic risk areas [5–10]. This increase in use is because of the significant advantages that the CFST column system offers over conventional steel or reinforced concrete systems, such as high axial load capacity [4], good plasticity and toughness [6], larger energy absorption capacity [7], convenient construction [11], economy of materials [12–14], and excellent seismic and refractory performance [15]. In particular, this type of structure can reduce the environmental burden by removing formwork [16], reusing steel pipes, and using high quality concrete with recycled aggregate [17]. The characteristics of CFST are that the steel material is located far from the central axis so the rigidity of the column is very large, and thus it also contributes to increasing the moment of inertia of the structure [5,18]. The ideal form of concrete core works against the compressive load and hinders the local buckling state of the steel pipe. Therefore, the CFST structures are often used in locations subject to large compressive loads [9,15,19]. The CFST columns are mainly divided into square columns, round columns, and rectangular columns, based on different cross-sectional forms [15]. In particular, the square and rectangular CFST columns have the advantage of easy connection and reliable work with other structural members such as beams, walls, and panels [20]. Compared with square CFST columns, rectangular columns have irregular bending stiffness along different axes, so this type of column is suitable for the mechanical behavior of members including arch ribs, pillars, abutments, and piers, and other structural members under load actions vary greatly from vertical to horizontal [6]. Because the scope of application of rectangular CFST columns is quite wide and this column is mainly subjected to compression, the main purpose of the paper is to analyze and evaluate the ultimate bearing capacity of rectangular columns.

In recent decades, the regulations for calculating the CFST column type have been proposed in design standards such as AISC-LRFD [21], ACI 318-05 [22], Japan Institute of Architecture [17], European Code EC 4, British Standard BS 5400 [23], and Australian Standard AS-5100.6 [24]. In addition, numerous experimental and numerical studies were conducted to analyze the mechanical properties of rectangular CFST columns under axial compression. As an example, Hatzigeorgiou [25] has proposed a theoretical analysis for modeling the behavior of CFST under extreme loading conditions. Later, the verification of such an approach against experimental and analytical results has also been reported in the work of Hatzigeorgiou [26]. In the work of Liu et al. [4], 26 rectangular CFST column samples were experimented under concentric compression with the main parameters such as strength and aspect ratio. In Chitawadagi et al. [8], the load capacity of CFST columns depended on the variation of CFST properties such as the wall thickness of pipes, strength of in-filled concrete, area of cross section of steel pipes, and pipe length. In this study, 243 rectangular CFST samples were investigated; the experimental results were compared with the predicted column strength, which was performed according to design codes such as EC4-1994 and AISC-LRFD-1994. In addition, there are many other test methods dealing with factors that affect the bearing capacity of rectangular CFST columns such as the effect of concrete compaction [27], load conditions, and boundary conditions [16]. The addition of steel fibers in core concrete had a significant effect on the performance of concrete steel pipes [28] and many other tests [9,13,29–32]. Finite element analysis is now also frequently used for design and research issues thanks to the existence of many commercial software such as ABAQUS [33] and ANSYS [34]. Tort et al. [35] carried out computational research to analyze the nonlinear response of composite frames including rectangular concrete pipe beams and steel frames subjected to static and dynamic loads. On the basis of the Drucker–Prager model, Wang et al. [36] developed a finite element model that can predict the axial compression behavior of a composite column with a fibrous reinforced concrete core. Collecting 340 test data of circular, square, and rectangular CFST columns, Tao et al. [37] developed new finite element models for simulating CFST stub columns under compression mode along the axis. The new model was more flexible and accurate for modeling the CFST stub columns. However, the design standards were limited by the scope of use and were not suitable for high-strength materials, and testing methods were often costly and time-consuming. The accuracy of finite element models was greatly affected by the input parameters, especially the suitable selection of the concrete model. Therefore, it is necessary to propose a uniform and effective approach to design rectangular CFST columns.

In recent years, artificial intelligence (AI) based on computer science has gradually become popular and applied in many different fields [38–41]. Artificial neural network (ANN) is a branch of AI techniques; different ANN-based modeling methods have been used by scientists in many construction engineering applications [42]. Sanad et al. [43] used ANN to estimate the reinforced concrete deep beams ultimate shear strength. Lima et al. [44] predicted the bending resistance and initial stiffness of steel beam connection using a back-propagation algorithm. Seleemah et al. [45] applied ANN to predict the maximum shear strength of concrete beams without horizontal reinforcement. Blachowski and Pnevmatikos [46] have developed a vibration control system based on the ANN method, for application in earthquake engineering. As an example for structural engineering, Kiani et al. [47] have applied AI techniques including support vector machines (SVM) and ANN for deriving seismic fragility curves. It is worth noticing that significant studies have been carried out to explore the prediction of damage using AI techniques. In a series of papers, Mangalathu et al. [48] have proposed various AI methods such as ANN and random forest for tracking damage of bridge portfolios [48] as well as assessing the seismic risk of skewed bridges [49]. In terms of structural failure, typical failure modes of reinforced concrete columns such as flexure, flexure–shear, and shear were investigated by Mangalathu et al. [50,51] using decision trees (DT), SVM, and ANN. Guo et al. [52,53] employed the ANN model for the identification of damage in different structures such as suspended-dome and offshore jacket platforms. Regarding structural uncertainty analysis, various published works by E. Zio should be consulted [54–56]. With rectangular CFST columns, the use of ANN has also been proposed. For example, Sadoon et al. [57] proposed an ANN model for predicting the final strength of rectangular concrete steel beam girder (RCFST) under eccentric shaft load. The results showed that the ANN model was more accurate than the AISC and Eurocode 4 standard. Du et al. [10] formulated an ANN model with different input parameters to determine the axial bearing capacity of rectangular CFST column. The results of the model were compared with the results calculated according to European Code EC 4 [23], ACI [22], and AISC360-10 [21], and found that the ANN model was accurate. However, in the above studies, the mentioned correlation coefficient (R) was less than 0.98. Therefore, in this paper, we tried to create a bulk sample set and proposed an algorithm to increase the accuracy of the prediction of the axial load bearing capacity of the CFST column.

In short, the aim of this paper is dedicated to the development and optimization of an AI-based model, namely the feedforward neural network (FNN), to predict the P_u of CFST. An optimization algorithm, invasive weed optimization (IWO), was used to finely tune the FNN parameters (i.e., weights and biases) to develop a hybrid model, namely FNN–IWO, and to improve the prediction performance. With respect to the CFST database, 99 samples were collected from the available literature and used for the training and testing phases of the FNN–IWO algorithm. Criteria such as coefficient of determination (R^2), standard deviation error (ErrorStD), root mean square error (RMSE), mean absolute error (MAE), and slope were used to evaluate the performance of FNN–IWO. Finally, an investigation of the prediction capability in the function of different structural parameters was conducted.

2. Materials and Methods

2.1. Feedforward Neural Network (FNN)

An artificial or neural network (also known as an artificial neural network (ANN)) is a biological neural network based a computational or mathematical model. It includes a number of artificial neurons (nodes) that are linked to each other and processes information by transmitting along the connections and calculating new values at the nodes (connection method for calculation) [58,59]. The ANN models are made up of three or more layers, including an input layer that is the leftmost layer of the network representing the inputs, an output layer that is the rightmost layer of the network representing the results achieved, and one or more hidden layers representing the logical reasoning of the network [60–62]. The neurons in each layer are linked to the front and rear neurons with each associated weight. A training algorithm is often used to repeat minimizing the cost function relative to the link weight and neuron threshold. Networks are usually divided into two categories based on how the units are connected, including the feedforward neural network (FNN) and the recurrent neural network. To date, FNN is the most popular architecture owing to its structural flexibility, good performance, and the availability of many training algorithms [63]. Currently, the most widely used training algorithm for multi-layer feedforward networks is the backpropagation algorithm (BP). In BP, network training is achieved by adjusting weights and is done through numerous training sets and training cycles [64]. With the ability to approximate the functions, FNNs have been successfully applied in a number of civil engineering and structural fields [65] such as predicting the compression strength of concrete [66], investigating the fire resistance of calves [67], determining the axial strength of cylindrical concrete pillars [58], and predicting the fire resistance of concrete tubular steel columns [65]. Therefore, in this study, FNN was selected and used to predict the axial capacity of CFST.

2.2. Invasive Weed Optimization (IWO)

IWO is a new random number optimization method inspired by a popular phenomenon in agriculture. The term of weed invasion was first introduced by Mehrabian and Lucas in 2006 [68]. This technique is based on a number of interesting features of invasive weed plants that reproduce and distribute fast and vigorously, and adapt themselves to changes in climatic conditions [69]. Therefore, capturing their characteristics will lead to a powerful optimization algorithm [70]. The advantages of IWO algorithm compared with other evolutionary algorithms are few parameters, simple structure, easy to understand, and easy to program features [71]. Up to now, the IWO algorithm has become more and more popular and has been successfully applied in areas such as antenna system design [72] and design of coding chains for DNA [73], as well as inter-related problems regarding economic [74], tourism [75], and construction techniques [76]. The IWO algorithm is implemented by the following steps:

Step 1. Initialization: Weeds are randomly scattered over a D-dimensional target area as the primary solution.

Step 2. Reproduction: During reproduction, each weed produces seed depending on the physical strength and colony. Weeds that acquire more resources have a better chance of producing seeds and plants that are less adapted to fields are not able to reproduce, and thus produce fewer seeds. The number of seeds increases linearly from the minimum value for the worst weed to the maximum value for the best weed.

Step 3. Spatial dispersal: The seeds generated from step 2 are randomly dispersed in the search space by means of normally distributed random numbers with an average of zero, but with different variances to ensure that the seeds are located around the main factory.

Step 4. Competitive exclusion: The spawning and dispersal process randomly create a new population for the next generation of weeds and their seeds. When the size of this new population is greater than a certain maximum value, the lower-strength weeds will be eliminated through competition and only some of the weeds will be equal to the dark weed population.

Step 5. Termination conditions: The process continues again from step 2 to step 4 until the maximum number of iterations is reached and the best physical tree is nearest to the optimized solution.

2.3. Quality Assessment Criteria

Evaluation of the AI model was performed using statistical measurements such as mean absolute error (MAE), coefficient of determination (R^2), and root mean square error (RMSE). In general, these criteria are popular methods to quantify the performance of AI algorithms [76,77]. More specifically, the mean squared difference between actual values and estimated values defines RMSE, whereas the mean magnitude of the errors defines MAE. The R^2 evaluates the correlation between actual and estimated values [78–80]. Quantitatively, lower RMSE and MAE show better performance of the models. In contrast, a higher R^2 shows better performance of the model [81,82]. MAE, RMSE, and R^2 are expressed as follows [83,84]:

$$\text{MAE} = \frac{1}{N}\sum_{i=1}^{N}(a_i - \bar{a}_i) \tag{1}$$

$$\text{RMSE} = \sqrt{\frac{1}{N}\sum_{i=1}^{N}(a_i - \bar{a}_i)^2} \tag{2}$$

$$R^2 = 1 - \frac{\sum_{i=1}^{N}(a_i - \bar{a}_i)^2}{\sum_{i=1}^{N}(a_i - \bar{a})^2} \tag{3}$$

where a_i is the actual output, $\bar{a}_i$ infers the predicted output, $\bar{a}$ infers the mean of the a_i, and N infers the number of used samples.

2.4. Data Used and Selection of Variables

In this study, a total of 99 compression tests of rectangular CFST columns (Figure 1) were extracted from the available literature: Bridge [85], Du et al. [86], Du et al. [87], Ghannam et al. [88], Han [89], Han & Yang [90], Han & Yao [91], Lin [92], Schneider [93], Shakir-Khalil & Mouli [94], and Shakir-Khalil & Zeghiche [95]. Information of the database is summarized in Table 1, including the number of data and the percentage of proportion, whereas Table 2 presents the initial statistical analysis of the corresponding database.

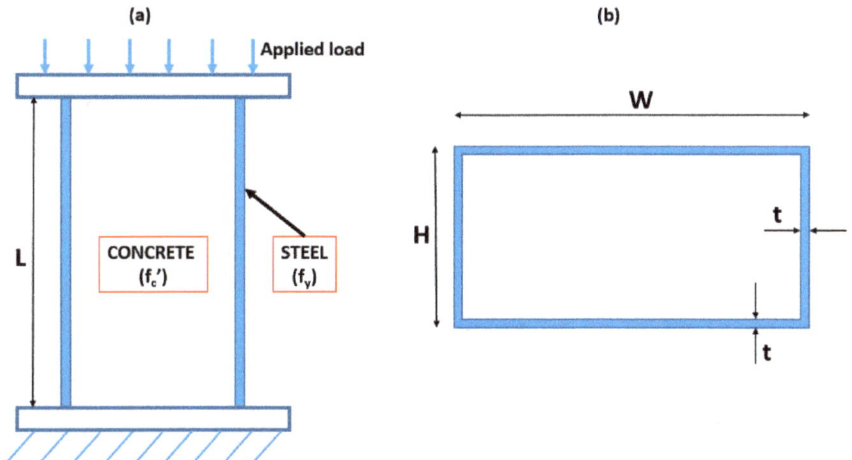

Figure 1. Schematic diagram of the compression test for concrete filled steel tubes (CFSTs): (a) front view; (b) cross-section view of the sample.

The experimental tests were carried out considering the following steps: design, processing of steel tube, production of concrete, curing of specimens, and loading measurement [15,86]. As proposed by Sarir et al. [96] and Ren et al. [15] in investigating CFST columns, initial geometric imperfections as well as residual stress exhibited a negligible effect on the behavior of columns under axial loading. Consequently, input variables affecting the axial capacity of rectangular CFST are from two main groups: geometry of columns and mechanical properties of constituent materials. Therefore, six independent variables were selected as inputs of the problem, such as depth of cross section (H), width of cross section (W), thickness of steel tube (t), length of column (L), yield stress of steel (f_y), and compressive strength of concrete (f_c'). It is seen in Table 2 of the initial statistical analysis that all input variables cover a wide range of values. More precisely, H varies from 90 to 360 mm with an average value of 163 mm and a coefficient of variation of 32%. W ranges from 60 to 240 mm with an average value of 111 mm and a coefficient of variation of 32%. t ranges from 0.7 to 10 mm with an average value of 4 mm and a coefficient of variation of 48%. L varies from 100 to 3050 mm with an average value of 869 mm and a coefficient of variation of 89%. f_y ranges from 194 to 515 MPa with an average value of 329 MPa and a coefficient of variation of 24%. f_c' varies from 8 to 47 MPa with an average value of 31 MPa and a coefficient of variation of 39%.

It should be pointed out that the steel tube of 43 specimens was cold-formed, whereas welded built-up was done in the other 56 configurations. In terms of failure modality, local outward buckling failure of the external steel was observed in all specimens, as shown in Figure 2a. This is the same as that observed by other investigations such as Han and Yao [91], Lyu et al. [97], Ding et al. [98], and Yan et al. [99]. Depending on the dimension of the cross section, the locations of the external folding of the steel tube are not the same. Such local buckling of the steel tube occurred mostly at the ends or in the center along the axis of the specimens, as seen in Figure 2a. In addition to outward buckling failure, fracture at the welding seam also occurred in welded specimens, as shown in Figure 2b. Such tensile fracture is the result of too much growth of the concrete in the core [99]. However, the tensile fracture of the steel tube generally occurred after the peak load [98]. Last, but not least, for all specimens, concrete in the core was damaged in most of specimens following a shear failure mode, as shown in Figure 2c [97,98]. Besides, the influence of temperature on the failure modality of stub CFST structural members could be referred to in Yan et al. [99] (low temperature) and Lyu et al. [97] (high temperature). Finally, Angelo et al. [100] and Kulkarni et al. [101] have tested and discussed about the failure of rectangular CFST structural members in junction with wide beam for earthquake engineering application.

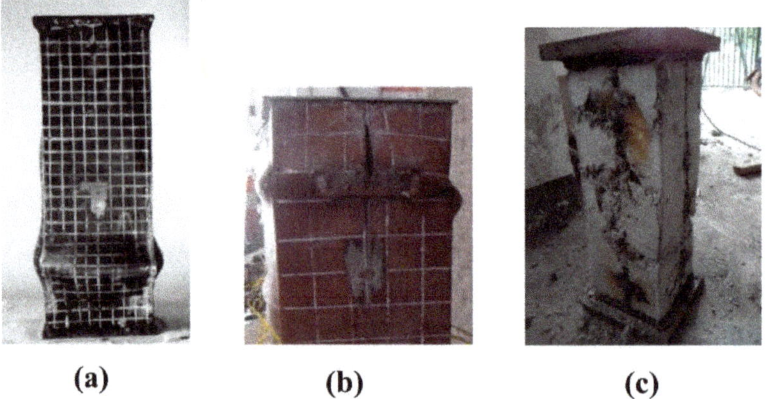

(a)   **(b)**   **(c)**

Figure 2. Failure of rectangular CFST specimens: (**a**) local outward buckling of steel tube (reproduced with permission from Han [89]), (**b**) tensile fracture at the welding seam of steel tube (reproduced with permission from Ding et al. [98]), (**c**) damage of concrete core (reproduced with permission from Lyu et al. [97]).

It is worth mentioning that only rectangular CFST columns (i.e., depth/width ratio greater than 1) were collected for investigation. As indicated in Table 2, the depth/width ratio ranges from 1 to 2, allowing for exploring the axial failure of CFST around the weak axis. In addition, as the depth/width ratio differs than 1, the stress of confined concrete applied to the steel wall is not the same along the weak and strong axes, while the thickness of the steel tube was constant. Consequently, the consideration of only rectangular CFST columns could strongly reveal the influence of both the structural geometry and mechanical properties of constituent materials.

Table 1. Information of the database used in this study.

No.	Reference	Number of Data	% of Proportion
1	Bridge [85]	1	1.0
2	Du et al. [86]	5	5.1
3	Du et al. [87]	8	8.1
4	Ghannam et al. [88]	12	12.1
5	Han [89]	20	20.2
6	Han & Yang [90]	4	4.0
7	Han & Yao [91]	19	19.2
8	Lin [92]	6	6.1
9	Schneider [93]	9	9.1
10	Shakir-Khalil & Mouli [94]	14	14.1
11	Shakir-Khalil & Zeghiche [95]	1	1.0
	Total	99	100

Table 2. Initial statistical analysis of database.

Parameters	Symbol	Unit	Role	Min	Q_{25}	Median	Q_{75}	Max	Mean	StD	Coefficient of Variation (%)
Depth of cross section	H	mm	Input	90	127.9	150	195	360	163.38	53.01	32.45
Width of cross section	W	mm	Input	60	90	100	124.48	240	110.94	35.63	32.12
Thickness of steel tube	t	mm	Input	0.7	2.7	3	5	10.01	4.12	1.97	47.84
Length of column	L	mm	Input	100	369.75	545	800	3050	869.23	772.12	88.83
Yield stress of steel	f_y	MPa	Input	194	245.18	340.1	357.88	514.53	329.09	78.73	23.92
Compressive strength of concrete	f_c'	MPa	Input	7.9	18.67	33.74	43.69	46.85	31.12	12.21	39.23
Axial capacity	P_u	kN	Output	490	760	1006	1340	3575	1267.61	768.72	60.64

The dataset was randomly divided into two sub-datasets including the training part (60%) and testing part (40%) part. All data were scaled into the range of [0,1] in order to reduce numerical biases while treating with the AI algorithms, as recommended by various studies in the literature [102–104]. Such a scaling process is expressed using Equation (4) between raw and scaled data [105–107]:

$$x^{scaled} = \frac{(x^{raw} - \beta)}{\alpha - \beta} \tag{4}$$

where α and β are the maximum and minimum values of the considered variable x, respectively. It should be noticed that a reverse transformation could be used for converting data from the scaling space to the raw one using Equation (4). Besides, a correlation analysis between the input and output variables is performed and plotted in Figure 3.

Figure 3 was generated in order to explore the linear statistical correlation between variables in the database. Therefore, a 7×7 matrix was generated, in which the upper triangular part indicates the value of the correlation coefficient, whereas the lower triangular part shows the scatter plot between two associated variables. The diagonal of the matrix indicates the name of the variable (i.e., as the correlation coefficient of a variable itself is equal to 1). For interpretation purpose, the correlation coefficient between H and W is indicated as 0.86, whereas the corresponding scatter plot between H and W is shown on the left side of W (row 2, column 1). It is seen that a high and positive value of statistical correlation was obtained in this case, confirmed by most of the data points being located around the diagonal in the scatter plot.

It can be seen that no direct correlation was observed between each input and output (P_u). The maximum value of the Pearson correlation coefficient (r) compared with P_u was calculated as 0.78 (for variable t), followed by 0.60 (for variable f_y), 0.39 (for variable W), 0.30 (for variable H), 0.27 (for variable f_c'), and 0.18 (for variable L). Besides, the correlation between H and W was highest (r = 0.86).

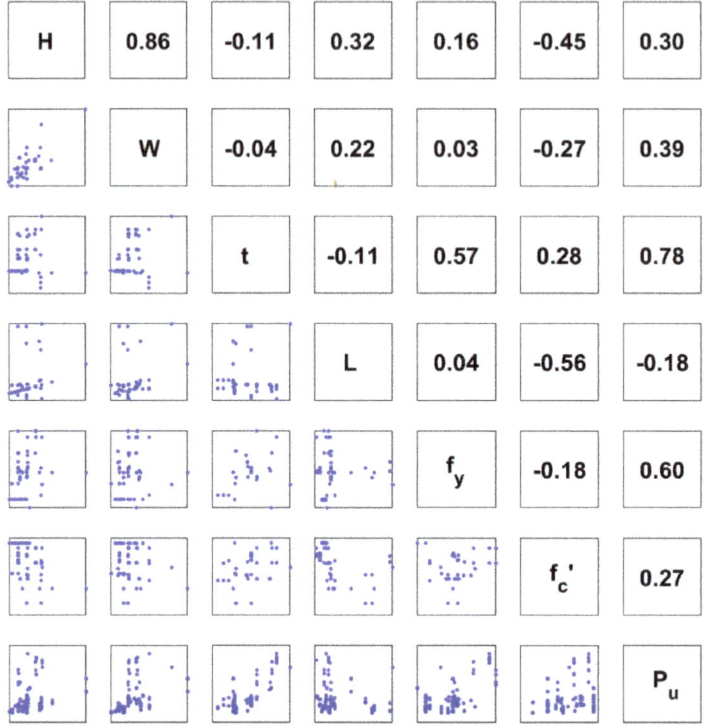

Figure 3. Correlation analysis between the depth of cross section (H), width of cross section (W), thickness of steel tube (t), length of column (L), yield stress of steel (f_y), concrete compressive strength (f_c'), and axial capacity (P_u).

3. Results and Discussion

3.1. Optimization of Weight Parameters of FNN using the IWO Technique

In this section, the optimization of weight parameters of FNN is presented using the IWO algorithm. It is not worth noticing that the architecture of the FNN model is very important. Depending on the problem of interest, the prediction results could exhibit significant variation from using one architecture to another [96,107,108]. As the numbers of inputs and outputs are fixed, the undetermined parameters of the architecture are the number of hidden layer(s) and the number of neurons in each hidden layer(s) [109]. As proved by many investigations in the literature, the FNN model involving only one hidden layer could be sufficient for exploring successfully complex nonlinear relationship between inputs and outputs. For instance, Mohamad et al. [110] have used one hidden layer architecture model for predicting ripping production, as have Singh et al. [111] for predicting cadmium removal. In civil engineering application, a prediction model involving one hidden layer has also been widely applied in many works, for instance, Gordan et al. [112] for earthquake slope stability or Sarir et al. [96] for bearing capacity of circular concrete-filled steel tube columns. Therefore, the one hidden layer FNN model was finally adopted in this work, also saving cost, processing time, and limitation of instruments. On the other hand, the number of neurons in the hidden layer was recommended to be equal to the sum of the number of inputs and outputs [109,113,114]. Consequently, the FNN model exhibits one hidden layer and seven neurons in the hidden layer. The activation function for the hidden layer was chosen as a sigmoid function, whereas the activation function for the output layer was a linear function [115]. The cost function was chosen such as the mean square error function [116]. Finally, Table 3 indicates the information of the FNN model.

As revealed in the literature, a key aspect of using evolutionary algorithms for optimizing AI models is to study the relationship between population size and problem dimensionality [117–120]. In many other evolutionary algorithms such as differential evolution, the number in the population is recommended to be 7–10 times the number of inputs [121,122]. In this study, the population size of the IWO technique was chosen as 50. Other parameters include the variance reduction exponent, chosen as 2; initial value of standard deviation, chosen as 0.01; final value of standard deviation, chosen as 0.001; and maximum iteration, chosen as 800. It is worth noticing that such ranges of parameters are commonly employed for training AI models using IWO algorithm, for instance, Huang et al. [76] and Mishagi et al. [123]. It should also be noticed that a large population size cannot be useful in evolutionary algorithms and affects the optimization results [124]. Information of the IWO algorithm is presented in Table 3.

Table 3. Values and description of feedforward neural network (FNN) and invasive weed optimization (IWO) parameters in this study.

Methods	Parameter	Values and Description
FNN	Number of neurons in input layer	6
	Number of neurons in output layer	1
	Number of hidden layers	1
	Number of neurons in hidden layer	7
	Size of weight matrix of hidden layer	42
	Size of weight matrix of output layer	7
	Size of bias vector of hidden layer	7
	Size of bias vector of output layer	1
	Dimension of optimization problem	57
	Activation function for hidden layer	Sigmoid
	Activation function for output layer	Linear
	Training algorithm	IWO
	Cost function	Mean square error
IWO	Population size	50
	Variance reduction exponent	2
	Initial value of standard deviation	0.01
	Final value of standard deviation	0.001
	Maximum iteration	800

Figure 4a presents the evolution of 42 weight parameters of the hidden layer, whereas Figure 4b shows such evolution of 7 weight parameters of the output layer. It is seen that, at the 300 first iterations, fluctuations were observed for all weight parameters, as the IWO algorithm imitated the colonizing behavior of weed plants. After about 500–600 iterations, stabilization was achieved for weight parameters for the 57-dimensional optimization problem. Consequently, at least 700–800 iterations are needed in order to ensure the stabilization of the process.

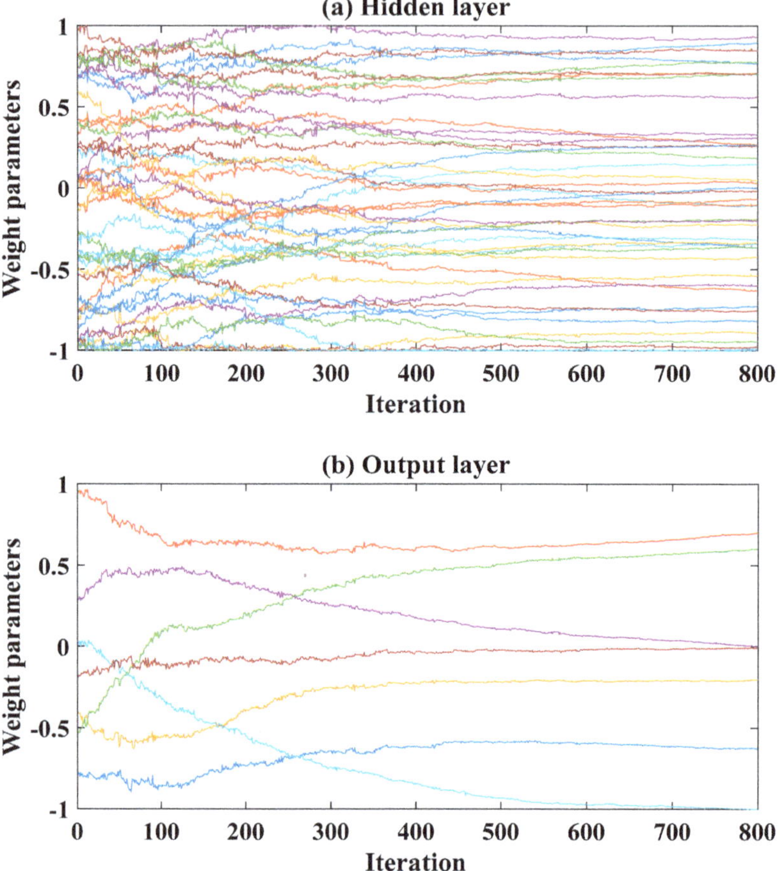

Figure 4. Evolution of weight parameters over 800 iterations: (**a**) weight parameters of input layer (42 parameters); (**b**) weight parameters of hidden layer (7 parameters).

Weight parameters at iteration 800 were extracted for constructing the final FNN–IWO model (a combination of FNN and IWO). This model was then used as a numerical prediction function for parametrically investigating the deviation of quality assessment criteria in function weight parameters. The parametric study could be helpful to verify if the results provided by the IWO were unique, that is, the IWO allowed reaching the global optimum of the problem. For illustration purposes, only three first weight parameters were plotted. Figure 5a presents the evolution of RMSE while varying weight parameters N°1 and N°2 from their lowest to highest values. In the same context, Figure 5b presents the evolution of RMSE while varying weight parameters N°1 and N°3 from their lowest to highest values. It is seen from Figure 5a,b that the global optimum of the two RMSE surfaces matched the final set of weight parameters provided by the IWO algorithm. This remark confirmed that the IWO technique allowed calibrating the global optimum of the optimization problem, thus providing the final FNN–IWO model.

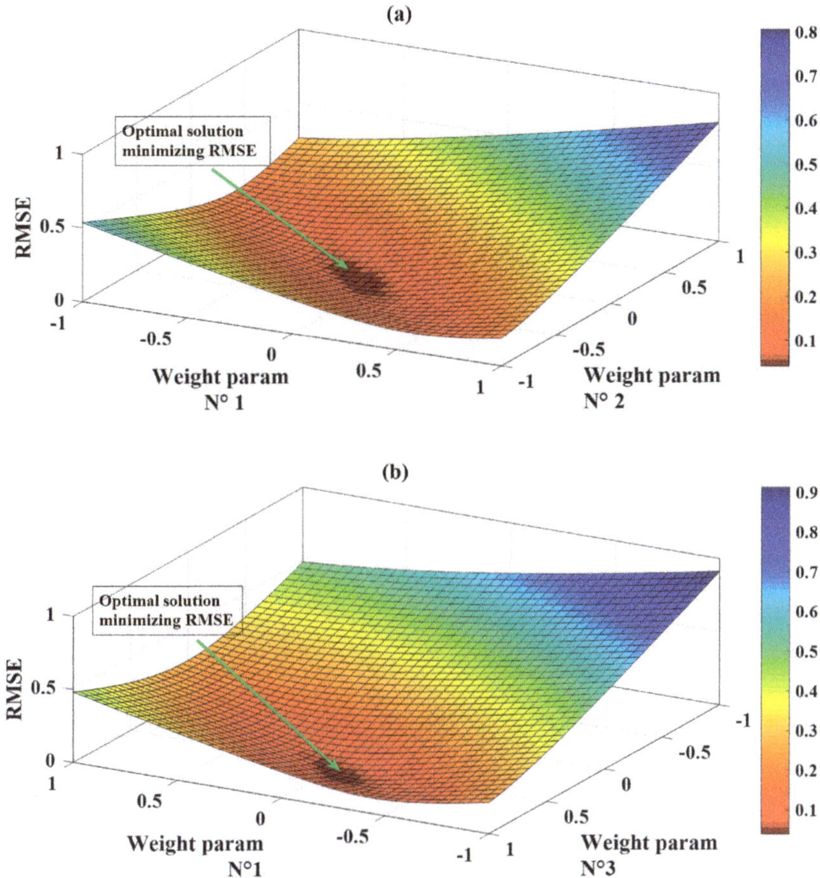

Figure 5. Verification of global optimum provided by the invasive weed optimization (IWO). The surfaces of root mean square error (RMSE) show unique optimal solution, which minimizes the value of RMSE: (**a**) between weight parameters N°1 and N°2, (**b**) between weight parameters N°1 and N°3.

Figure 6a–c present the evolution of RMSE, MAE, and R^2 during the optimization process of FNN weight parameters, for both training and testing data. It is seen that during the optimization using the training data, good results of RMSE, MAE, and R^2 for the testing data were obtained. It is not worth noting that the testing data were totally new when applying. This remark allows exploring that no overfitting occurred during the training phase (i.e., performance indicators of testing data go in a bad direction). The efficiency and robustness of the IWO technique are then confirmed.

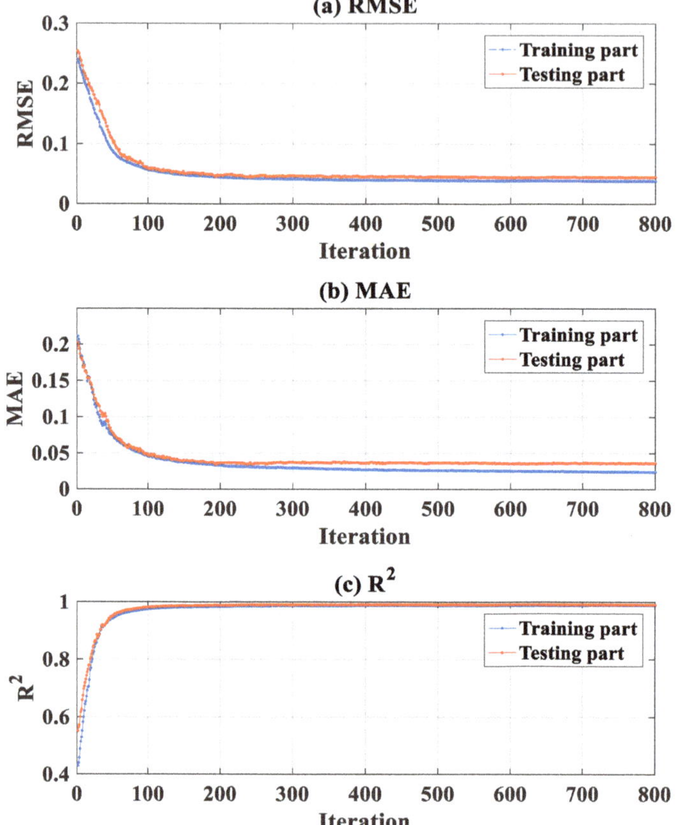

Figure 6. Evaluation of the performance indicators during optimization: (**a**) RMSE, (**b**) mean absolute error (MAE), and (**c**) R^2, for training and testing data, respectively.

3.2. Influence of the Training Set Size

In this section, the influence of training set size (in %) on the prediction results is presented. The training dataset was varied from 10% to 90% of the total data (with a resolution of 10%). Figure 7 illustrates the influence of training set size, with respect to R^2 (Figure 7a), RMSE (Figure 7b), MAE (Figure 7c), ErrorStD (Figure 7d), and slope (Figure 7e). All relevant values are also highlighted in Table 4.

As seen in Figure 7a,e for R^2 and slope, the performance of the prediction model progressively increased during the increasing of the training set size from 10% to 90%. For instance, for the testing part, $R^2 = 0.387$ when the training set size was 10%, which was increased to 0.987 when the training set size was 90%. The same remark was also obtained when regarding Figure 7b,c,d for RMSE, MAE, and ErrorStD, respectively. Moreover, the performance of the prediction model for both training and testing parts became stable from 60% of the training set size (Figure 7a). This observation indicates that no over-fitting occurred when the training set size surpassed a high percentage, for instance, 80%. This point proves that the prediction model is robust, exhibiting a strong capability in tracking relevant information in the testing part even it is small. Finally, yet importantly, the prediction model is promising in the case in which more data are available.

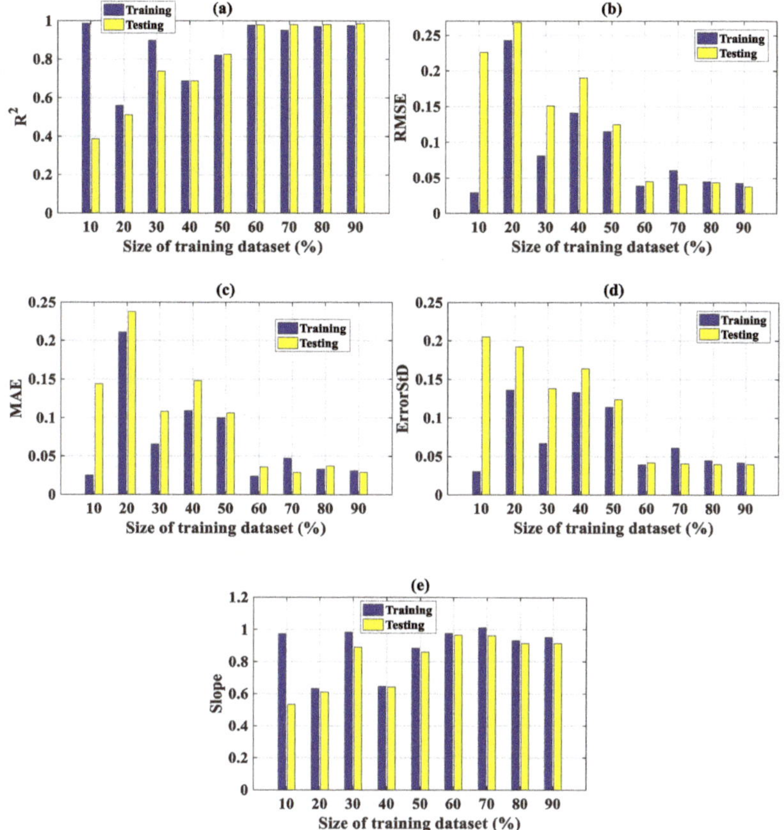

Figure 7. Influence of training set size with respect to (**a**) R^2, (**b**) RMSE, (**c**) MAE, (**d**) ErrorStD, and (**e**) slope.

Table 4. Summary of influence of training set size on the prediction results. RMSE, root mean square error; MAE, mean absolute error.

Dataset	Size of Training Dataset (%)	Size of Testing Dataset (%)	R^2	RMSE	MAE	ErrorStD	Slope
	10	90	0.987	0.029	0.025	0.030	0.974
	20	80	0.559	0.243	0.211	0.136	0.633
	30	70	0.897	0.081	0.066	0.067	0.984
	40	60	0.687	0.141	0.109	0.133	0.646
Training	50	50	0.821	0.115	0.100	0.114	0.883
	60	40	0.978	0.039	0.024	0.039	0.976
	70	30	0.952	0.061	0.047	0.061	1.011
	80	20	0.971	0.045	0.033	0.045	0.931
	90	10	0.977	0.043	0.031	0.042	0.952
	10	90	0.387	0.226	0.144	0.205	0.535
	20	80	0.511	0.268	0.238	0.192	0.610
	30	70	0.737	0.151	0.108	0.138	0.890
	40	60	0.688	0.190	0.148	0.164	0.643
Testing	50	50	0.826	0.125	0.106	0.124	0.860
	60	40	0.979	0.045	0.036	0.042	0.966
	70	30	0.982	0.041	0.029	0.041	0.961
	80	20	0.982	0.044	0.037	0.040	0.914
	90	10	0.987	0.038	0.029	0.040	0.914

3.3. Prediction Capability of the FNN–IWO Model

In this section, the performance of FNN–IWO in predicting the P_u of CFST is investigated. The predicted outputs versus the corresponding experimental results associated with the training, testing, and all datasets are presented in Figure 8. The fitted linear lines are also plotted (red lines) in each graph to show the performance of the algorithm. R^2 values with respect to the training, testing, and all datasets were estimated at 0.978, 0.979, and 0.978, respectively, showing an excellent prediction capability of FNN–IWO. Furthermore, three linear equations representing the relationships between actual and predicted data were also given in each graph, including the intercepts and slopes. It is observed that the FNN–IWO algorithm possessed a strong linear correlation between actual and predicted P_u values.

The detailed performance of the proposed FNN–IWO algorithm is summarized in Table 5, including R^2, RMSE, MAE, standard deviation error (ErrorStD), slope, and slope angle. Regarding the results of quality assessment and error analysis, FNN–IWO exhibited a strong capability in predicting the critical compression capacity of the rectangular section.

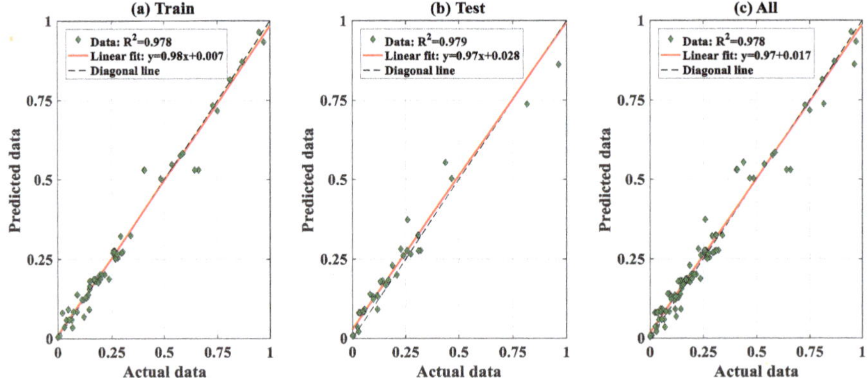

Figure 8. Comparison between actual and predicted data in regression scatter mode for (**a**) training data, (**b**) testing data, and (**c**) all data.

Table 5. Performance indicators of the optimal FNN–IWO model.

Indicator	R^2	RMSE	MAE	ErrorStD	Slope	Slope Angle
Training part	0.978	0.039	0.024	0.039	0.976	44.296°
Testing part	0.979	0.045	0.036	0.042	0.966	44.015°
All data	0.978	0.042	0.029	0.041	0.969	44.101°

For further assessment of the performance of the FNN–IWO algorithm, comparison between the experimental and predicted results was performed at different quantile levels. For this purpose, quantiles from 10% to 90% were computed to track the behavior of the distribution of the data, with a focus on the most important statistical distribution. The results are presented (Figure 9a–c) for the training, testing, and all data, respectively, whereas the percentage of error (%) between the predicted and actual values at each quantile level is displayed in Figure 10.

It is seen that, for the training dataset, the actual and predicted data were highly correlated, whereas a small difference was observed at each level of quantile for the testing part. With respect to the whole dataset, the highest error ratio was observed at Q80, followed by Q90 and Q10. For the values of error, it was seen that the FNN–IWO model exhibited a strong efficiency in predicting P_u within the Q10–Q70 range (error < 5%) and from Q80 to Q90 (with error in the 5%–10% range).

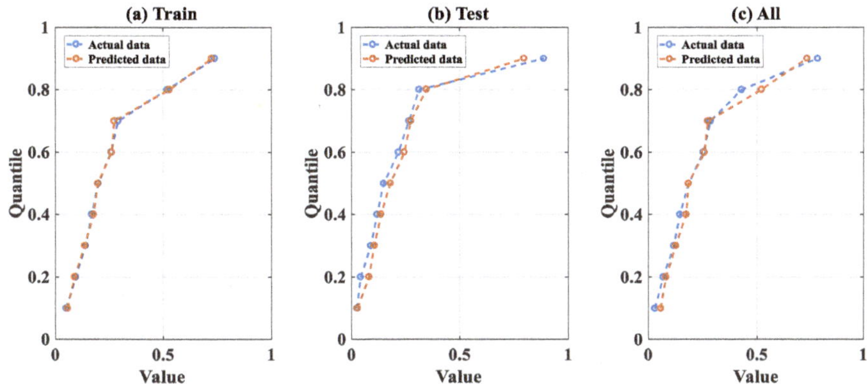

Figure 9. Comparison between actual and predicted data at different quantile levels of the distributions for (**a**) training data, (**b**) testing data, and (**c**) all data.

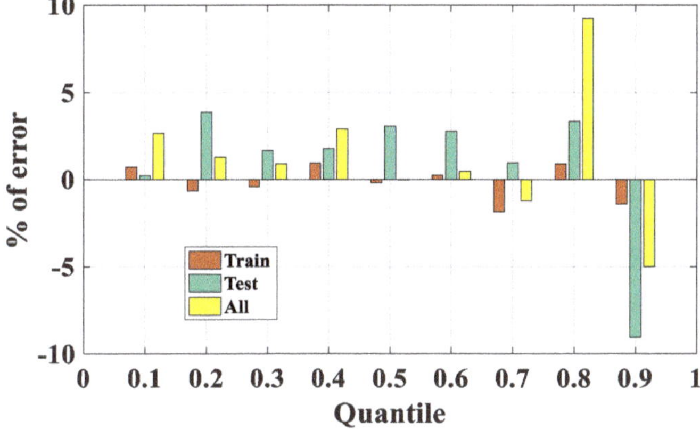

Figure 10. Percentage of error between quantile estimation for training, testing, and all data.

3.4. Prediction Accuracy in Function of Structural Parameters of FNN–IWO

In this section, the prediction accuracy of FNN–IWO with respect to different ranges of structural parameters is presented. The actual and predicted P_u in function of the depth /width ratio, t, f_y, f_c', and slenderness ratio are displayed in Figure 11a–e, respectively. Besides, error analysis in terms of R^2, RMSE, and MAE for several intervals of the depth/width ratio, t, f_y, f_c', and slenderness ratio, respectively, is also indicated in Table 6 and Figure 11, together with the associated number of data.

In the case of the depth/width ratio, 11 configurations were found between 1 and 1.2, exhibiting $R^2 = 0.98$, RMSE = 137.57 kN, and MAE = 95.25 kN; 22 configurations were found between 1.2 and 1.4, showing $R^2 = 0.98$, RMSE = 71.07 kN, and MAE = 56.01 kN; 43 configurations were found between 1.4 and 1.6, exhibiting $R^2 = 0.97$, RMSE = 144.71 kN, and MAE = 109.65 kN; 11 configurations were found between 1.6 and 1.8, exhibiting $R^2 = 0.89$, RMSE = 56.16 kN, and MAE = 38.91 kN; and only 3 configurations were found between 1.8 and 2, exhibiting $R^2 = 1.00$, RMSE = 24.75 kN, and MAE = 21.81 kN. Such an analysis allowed confirming that the FNN–IWO model is efficient in predicting P_u from nearly square to highly rectangular columns.

In the case of slenderness, 78 configurations were found between 0 and 20 of slenderness, exhibiting $R^2 = 0.98$, RMSE = 123.29 kN, and MAE = 86.64 kN; 6 configurations were found between 20 and 40 of slenderness, showing $R^2 = 0.98$, RMSE = 42.80 kN, and MAE = 32.09 kN; 13 configurations were found between 40 and 60 of slenderness, exhibiting $R^2 = 0.99$, RMSE = 72.25 kN, and MAE = 55.32 kN. Although the number of data is small for large slenderness, such an analysis allowed remarking that the FNN–IWO model is efficient in predicting P_u for short, medium, and long columns.

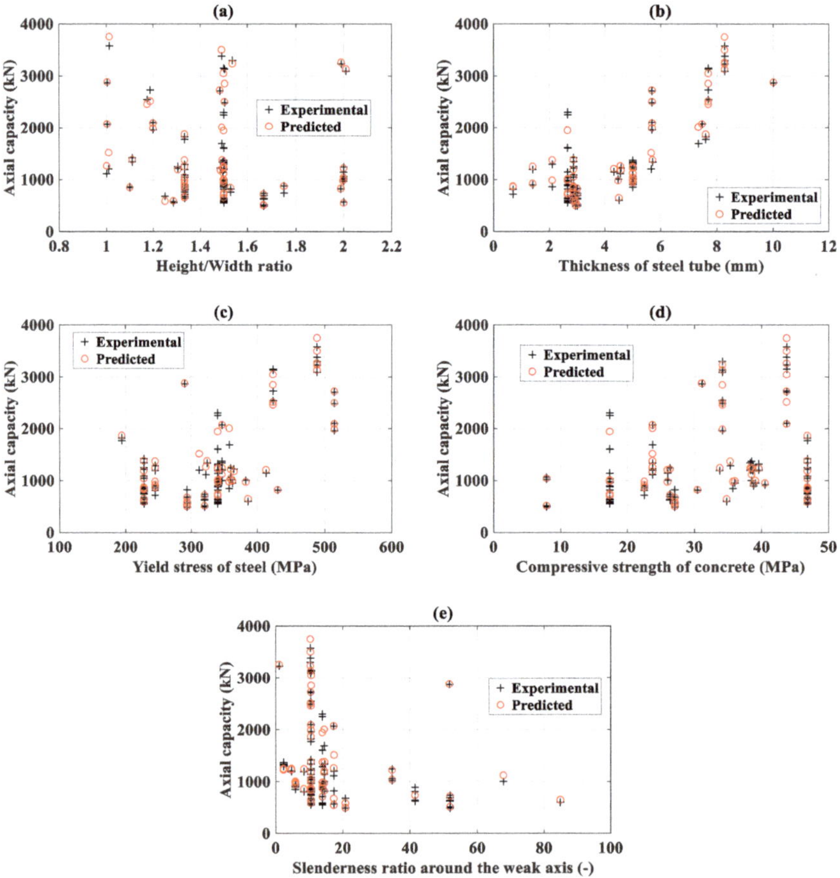

Figure 11. Evaluation of axial capacity in function of the (**a**) depth/width ratio, (**b**) thickness, (**c**) yield stress, (**d**) compressive strength, and (**e**) slenderness ratio.

Table 6. Error analysis of prediction performance with respect to different ranges of values of structural variables.

Structural Parameter	Lower Bound	Upper Bound	Number of Data	R^2	RMSE (kN)	MAE (kN)
	1	1.2	11	0.98	137.57	95.25
	1.2	1.4	22	0.98	71.07	56.01
Depth/width ratio (-)	1.4	1.6	43	0.97	144.71	109.65
	1.6	1.8	11	0.89	56.16	38.91
	1.8	2	3	1.00	24.75	21.81
	0	2	4	0.91	87.07	70.74
Thickness of steel	2	4	52	0.91	118.58	80.36
tube (mm)	4	6	29	0.97	85.70	61.60
	6	8	8	0.91	178.54	143.84
	8	10	6	0.89	92.27	72.82
	190	260	26	0.97	64.45	50.11
Yield stress of steel	260	320	6	0.97	146.68	99.40
(MPa)	320	380	50	0.91	129.64	92.09
	380	440	8	0.99	137.34	104.47
	440	515	9	0.99	76.28	55.16
Compressive	5	20	25	0.90	157.25	113.71
strength of concrete	20	30	22	0.93	120.57	84.45
(MPa)	30	40	24	0.99	87.44	67.86
	40	50	28	0.99	75.40	53.80
	0	20	78	0.98	123.29	86.64
	20	40	6	0.98	42.80	32.09
Slenderness ratio (-)	40	60	13	0.99	72.25	55.32
	60	80	1	-	116.74	116.74
	80	100	1	-	49.65	49.65

3.5. Comparison of the Hybrid Model of FNN–IWO and the Single FNN Model

In order to highlight the efficiency of the evolutionary IWO algorithm, comparisons between FNN–IWO and the individual FNN were performed, using a similar training algorithm (scaled conjugate gradient (SCG)), FNN architecture, and dataset.

Considering RMSE, MAE, and standard deviation error (ErrorStD), Figure 12 identifies the values of the two algorithms for the training part (Figure 12a) and testing part (Figure 12b). It can be clearly seen that FNN–IWO is more accurate than the single FNN, represented by a reduction of error for RMSE (2 times), MAE (3 times), or ErrorStD (2 times). Improvement of the accuracy is more pronounced in the training part than the testing part. Considering R^2 and slope as error criteria, FNN–IWO also exhibited an advantage compared with FNN without optimization, for both the training and testing datasets (Figure 11c,d).

For the sake of comparison, Table 7 indicates the exact values and gains (in %) while using FNN–IWO with FNN for five error criteria. With a focus on the testing part, the gains reached 47.9%, 49.2%, 41.3%, 6.5%, and 1.5% for RMSE, MAE, ErrorStD, R^2, and slope, respectively. As a conclusion, using IWO to tune the weights and bias of FNN strongly enhanced the accuracy in predicting P_u.

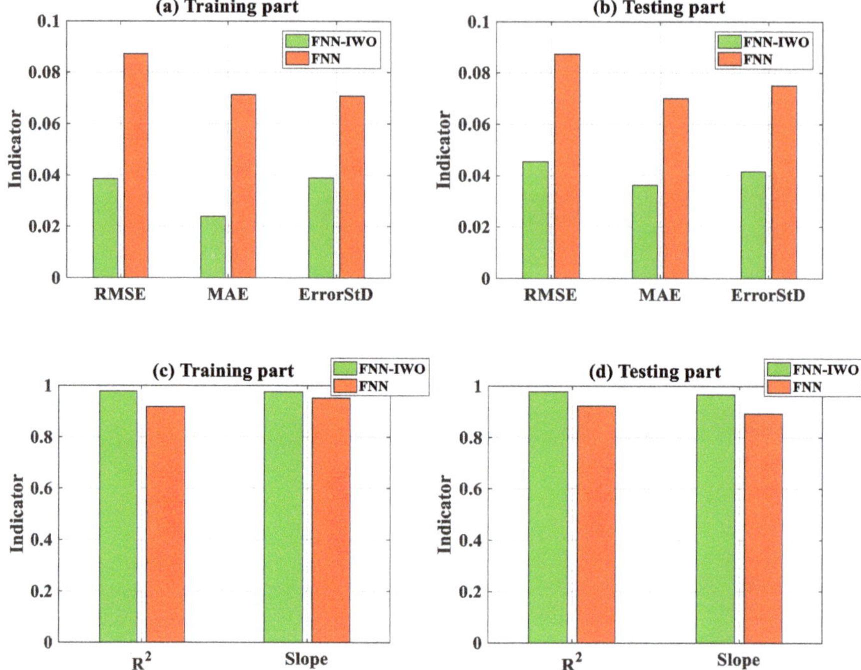

Figure 12. Comparison of performance indicators between the individual FNN and FNN–IWO model: (**a**) RMSE, MAE, and ErrorStD for training data; (**b**) RMSE, MAE, and ErrorStD for testing data; (**c**) R^2 and slope for training data; and (**d**) R^2 and slope for testing data.

Table 7. Comparison of performance indicators between FNN–IWO and individual FNN.

Data	Model Used	RMSE	MAE	ErrorStD	R^2	Slope
	FNN–IWO	0.039	0.024	0.039	0.978	0.976
Training	FNN	0.087	0.070	0.075	0.923	0.893
	% Gain	+55.8	+65.9	+48.1	+6.0	+9.2
	FNN–IWO	0.045	0.036	0.042	0.979	0.966
Testing	FNN	0.087	0.071	0.071	0.919	0.952
	% Gain	+47.9	+49.2	+41.3	+6.5	+1.5

4. Conclusions and Outlook

Even though many studies attempted to predict the P_u of CFST with different AI algorithms, the accuracy and robustness of these algorithms still need further comprehensive investigation. In this study, a novel hybrid approach of FNN–IWO was proposed and improved for the prediction of P_u of CFST, of which IWO was used for tuning and optimizing the FNN weights and biases to improve the prediction performance.

The results showed that the FNN–IWO algorithm is an excellent predictor of P_u, with a value of R^2 of up to 0.979. The performance of FNN–IWO in predicting P_u function of structural parameters such as depth/width ratio, thickness of steel tube, yield stress of steel, concrete compressive strength, and slenderness ratio was investigated and the results showed that FNN–IWO is efficient in predicting P_u from nearly square to highly rectangular columns, as well as for short, medium, and long columns. Better performance of FNN–IWO was also pointed out with the gains in accuracy of 47.9%, 49.2%, and 6.5% for RMSE, MAE, and R^2, respectively, compared with the simulation using the single FNN. This study may help in quick and accurate prediction of P_u of CFST for better practice purposes.

In general, the main advantage of AI-based methods is its efficient capability to model the macroscopic mechanical behavior of the structural members without any prior assumptions or constraints. Therefore, the developed AI model in this study could be applied to the pre-design phase of the design process. Indeed, such quick numerical estimation is helpful to explore some initial evaluations of the outcome before conducting any extensive laboratory experiments. To this aim, a graphical user interface application should be compiled for facilitating the application by engineers/researchers.

On the other hand, empirical formulae should be derived based on the "black-box" AI-based model developed in this study for estimating the axial behavior of rectangular CFST columns. In addition, the performance of such empirical formulae should be compared with other existing equations in the literature such as Ding et al. [98], Wang et al. [125], and Han et al. [126]. Besides, numerical finite element scheme should also be studied, especially for investigating the mechanical behaviors of composite columns at both the micro and macro levels. Finally, improvement for current designs (such as Eurocode-4 [127], AISC [128], and ACI [129]), if it exists, should be proposed.

The axial behavior of CFST composite columns is a complex problem, involving various variables such as geometry and mechanical properties of constituent materials. Consequently, experimental databases are crucial for studying this problem. In further studies, a larger database should be considered, in order to cover more material strengths and geometric dimension ranges.

The methodology modeling of this work could be extended for predicting other macroscopic properties such as bending, compression, or tension strength of not only composite members, but also members made of a single material (i.e., concrete or steel members). Besides, an investigation based on homogenization and de-homogenization approaches [130–134] could also be useful for studying structural members under different boundary conditions and loadings. Such a framework, including the finite element scheme, could also be coupled with AI-based prediction in order to better understand the micro and macro behaviors of structural members.

Author Contributions: Conceptualization, H.Q.N., T.-T.L. and H.-B.L.; methodology, H.-B.L., T.-T.L., and B.T.P.; validation, H.-B.L., H.Q.N., and B.T.P.; formal analysis, H.Q.N., T.-A.N, T.-T.L., V.Q.T. and H.-B.L.; data curation, V.Q.T. and T.-A.N.; writing—original draft preparation, all authors; writing—review and editing, H.-B.L., T.-T.L., and B.T.P.; project administration, H.-B.L.; funding acquisition, H.Q.N. All authors have read and agreed to the published version of the manuscript.

Funding: This research received no external funding.

Conflicts of Interest: The authors declare no conflict of interest.

References

1. Giakoumelis, G.; Lam, D. Axial capacity of circular concrete-filled tube columns. *J. Constr. Steel Res.* **2004**, *60*, 1049–1068. [CrossRef]
2. Tran, V.Q.; Nguyen, H.L.; Dao, V.D.; Hilloulin, B.; Nguyen, L.K.; Nguyen, Q.H.; Le, T.-T.; Ly, H.-B. Temperature effects on chloride binding capacity of cementitious materials. *Mag. Concr. Res.* **2019**, 1–39. [CrossRef]
3. Ly, H.-B.; Pham, B.T.; Dao, D.V.; Le, V.M.; Le, L.M.; Le, T.-T. Improvement of ANFIS Model for Prediction of Compressive Strength of Manufactured Sand Concrete. *Appl. Sci.* **2019**, *9*, 3841. [CrossRef]
4. Liu, D.; Gho, W.-M. Axial load behaviour of high-strength rectangular concrete-filled steel tubular stub columns. *Thin-Walled Struct.* **2005**, *43*, 1131–1142.
5. Tao, Z.; Han, L.-H. Behaviour of concrete-filled double skin rectangular steel tubular beam–columns. *J. Constr. Steel Res.* **2006**, *62*, 631–646. [CrossRef]
6. Liu, S.; Ding, X.; Li, X.; Liu, Y.; Zhao, S. Behavior of Rectangular-Sectional Steel Tubular Columns Filled with High-Strength Steel Fiber Reinforced Concrete under Axial Compression. *Materials* **2019**, *12*, 2716. [CrossRef]
7. Liu, D.; Gho, W.-M.; Yuan, J. Ultimate capacity of high-strength rectangular concrete-filled steel hollow section stub columns. *J. Constr. Steel Res.* **2003**, *59*, 1499–1515. [CrossRef]

8. Chitawadagi, M.V.; Narasimhan, M.C.; Kulkarni, S.M. Axial capacity of rectangular concrete-filled steel tube columns–DOE approach. *Constr. Build. Mater.* **2010**, *24*, 585–595. [CrossRef]

9. Yang, Y.-F.; Han, L.-H. Experiments on rectangular concrete-filled steel tubes loaded axially on a partially stressed cross-sectional area. *J. Constr. Steel Res.* **2009**, *65*, 1617–1630. [CrossRef]

10. Du, Y.; Chen, Z.; Zhang, C.; Cao, X. Research on axial bearing capacity of rectangular concrete-filled steel tubular columns based on artificial neural networks. *Front. Comput. Sci.* **2017**, *11*, 863–873. [CrossRef]

11. Evirgen, B.; Tuncan, A.; Taskin, K. Structural behavior of concrete filled steel tubular sections (CFT/CFSt) under axial compression. *Thin-Walled Struct.*.

12. Liu, D. Behaviour of eccentrically loaded high-strength rectangular concrete-filled steel tubular columns. *J. Constr. Steel Res.* **2006**, *62*, 839–846. [CrossRef]

13. Gho, W.-M.; Liu, D. Flexural behaviour of high-strength rectangular concrete-filled steel hollow sections. *J. Constr. Steel Res.* **2004**, *60*, 1681–1696. [CrossRef]

14. Lai, Z.; Varma, A.H.; Zhang, K. Noncompact and slender rectangular CFT members: Experimental database, analysis, and design. *J. Constr. Steel Res.* **2014**, *101*, 455–468. [CrossRef]

15. Ren, Q.; Li, M.; Zhang, M.; Shen, Y.; Si, W. Prediction of Ultimate Axial Capacity of Square Concrete-Filled Steel Tubular Short Columns Using a Hybrid Intelligent Algorithm. *Appl. Sci.* **2019**, *9*, 2802. [CrossRef]

16. Zeghiche, J.; Chaoui, K. An experimental behaviour of concrete-filled steel tubular columns. *J. Constr. Steel Res.* **2005**, *61*, 53–66. [CrossRef]

17. Morino, S.; Tsuda, K. Design and construction of concrete-filled steel tube column system in Japan. *Earthq. Eng. Eng. Seismol.* **2003**, *4*, 51–73.

18. Sherman, D.R. Designing with structural tubing. *Eng. J. Am. Inst. Steel Constr.* **1996**, *33*, 101–109.

19. Kwon, Y.B.; Jeong, I.K. Resistance of rectangular concrete-filled tubular (CFT) sections to the axial load and combined axial compression and bending. *Thin-Walled Struct.* **2014**, *79*, 178–186. [CrossRef]

20. Tian, Z.; Liu, Y.; Jiang, L.; Zhu, W.; Ma, Y. A review on application of composite truss bridges composed of hollow structural section members. *J. Traffic Transp. Eng.* **2019**, *6*, 94–108. [CrossRef]

21. American Institute of Steel Construction. *Load and Resistance Factor Design Specification for Structural Steel Buildings*; American Institute for Steel Construction Inc.: Chicago, IL, USA, 1993.

22. *American Concrete Institute. Building Code Requirements for Structural Concrete (ACI 318-05) and Commentary (ACI 318R-05)*; American Concrete Inst.: Farmington Hills, MI, USA, 2004.

23. Arya, C. *Design of Structural Elements: Concrete, Steelwork, Masonry and Timber Designs to British Standards and Eurocodes*; CRC Press: Boca Raton, FL, USA, 2009.

24. Hicks, S.; Uy, B.; Kang, W.H. AS/NZS 5100.6, Design of steel and composite bridges. In Proceedings of the Austroads Bridge Conference, Melbourne, Australia, 3–6 April 2017.

25. Hatzigeorgiou, G.D. Numerical model for the behavior and capacity of circular CFT columns, Part I: Theory. *Eng. Struct.* **2008**, *30*, 1573–1578. [CrossRef]

26. Hatzigeorgiou, G.D. Numerical model for the behavior and capacity of circular CFT columns, Part II: Verification and extension. *Eng. Struct.* **2008**, *30*, 1579–1589. [CrossRef]

27. Han, L.-H.; Yang, Y.-F. Influence of concrete compaction on the behavior of concrete filled steel tubes with rectangular sections. *Adv. Struct. Eng.* **2001**, *4*, 93–100. [CrossRef]

28. Tokgoz, S.; Dundar, C. Experimental study on steel tubular columns in-filled with plain and steel fiber reinforced concrete. *Thin-Walled Struct.* **2010**, *48*, 414–422. [CrossRef]

29. Fam, A.; Schnerch, D.; Rizkalla, S. Rectangular filament-wound glass fiber reinforced polymer tubes filled with concrete under flexural and axial loading: Experimental investigation. *J. Compos. Constr.* **2005**, *9*, 25–33. [CrossRef]

30. Mouli, M.; Khelafi, H. Strength of short composite rectangular hollow section columns filled with lightweight aggregate concrete. *Eng. Struct.* **2007**, *29*, 1791–1797. [CrossRef]

31. Espinos, A.; Romero, M.L.; Serra, E.; Hospitaler, A. Experimental investigation on the fire behaviour of rectangular and elliptical slender concrete-filled tubular columns. *Thin-Walled Struct.* **2015**, *93*, 137–148. [CrossRef]

32. Lue, D.M.; Liu, J.-L.; Yen, T. Experimental study on rectangular CFT columns with high-strength concrete. *J. Constr. Steel Res.* **2007**, *63*, 37–44. [CrossRef]

33. Smith, M. *ABAQUS/Standard User's Manual*; Version 6.9; Dassault Systemes Simulia Corp: Providence, RI, USA, 2009.

34. DeSalvo, G.J.; Swanson, J.A. *Swanson ANSYS Engineering Analysis System User's Manual*; Houston, P., Ed.; Swanson Analysis Systems: Canonsburg, PA, USA, 1985.

35. Tort, C.; Hajjar, J.F. Mixed finite-element modeling of rectangular concrete-filled steel tube members and frames under static and dynamic loads. *J. Struct. Eng.* **2009**, *136*, 654–664. [CrossRef]

36. Wang, X.; Qi, Y.; Sun, Y.; Xie, Z.; Liu, W. Compressive Behavior of Composite Concrete Columns with Encased FRP Confined Concrete Cores. *Sensors* **2019**, *19*, 1792. [CrossRef]

37. Tao, Z.; Wang, Z.-B.; Yu, Q. Finite element modelling of concrete-filled steel stub columns under axial compression. *J. Constr. Steel Res.* **2013**, *89*, 121–131. [CrossRef]

38. Dao, D.V.; Trinh, S.H.; Ly, H.-B.; Pham, B.T. Prediction of Compressive Strength of Geopolymer Concrete Using Entirely Steel Slag Aggregates: Novel Hybrid Artificial Intelligence Approaches. *Appl. Sci.* **2019**, *9*, 1113. [CrossRef]

39. Ly, H.-B.; Monteiro, E.; Le, T.-T.; Le, V.M.; Dal, M.; Regnier, G.; Pham, B.T. Prediction and Sensitivity Analysis of Bubble Dissolution Time in 3D Selective Laser Sintering Using Ensemble Decision Trees. *Materials* **2019**, *12*, 1544. [CrossRef] [PubMed]

40. Qi, C.; Ly, H.-B.; Chen, Q.; Le, T.-T.; Le, V.M.; Pham, B.T. Flocculation-dewatering prediction of fine mineral tailings using a hybrid machine learning approach. *Chemosphere* **2020**, *244*, 125450. [CrossRef] [PubMed]

41. Dao, D.V.; Ly, H.-B.; Vu, H.-L.T.; Le, T.-T.; Pham, B.T. Investigation and Optimization of the C-ANN Structure in Predicting the Compressive Strength of Foamed Concrete. *Materials* **2020**, *13*, 1072. [CrossRef] [PubMed]

42. Dao, D.V.; Ly, H.-B.; Trinh, S.H.; Le, T.-T.; Pham, B.T. Artificial Intelligence Approaches for Prediction of Compressive Strength of Geopolymer Concrete. *Materials* **2019**, *12*, 983. [CrossRef] [PubMed]

43. Sanad, A.; Saka, M.P. Prediction of ultimate shear strength of reinforced-concrete deep beams using neural networks. *J. Struct. Eng.* **2001**, *127*, 818–828. [CrossRef]

44. De Lima, L.R.; Vellasco, P.D.; De Andrade, S.A.; Da Silva, J.G.; Vellasco, M.M. Neural networks assessment of beam-to-column joints. *J. Braz. Soc. Mech. Sci. Eng.* **2005**, *27*, 314–324. [CrossRef]

45. Seleemah, A.A. A neural network model for predicting maximum shear capacity of concrete beams without transverse reinforcement. *Can. J. Civ. Eng.* **2005**, *32*, 644–657. [CrossRef]

46. Blachowski, B.; Pnevmatikos, N. Neural Network Based Vibration Control of Seismically Excited Civil Structures. *Period. Polytech. Civ. Eng.* **2018**, *62*, 620–628. [CrossRef]

47. Kiani, J.; Camp, C.; Pezeshk, S. On the application of machine learning techniques to derive seismic fragility curves. *Comput. Struct.* **2019**, *218*, 108–122. [CrossRef]

48. Mangalathu, S.; Hwang, S.-H.; Choi, E.; Jeon, J.-S. Rapid seismic damage evaluation of bridge portfolios using machine learning techniques. *Eng. Struct.* **2019**, *201*, 109785. [CrossRef]

49. Mangalathu, S.; Heo, G.; Jeon, J.-S. Artificial neural network based multi-dimensional fragility development of skewed concrete bridge classes. *Eng. Struct.* **2018**, *162*, 166–176. [CrossRef]

50. Mangalathu Sujith; Jeon Jong-Su Machine Learning–Based Failure Mode Recognition of Circular Reinforced Concrete Bridge Columns: Comparative Study. *J. Struct. Eng.* **2019**, *145*, 04019104. [CrossRef]

51. Mangalathu, S.; Jeon, J.-S. Classification of failure mode and prediction of shear strength for reinforced concrete beam-column joints using machine learning techniques. *Eng. Struct.* **2018**, *160*, 85–94. [CrossRef]

52. Guo, J.; Zhao, X.; Guo, J.; Yuan, X.; Dong, S.; Xiong, Z. Model updating of suspended-dome using artificial neural networks. *Adv. Struct. Eng.* **2017**, *20*, 1727–1743. [CrossRef]

53. Guo, J.; Wu, J.; Guo, J.; Jiang, Z. A Damage Identification Approach for Offshore Jacket Platforms Using Partial Modal Results and Artificial Neural Networks. *Appl. Sci.* **2018**, *8*, 2173. [CrossRef]

54. Pedroni, N.; Zio, E.; Apostolakis, G.E. Comparison of bootstrapped artificial neural networks and quadratic response surfaces for the estimation of the functional failure probability of a thermal–hydraulic passive system. *Reliab. Eng. Syst. Saf.* **2010**, *95*, 386–395. [CrossRef]

55. Ricotti, M.E.; Zio, E. Neural network approach to sensitivity and uncertainty analysis. *Reliab. Eng. Syst. Saf.* **1999**, *64*, 59–71. [CrossRef]

56. Ak, R.; Li, Y.; Vitelli, V.; Zio, E.; López Droguett, E.; Magno Couto Jacinto, C. NSGA-II-trained neural network approach to the estimation of prediction intervals of scale deposition rate in oil & gas equipment. *Expert Syst. Appl.* **2013**, *40*, 1205–1212.

57. Saadoon, A.S.; Nasser, K.Z.; Mohamed, I.Q. A Neural Network Model to Predict Ultimate Strength of Rectangular Concrete Filled Steel Tube Beam–Columns. *Eng. Technol. J.* **2012**, *30*, 3328–3340.

58. Jayalekshmi, S.; Jegadesh, J.S.; Goel, A. Empirical Approach for Determining Axial Strength of Circular Concrete Filled Steel Tubular Columns. *J. Inst. Eng. Ser. A* **2018**, *99*, 257–268. [CrossRef]

59. Pham, B.T.; Nguyen, M.D.; Ly, H.-B.; Pham, T.A.; Hoang, V.; Van Le, H.; Le, T.-T.; Nguyen, H.Q.; Bui, G.L. Development of Artificial Neural Networks for Prediction of Compression Coefficient of Soft Soil. In Proceedings of the CIGOS 2019, Innovation for Sustainable Infrastructure, Hanoi, Vietnam, 11 October 2019; Ha-Minh, C., Dao, D.V., Benboudjema, F., Derrible, S., Huynh, D.V.K., Tang, A.M., Eds.; Springer: Singapore, 2020; pp. 1167–1172.

60. Asteris, P.G.; Armaghani, D.J.; Hatzigeorgiou, G.D.; Karayannis, C.G.; Pilakoutas, K. Predicting the shear strength of reinforced concrete beams using Artificial Neural Networks. *Comput. Concr.* **2019**, *24*, 469–488.

61. Asteris, P.G.; Apostolopoulou, M.; Skentou, A.D.; Moropoulou, A. Application of artificial neural networks for the prediction of the compressive strength of cement-based mortars. *Comput. Concr.* **2019**, *24*, 329–345.

62. Asteris, P.G.; Roussis, P.C.; Douvika, M.G. Feed-forward neural network prediction of the mechanical properties of sandcrete materials. *Sensors* **2017**, *17*, 1344. [CrossRef] [PubMed]

63. Montana, D.J.; Davis, L. Training Feedforward Neural Networks Using Genetic Algorithms. *Proc. IJCAI* **1989**, *89*, 762–767.

64. Caglar, N. Neural network based approach for determining the shear strength of circular reinforced concrete columns. *Constr. Build. Mater.* **2009**, *23*, 3225–3232. [CrossRef]

65. Al-Khaleefi, A.M.; Terro, M.J.; Alex, A.P.; Wang, Y. Prediction of fire resistance of concrete filled tubular steel columns using neural networks. *Fire Saf. J.* **2002**, *37*, 339–352. [CrossRef]

66. Ni, H.-G.; Wang, J.-Z. Prediction of compressive strength of concrete by neural networks. *Cem. Concr. Res.* **2000**, *30*, 1245–1250. [CrossRef]

67. Chan, Y.N.; Jin, P.; Anson, M.; Wang, J.S. Fire resistance of concrete: Prediction using artificial neural networks. *Mag. Concr. Res.* **1998**, *50*, 353–358. [CrossRef]

68. Mehrabian, A.R.; Lucas, C. A novel numerical optimization algorithm inspired from weed colonization. *Ecol. Inform.* **2006**, *1*, 355–366. [CrossRef]

69. Barisal, A.K.; Prusty, R.C. Large scale economic dispatch of power systems using oppositional invasive weed optimization. *Appl. Soft Comput.* **2015**, *29*, 122–137. [CrossRef]

70. Pourjafari, E.; Mojallali, H. Solving nonlinear equations systems with a new approach based on invasive weed optimization algorithm and clustering. *Swarm Evol. Comput.* **2012**, *4*, 33–43. [CrossRef]

71. Ahmadi, M.; Mojallali, H. Chaotic invasive weed optimization algorithm with application to parameter estimation of chaotic systems. *Chaos Solitons Fractals* **2012**, *45*, 1108–1120. [CrossRef]

72. Mallahzadeh, A.R.R.; Oraizi, H.; Davoodi-Rad, Z. Application of the invasive weed optimization technique for antenna configurations. *Prog. Electromagn. Res.* **2008**, *79*, 137–150. [CrossRef]

73. Zhang, X.; Wang, Y.; Cui, G.; Niu, Y.; Xu, J. Application of a novel IWO to the design of encoding sequences for DNA computing. *Comput. Math. Appl.* **2009**, *57*, 2001–2008. [CrossRef]

74. Saravanan, B.; Vasudevan, E.R.; Kothari, D.P. Unit commitment problem solution using invasive weed optimization algorithm. *Int. J. Electr. Power Energy Syst.* **2014**, *55*, 21–28. [CrossRef]

75. Zhou, Y.; Luo, Q.; Chen, H.; He, A.; Wu, J. A discrete invasive weed optimization algorithm for solving traveling salesman problem. *Neurocomputing* **2015**, *151*, 1227–1236. [CrossRef]

76. Huang, L.; Asteris, P.G.; Koopialipoor, M.; Armaghani, D.J.; Tahir, M.M. Invasive weed optimization technique-based ANN to the prediction of rock tensile strength. *Appl. Sci.* **2019**, *9*, 5372. [CrossRef]

77. Asteris, P.G.; Mokos, V.G. Concrete compressive strength using artificial neural networks. *Neural Comput. Appl.* **2019**, 1–20. [CrossRef]

78. Guilleminot, J.; Le, T.T.; Soize, C. Stochastic framework for modeling the linear apparent behavior of complex materials: Application to random porous materials with interphases. *Acta Mech. Sin.* **2013**, *29*, 773–782. [CrossRef]

79. Ly, H.-B.; Desceliers, C.; Le, L.M.; Le, T.-T.; Pham, B.T.; Nguyen-Ngoc, L.; Doan, V.T.; Le, M. Quantification of Uncertainties on the Critical Buckling Load of Columns under Axial Compression with Uncertain Random Materials. *Materials* **2019**, *12*, 1828. [CrossRef]

80. Pham, B.T.; Nguyen, M.D.; Dao, D.V.; Prakash, I.; Ly, H.-B.; Le, T.-T.; Ho, L.S.; Nguyen, K.T.; Ngo, T.Q.; Hoang, V.; et al. Development of artificial intelligence models for the prediction of Compression Coefficient of soil: An application of Monte Carlo sensitivity analysis. *Sci. Total Environ.* **2019**, *679*, 172–184. [CrossRef] [PubMed]

81. Le, L.M.; Ly, H.-B.; Pham, B.T.; Le, V.M.; Pham, T.A.; Nguyen, D.-H.; Tran, X.-T.; Le, T.-T. Hybrid Artificial Intelligence Approaches for Predicting Buckling Damage of Steel Columns Under Axial Compression. *Materials* **2019**, *12*, 1670. [CrossRef] [PubMed]

82. Nguyen, H.-L.; Le, T.-H.; Pham, C.-T.; Le, T.-T.; Ho, L.S.; Le, V.M.; Pham, B.T.; Ly, H.-B. Development of Hybrid Artificial Intelligence Approaches and a Support Vector Machine Algorithm for Predicting the Marshall Parameters of Stone Matrix Asphalt. *Appl. Sci.* **2019**, *9*, 3172. [CrossRef]

83. Pham, B.T.; Le, L.M.; Le, T.-T.; Bui, K.-T.T.; Le, V.M.; Ly, H.-B.; Prakash, I. Development of advanced artificial intelligence models for daily rainfall prediction. *Atmos. Res.* **2020**, *237*, 104845. [CrossRef]

84. Le, T.-T.; Pham, B.T.; Ly, H.-B.; Shirzadi, A.; Le, L.M. Development of 48-hour Precipitation Forecasting Model using Nonlinear Autoregressive Neural Network. In Proceedings of the CIGOS 2019, Innovation for Sustainable Infrastructure, Hanoi, Vietnam, 11 October 2019; Ha-Minh, C., Dao, D.V., Benboudjema, F., Derrible, S., Huynh, D.V.K., Tang, A.M., Eds.; Springer: Singapore, 2020; pp. 1191–1196.

85. Bridge, R.Q. *Concrete Filled Steel Tubular Columns/by R.Q. Bridge*; School of Civil Engineering, University of Sydney: Sydney, Australia, 1976.

86. Du, Y.; Chen, Z.; Xiong, M.-X. Experimental behavior and design method of rectangular concrete-filled tubular columns using Q460 high-strength steel. *Constr. Build. Mater.* **2016**, *125*, 856–872. [CrossRef]

87. Du, Y.; Chen, Z.; Yu, Y. Behavior of rectangular concrete-filled high-strength steel tubular columns with different aspect ratio. *Thin-Walled Struct.* **2016**, *109*, 304–318. [CrossRef]

88. Ghannam, S.; Jawad, Y.A.; Hunaiti, Y. Failure of lightweight aggregate concrete-filled steel tubular columns. *Steel Compos. Struct.* **2004**, *4*, 1–8. [CrossRef]

89. Han, L.-H. Tests on stub columns of concrete-filled RHS sections. *J. Constr. Steel Res.* **2002**, *58*, 353–372. [CrossRef]

90. Han, L.-H.; Yang, Y.-F. Analysis of thin-walled steel RHS columns filled with concrete under long-term sustained loads. *Thin-Walled Struct.* **2003**, *41*, 849–870. [CrossRef]

91. Han, L.-H.; Yao, G.-H. Influence of concrete compaction on the strength of concrete-filled steel RHS columns. *J. Constr. Steel Res.* **2003**, *59*, 751–767. [CrossRef]

92. Lin, C.Y. Axial Capacity of Concrete Infilled Cold-formed Steel Columns. In Proceedings of the Ninth International Specialty Conference on Cold-Formed Steel Structures; St. Louis, MO, USA, 8–9 November 1988; pp. 443–457.

93. Schneider Stephen, P. Axially Loaded Concrete-Filled Steel Tubes. *J. Struct. Eng.* **1998**, *124*, 1125–1138. [CrossRef]

94. Shakir-Khalil, H.; Mouli, M. Further Tests on Concrete-Filled Rectangular Hollow-Section Columns. *Struct. Eng.* **1990**, *68*, 405–413.

95. Shakir-Khalil, H.; Zeghiche, J. Experimental Behaviour of Concrete-Filled Rolled Rectangular Hollow-Section Columns. *Struct. Eng.* **1989**, *67*, 346–353.

96. Sarir, P.; Chen, J.; Asteris, P.G.; Armaghani, D.J.; Tahir, M.M. Developing GEP tree-based, neuro-swarm, and whale optimization models for evaluation of bearing capacity of concrete-filled steel tube columns. *Eng. Comput.* **2019**, 1–19. [CrossRef]

97. Lyu, X.; Xu, Y.; Xu, Q.; Yu, Y. Axial Compression Performance of Square Thin Walled Concrete-Filled Steel Tube Stub Columns with Reinforcement Stiffener under Constant High-Temperature. *Materials* **2019**, *12*, 1098. [CrossRef] [PubMed]

98. Ding, F.; Fang, C.; Bai, Y.; Gong, Y. Mechanical performance of stirrup-confined concrete-filled steel tubular stub columns under axial loading. *J. Constr. Steel Res.* **2014**, *98*, 146–157. [CrossRef]

99. Yan, J.-B.; Dong, X.; Wang, T. Axial compressive behaviours of square CFST stub columns at low temperatures. *J. Constr. Steel Res.* **2020**, *164*, 105812. [CrossRef]

100. Masi, A.; Santarsiero, G.; Mossucca, A.; Nigro, D. Influence of Axial Load on the Seismic Behavior of RC Beam-Column Joints with Wide Beam. *Appl. Mech. Mater.* **2014**, *508*, 208–214. [CrossRef]

101. Kulkarni, S.A.; Li, B. Seismic Behavior of Reinforced Concrete Interior Wide-Beam Column Joints. *J. Earthq. Eng.* **2008**, *13*, 80–99. [CrossRef]

102. Ly, H.-B.; Le, L.M.; Duong, H.T.; Nguyen, T.C.; Pham, T.A.; Le, T.-T.; Le, V.M.; Nguyen-Ngoc, L.; Pham, B.T. Hybrid Artificial Intelligence Approaches for Predicting Critical Buckling Load of Structural Members under Compression Considering the Influence of Initial Geometric Imperfections. *Appl. Sci.* **2019**, *9*, 2258. [CrossRef]

103. Ly, H.-B.; Le, L.M.; Phi, L.V.; Phan, V.-H.; Tran, V.Q.; Pham, B.T.; Le, T.-T.; Derrible, S. Development of an AI Model to Measure Traffic Air Pollution from Multisensor and Weather Data. *Sensors* **2019**, *19*, 4941. [CrossRef] [PubMed]

104. Nguyen, H.-L.; Pham, B.T.; Son, L.H.; Thang, N.T.; Ly, H.-B.; Le, T.-T.; Ho, L.S.; Le, T.-H.; Tien Bui, D. Adaptive Network Based Fuzzy Inference System with Meta-Heuristic Optimizations for International Roughness Index Prediction. *Appl. Sci.* **2019**, *9*, 4715. [CrossRef]

105. Ly, H.-B.; Le, T.-T.; Le, L.M.; Tran, V.Q.; Le, V.M.; Vu, H.-L.T.; Nguyen, Q.H.; Pham, B.T. Development of Hybrid Machine Learning Models for Predicting the Critical Buckling Load of I-Shaped Cellular Beams. *Appl. Sci.* **2019**, *9*, 5458. [CrossRef]

106. Dao, D.V.; Jaafari, A.; Bayat, M.; Mafi-Gholami, D.; Qi, C.; Moayedi, H.; Phong, T.V.; Ly, H.-B.; Le, T.-T.; Trinh, P.T.; et al. A spatially explicit deep learning neural network model for the prediction of landslide susceptibility. *Catena* **2020**, *188*, 104451. [CrossRef]

107. Dao, D.V.; Adeli, H.; Ly, H.-B.; Le, L.M.; Le, V.M.; Le, T.-T.; Pham, B.T. A Sensitivity and Robustness Analysis of GPR and ANN for High-Performance Concrete Compressive Strength Prediction Using a Monte Carlo Simulation. *Sustainability* **2020**, *12*, 830. [CrossRef]

108. Roshni, T.; Jha, M.K.; Deo, R.C.; Vandana, A. Development and Evaluation of Hybrid Artificial Neural Network Architectures for Modeling Spatio-Temporal Groundwater Fluctuations in a Complex Aquifer System. *Water Resour. Manag.* **2019**, *33*, 2381–2397. [CrossRef]

109. Sheela, K.G.; Deepa, S.N. Review on Methods to Fix Number of Hidden Neurons in Neural Networks. Available online: https://www.hindawi.com/journals/mpe/2013/425740/ (accessed on 8 December 2019).

110. Mohamad, E.T.; Faradonbeh, R.S.; Armaghani, D.J.; Monjezi, M.; Majid, M.Z.A. An optimized ANN model based on genetic algorithm for predicting ripping production. *Neural Comput. Appl.* **2017**, *28*, 393–406. [CrossRef]

111. Singh, T.N.; Singh, V.K.; Sinha, S. Prediction of Cadmium Removal Using an Artificial Neural Network and a Neuro-Fuzzy Technique. *Mine Water Environ.* **2006**, *25*, 214–219. [CrossRef]

112. Gordan, B.; Jahed Armaghani, D.; Hajihassani, M.; Monjezi, M. Prediction of seismic slope stability through combination of particle swarm optimization and neural network. *Eng. Comput.* **2016**, *32*, 85–97. [CrossRef]

113. Hornik, K. Approximation capabilities of multilayer feedforward networks. *Neural Netw.* **1991**, *4*, 251–257. [CrossRef]

114. Kuri-Morales, A. Closed determination of the number of neurons in the hidden layer of a multi-layered perceptron network. *Soft Comput.* **2017**, *21*, 597–609. [CrossRef]

115. Abambres, M.; Rajana, K.; Tsavdaridis, K.D.; Ribeiro, T.P. Neural Network-Based Formula for the Buckling Load Prediction of I-Section Cellular Steel Beams. *Computers* **2019**, *8*, 2. [CrossRef]

116. Bayat, M.; Ghorbanpour, M.; Zare, R.; Jaafari, A.; Thai Pham, B. Application of artificial neural networks for predicting tree survival and mortality in the Hyrcanian forest of Iran. *Comput. Electron. Agric.* **2019**, *164*, 104929. [CrossRef]

117. Freitas, A.A. A Review of evolutionary Algorithms for Data Mining. In *Soft Computing for Knowledge Discovery and Data Mining*; Maimon, O., Rokach, L., Eds.; Springer: Boston, MA, USA, 2008; pp. 79–111, ISBN 978-0-387-69935-6.

118. Baioletti, M.; Di Bari, G.; Milani, A.; Poggioni, V. Differential Evolution for Neural Networks Optimization. *Mathematics* **2020**, *8*, 69. [CrossRef]

119. Sun, Y.; Gao, Y. A Multi-Objective Particle Swarm Optimization Algorithm Based on Gaussian Mutation and an Improved Learning Strategy. *Mathematics* **2019**, *7*, 148. [CrossRef]

120. Javed, H.; Jan, M.A.; Tairan, N.; Mashwani, W.K.; Khanum, R.A.; Sulaiman, M.; Khan, H.U.; Shah, H. On the Efficacy of Ensemble of Constraint Handling Techniques in Self-Adaptive Differential Evolution. *Mathematics* **2019**, *7*, 635. [CrossRef]

121. Piotrowski, A.P. Review of Differential Evolution population size. *Swarm Evol. Comput.* **2017**, *32*, 1–24. [CrossRef]

122. Fu, Y.; Wang, H.; Yang, M.-Z. An Adaptive Population Size Differential Evolution with Novel Mutation Strategy for Constrained Optimization. *arXiv* **2018**, arXiv:1805.04217.

123. Misaghi, M.; Yaghoobi, M. Improved invasive weed optimization algorithm (IWO) based on chaos theory for optimal design of PID controller. *J. Comput. Des. Eng.* **2019**, *6*, 284–295. [CrossRef]

124. Chen, T.; Tang, K.; Chen, G.; Yao, X. A Large Population Size Can Be Unhelpful in Evolutionary Algorithms. *arXiv* **2012**, arXiv:1208.2345. [CrossRef]
125. Wang, Z.-B.; Tao, Z.; Han, L.-H.; Uy, B.; Lam, D.; Kang, W.-H. Strength, stiffness and ductility of concrete-filled steel columns under axial compression. *Eng. Struct.* **2017**, *135*, 209–221. [CrossRef]
126. Han, L.-H.; Yao, G.-H.; Zhao, X.-L. Tests and calculations for hollow structural steel (HSS) stub columns filled with self-consolidating concrete (SCC). *J. Constr. Steel Res.* **2005**, *61*, 1241–1269. [CrossRef]
127. Eurocode 4. Design of Composite Steel and Concrete Structures. In *Part 1.1, General Rules and Rules for Buildings*; European Committee for Standardization, British Standards Institution: London, UK, 2004.
128. *Specification for Structural Steel Buildings*; American Institute of Steel Construction: Chicago, IL, USA, 2010.
129. A.C.I. Committee. *Building Code Requirements for Structural Concrete (ACI 318-08) and Commentary*; American Concrete Institute: Farmington Hills, MI, USA, 2008.
130. Peng, B.; Yu, W. A micromechanics theory for homogenization and dehomogenization of aperiodic heterogeneous materials. *Compos. Struct.* **2018**, *199*, 53–62. [CrossRef]
131. Le, T.-T. Stochastic Modeling, in Continuum Mechanics, of the Inclusion-Matrix Interphase from Molecular Dynamics Simulations. PhD Thesis, University of Paris-Est Marne-la-Vallée, Paris, France, 2015.
132. Yvonnet, J.; Bonnet, G. A consistent nonlocal scheme based on filters for the homogenization of heterogeneous linear materials with non-separated scales. *Int. J. Solids Struct.* **2014**, *51*, 196–209. [CrossRef]
133. Soize, C.; Desceliers, C.; Guilleminot, J.; Le, T.-T.; Nguyen, M.-T.; Perrin, G.; Allain, J.-M.; Gharbi, H.; Duhamel, D.; Funfschilling, C. Stochastic Representations and Statistical Inverse Identification for Uncertainty Quantification in Computational Mechanics. In Proceedings of the 1st ECCOMAS Thematic Conference on Uncertainty Quantification in Computational Sciences and Engineering, Crete Island, Greece, 25–27 May 2015; pp. 1–26.
134. Cherkaev, E.; Ou, M.-J.Y. Dehomogenization: Reconstruction of moments of the spectral measure of the composite. *Inverse Probl.* **2008**, *24*, 065008. [CrossRef]

Article

Numerical Modelling of Ballistic Impact Response at Low Velocity in Aramid Fabrics

Norberto Feito [1,*] , **José Antonio Loya** [2] , **Ana Muñoz-Sánchez** [3] and **Raj Das** [4]

[1] Centre of Research in Mechanical Engineering—CIIM, Department of Mechanical and Materials Engineering, Universitat Politècnica de València, Camino de Vera, s/n, 46022 Valencia, Spain

[2] Department of Continuum Mechanics and Structural Analysis, University Carlos III of Madrid, Avda. Universidad 30, 28911 Leganés, Madrid, Spain

[3] Department of Mechanical Engineering, University Carlos III of Madrid, Avda. Universidad 30, 28911 Leganés, Madrid, Spain

[4] Sir Lawrence Wackett Research Centre, School of Engineering, RMIT University, 124 Latrobe Street, Melbourne, Victoria 3000, Australia

* Correspondence: norfeisa@upvnet.upv.es

Received: 4 June 2019; Accepted: 22 June 2019; Published: 28 June 2019

Abstract: In this study, the effect of the impact angle of a projectile during low-velocity impact on Kevlar fabrics has been investigated using a simplified numerical model. The implementation of mesoscale models is complex and usually involves long computation time, in contrast to the practical industry needs to obtain accurate results rapidly. In addition, when the simulation includes more than one layer of composite ply, the computational time increases even in the case of hybrid models. With the goal of providing useful and rapid prediction tools to the industry, a simplified model has been developed in this work. The model offers an advantage in the reduced computational time compared to a full 3D model (around a 90% faster). The proposed model has been validated against equivalent experimental and numerical results reported in the literature with acceptable deviations and accuracies for design requirements. The proposed numerical model allows the study of the influence of the geometry on the impact response of the composite. Finally, after a parametric study related to the number of layers and angle of impact, using a response surface methodology, a mechanistic model and a surface diagram have been presented in order to help with the calculation of the ballistic limit.

Keywords: aramid; impact; computational techniques; finite elements; mechanical analysis

---

1. Introduction

With the aim of damage reduction in person subjected to ballistic impact, polymeric, carbon, and glass fibers are commonly used to develop protective systems. However, during last decades it has been found that a combination of high strength yarns in different directions generates flexible woven fabrics which are light and highly resistant at the same time [1]. Therefore, aramid fibers are one of the most used protection materials nowadays [2], with a growing trend in the industry.

Experimental studies have proved that the ballistic performance of aramid fabrics depends on many factors, such as the projectile geometry [3,4], the impact velocity [3,5–7], the friction between yarns [8–10], the woven structure of the fabric [6,11–13] and material properties [14]. However, most research has found that each property individually does not control the ballistic performance [15,16], rather the combined effect of all of these properties play a key role.

New studies to predict and to reduce the damage are constantly under study [17,18]. The use of surrogate models is common in optimal design problems to approximate the objective functions, as neuronal networks [19,20] or genetic algorithms [21]. However, a big dataset is needed in most cases to obtain good predicted values what can be expensive.

A useful tool for obtaining improved understanding of the behavior of aramid materials under impact loads is the use of numerical models. The ballistic limit, V_{50}, the deformed shape of the woven fabric and the effect of the internal interlayer friction can be studied using the finite element method. However, the geometry of the fabric structure and the failure mechanisms of the yarns make the modelling of 3D fabrics complex [22–30]. To decrease the complexity of the finite element model, new simple models using shell elements [31–35] and truss elements [36] have been successfully implemented.

Duan et al. [22], Rao et al. [23] and Grujicic et al. [29,30] developed 3D numerical models with solid elements representing yarns as homogenous continua. These studies analyzed mainly the friction coefficient between the yarns and the clamping conditions. Modelling results showed that the fabric boundary condition is a primary factor that influenced the friction effect, when only two edges are clamped, fabric reduces the residual velocity of the projectile and absorbs energy more effectively. Fabrics with high stiffness decelerate the projectile relatively rapidly meanwhile fabrics with high-strength yarns need more time to initiate the failure. The material which combines both properties is the most favorable of all the examined ones under the imposed boundary conditions considered in the study.

Despite the accuracy of 3D models, some researchers observed that they have two limitations. The first is that these models do not consider the statistical variability in the fabric geometry and material properties. The second one is the high computational cost due to the realistic representation of the fabric that requires a fine mesh with many elements.

To address the first problem and improve the prediction of the V_{50} velocities, Nilakantan et al. [24,26,37] implemented a framework which incorporated the inherent statistical variability in the system. This framework can be used to calculate the probabilistic velocity response (PVR) curve for projectiles with different size and shape and for different clamped conditions. These studies analyzed the projectile geometry [26], the impact location sensitivity [27] and the fabric clamping conditions [28,38]. The findings showed that projectiles with a small impact area resulted in a higher residual velocity when the impact occurs in the gap between the yarns compared to when it occurs at the yarn crossover junction. The large cylindrical projectile, which had the flattest and the largest impact face, showed no sensitivity to impact location. The study also remarked that a circular shaped projectile shows lower dependency on the residual velocity with the projectile impact location than other configurations. This is mainly due to the fact that this configuration is not sensitive to in-plane fabric rotations, giving it a slight advantage over the other clamping configurations.

To address the second problem related to the computational efficiency, researchers presented different solutions. One option was the use of shell elements instead of solid elements [32–35]. Ha–Minh et al. [32,33] studied the effect of the number of elements used in the model (four and eight elements by yarn) and showed, in terms of velocity evolution, similar results in both cases. This is a significant conclusion because the computation time of the model with 8 elements is double than the model with 4 elements.

Chu et al. [35] studied mechanical properties of the yarns. It was concluded that the yarn density does not affect significantly the ballistic performance of the fabric. As opposed to this, it was observed that a high value of longitudinal Young's modulus produces a faster deceleration of the projectile.

The second solution to reduce the computational cost is to modify the original 3D model, generating a hybrid model, which is comprised of zones with different modelling resolutions and different finite element formulation, all coupled together with impedance matching interfaces [31,39–42]. Barauskas et al. [31] developed a model for a plain-woven single-ply. The fabric model presented three different zones: a zone close to the impact area including failure modelling, a second zone where the woven structure did not undergo failure, and a last zone (far from the impact area) implemented as orthotropic membrane. During this research, it was proved that it is not easy to validate this kind of approach due to the many different levels involved in the process.

The original model of Nilakantan et al. [26], was modified in later studies [39–41]. New models which combine solid and shell elements were tested under different distributions of both meso-level and macro-level in the fabric producing a low cost computational model.

Finally, Ha-Minh et al. [42] carried out a study comparing three different hybrid models, that was a modification of their initial macroscopic-model. The numerical analysis showed that the difference among the models was negligible for results such as the evolution of projectile velocity and the global impact fabric behavior.

With the objective of studying the friction coefficient in aramid woven, Das et al. [36] implemented a model with truss elements. The study analyzed the influence of the friction coefficient and the clamping conditions. It was reported that there is a limit for the friction between yarns since the shape of the projectile is negligible and affects in a negative way to the energy dissipation and the failure mechanism.

In the present study, a simplified finite element model is developed using truss elements with the aim of establishing the advantage that 1D element models present over solid element models. A validation analysis proved that the simplified model is efficient in terms of computational cost and accuracy. A parametric study has been carried out using the numerical model, with specific focus on the effect of the impact angle, a parameter that is not usually analyzed in the literature. Results are used to generate a response surface diagram able to predict the ballistic limit for low impact velocities. This methodology will be a useful tool for rapid impact response analysis in the industry.

2. Theoretical Study

The study of the physical mechanisms by which impact pressure waves reach the chest and cause injury are continuously under study in the research community. It is a goal to determine how the impact pressure waves are transmitted to the thorax or the brain in order to implement effective preventive measures and reduce the exposure risks [43,44].

The energy absorption rate of aramid fabric during impact depends on many variables. Among all of them, it should be mentioned the fiber modulus and the material modulus, which is related to the ability to brake the projectile [22,45].

The breaking energy of the yarn is determined by its characteristics, such as tensile strength, elongation or modulus. These factors affect the transmitting velocity of the stress wave that is generated by the impact of the projectile. The wave must be dispersed rapidly, and the breaking energy must be as high as possible to increase the impact resistance and the ballistic limit (velocity at which the projectile passes through the material) [46].

The characteristics of the material, such as the interlacing geometry of the fibers, the thickness and the number of layers, also affect the distance at which the impact disturbance will have moved in a given time. They also influence the breaking energy of the material (ability to resist breakage due to an external force), which depends on the tensile strength and elongation of the material.

3. Numerical Model

The numerical models implemented in this study were developed using the finite element explicit code ABAQUS. The following sections describe the methodology used to generate both models.

3.1. Projectile

All projectiles were implemented as rigid solid objects to reduce the computational time [28]. Four geometric shapes (blunt, hemispherical, conical and truncated conical) was defined to keep the mass constant to 7.5 g for all cases (see Figure 1). The friction coefficient between projectile and material was established to be 0.22, taken from the literature [27]. As a rigid 3D object contained in a 3D space, a projectile has six degrees of freedom for motion. All rotation degrees have been restrained.

The projectile can only traverse in the direction of the impact. The projectile velocity normal to the fabric is specified before each simulation and it ranged between 30–130 m/s based on experimental tests carried out in the same material by Yu et al. [47].

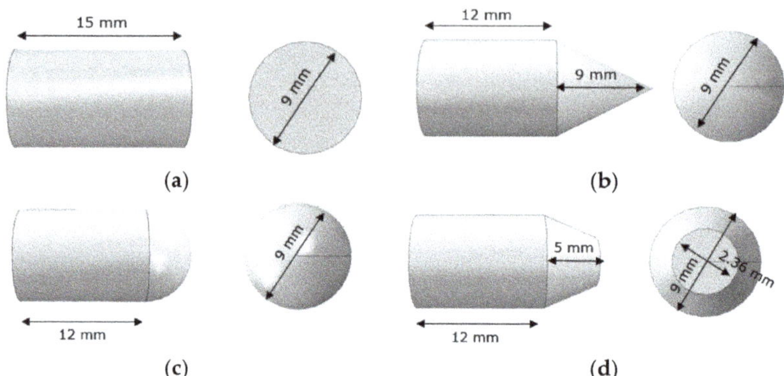

Figure 1. Dimensions of the projectiles: (**a**) blunt projectile, (**b**) conical point projectile, (**c**) hemispherical projectile and (**d**) truncated conical point projectile.

3.2. Specimen

The specimen was impacted by the projectile in the center. It was fully clamped, resulting an effective area of 101×101 mm^2 (Figure 2a). Each yarn had density of 1440 kg/m^3 and was modelled as a linear-elastic material until failure to capture the behavior of the woven fabric subjected to ballistic impact (Figure 2b) [23,24]. When the ultimate strength was reached, the element was assumed to be damaged and removed from the computational domain. The mechanical properties of the aramid fabric were taken from the literature [36], where no apparent plastic deformation before fracture was reported. Table 1 shows the most important parameters of the yarns in both the directions. The static and dynamic friction coefficient values between yarns are 0.186 and 0.17, respectively. Both coefficients were obtained from semi-analytical model based on yarn pull-out tests [36].

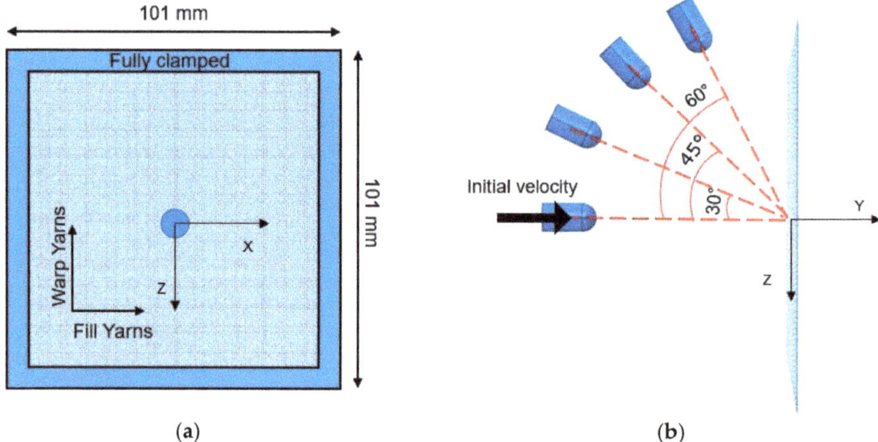

Figure 2. Fabric impact test setup with angular impact of the projectiles: (**a**) front view and (**b**) side view.

Table 1. Material properties of the Kevlar yarns [36].

Property	Warp Yarn	Fill Yarn
Young Module (GPa)	63.86	77.84
Poisson's ratio	0.01	0.01
Ultimate Strength (GPa)	2.28	2.76
Thickness (mm)	0.11	0.12

The difference between both numerical models proposed, presented in Figure 3, is the representation of the yarns. In the complete full model case, the yarn dimensions were taken from [36] and were modelled with 3D solid elements (Figure 3a). The 3D elements were the standard volume elements of Abaqus. This element can be composed of a single homogeneous or heterogeneous material; they are more accurate if not distorted, particularly for quadrilaterals and hexahedra. In the present study the element used was C3D8R (eight-node linear brick) with reduced integration and hourglass control. It has three degrees of freedom by node [48].

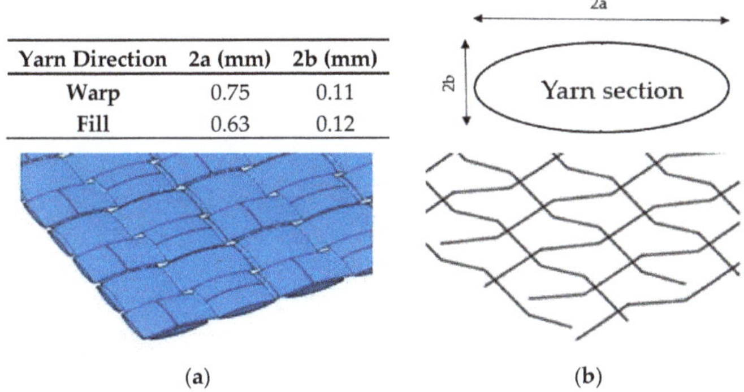

Yarn Direction	2a (mm)	2b (mm)
Warp	0.75	0.11
Fill	0.63	0.12

(a) (b)

Figure 3. Yarn geometrical parameters. (**a**) Solid elements model and (**b**) Truss elements model.

On the other hand, the simplified model presents the yarns as 3D one-dimensional truss elements, which use linear interpolation for position and displacement, and have a constant stress. In this study, the element T3D2 (two-node linear displacement) was chosen. It had three degrees of freedom by node. Truss elements are long, slender structural members that can transmit only axial force and do not transmit moments [48]. The cross section was considered to be made up of trusses of an approximate cross section of 0.064 mm^2 (Figure 3b). Dimensions of the yarns are also given in Figure 3.

4. Validation and Comparison between Models

The study encompasses evaluation of computational time, projectile geometry effects and ballistic curve prediction (initial and residual velocities). For the flat nose projectile case, with and impact angle of 0°, the ballistic curves obtained with the 1D and 3D models proposed, and the corresponding experimental values published in [36], are presented in Figure 4.

In the Figure 4, three different regions, typical from this kind of curves, are observed:

1. In the first one ($V_i < 44$ m/s), the negative residual velocities indicate that the projectile rebounds due to the impact. This phase was dominated by fiber elongation. The yarns returned to the original configuration and released most of the elastic energy stored, imparting the projectile almost its original velocity towards the opposite direction. Thus, only a small amount of energy was dissipated.

2. In the second phase (44 m/s $< V_i <$ 52 m/s) an abrupt increment from a negative to a positive value of the residual velocity was observed. This corresponds to the rupture failure of the fabric.

Some of the yarns surpassed their tensile yield stress and collapsed, dissipating energy in the process. The number of breaking (failed) yarns increased rapidly in a short range of velocities, allowing the projectile to pierce or penetrate the fabric.

3. If the initial velocity is further increased ($V_i > 52$ m/s), the projectile always pierced the fabric, but the initial velocity was reduced by the dissipated energy. That is because the yarns first suffered an elongation, and then they failed by rupture because of the impact, absorbing some portion of the impact energy from the projectile.

The results obtained accurately reproduced the experimental trends above and below the ballistic limit. Here, negative residual velocities mean projectile rebound (impact velocity below the ballistic limit). In particular, the ballistic limit error was close to 4%, as can be seen in Table 2.

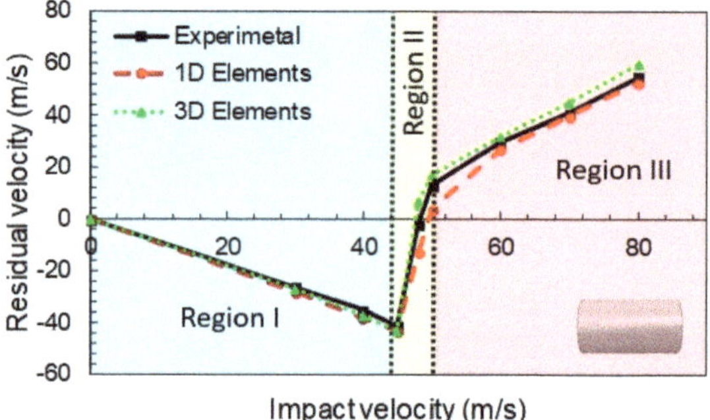

Figure 4. Calibrated ballistic curves for flat projectile. Experimental data taken from [36].

Table 2. Ballistic limit results for the truss elements model.

Projectile Type	Experimental (m/s)	1D Model (m/s)	Error (%)
Blunt projectile	48	50	4.1
Hemispherical projectile	43	40	−6.9

Once the model was calibrated for the flat nose projectile, the hemispherical point projectile case was analyzed. Is this case, the general trends were also quite well predicted, see Figure 5. The three different regions described for the Figure 4 were also observed here. Although the error for the ballistic limit derived from the change of projectile geometry was close to −7% (Table 2), it can be considered acceptable according to the simplifications of the model and the level of engineering accuracy desired. Figure 6 shows the stressed fibers (typical cross section) at different instants of the simulation (Figure 6a). The fiber separation and fiber breakage during perforation is shown in Figure 6b.

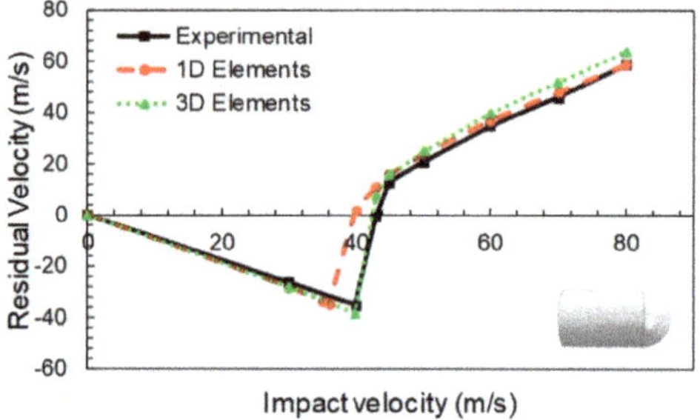

Figure 5. Validated ballistic curves for hemispherical point projectile. Experimental data taken from [36].

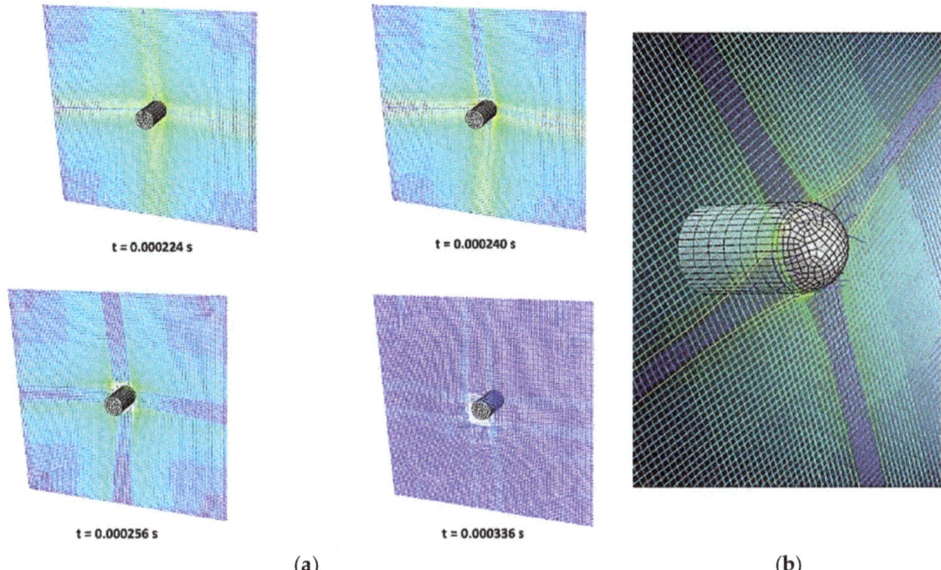

Figure 6. (**a**) Penetration of the hemispherical point projectile into a fully clamped layer at different time instants and (**b**) detail of the yarns failure.

In general, it can be observed that both 1D and 3D element models present reasonable accuracy when predicting the residual velocity of the projectile in terms of magnitude and trending. It is possible to observe the overestimation of residual velocity predicted with the solid model; in particular, it is slightly larger than that obtained with the simplified model, which involves that the solid model is conservative.

Despite an improved velocity estimation by the 3D model, the results obtained by the simplified model (which is around 99% faster, Table 3) lead to consideration it as an efficient tool in terms of computational requirement and accuracy level.

Table 3. Comparison of computational time for the case of 60 m/s between 1D and 3D models.

Projectile Type	1D Model	3D Model	Computational Cost Reduction
Blunt projectile	480 s	33,213 s	98.55%
Hemispherical projectile	270 s	29,058 s	99.07%

5. Results

After validating and comparing the results of both 1D and 3D element based models, a parametric study was carried out in order to analyze the influence of the projectile shape, the number of layers of the impacted panel and the impact angle. Even though both models are suitable to estimate the residual velocity, the 1D model was selected for the study because of its simplicity and low computational time. Results are discussed in the following sections.

5.1. Influence of the Projectile Geometry

In agreement with the literature related to low impact velocity in soft fabrics [47], the fabric area far from the impact zone, suffered a low deformation (a low displacement). Fixing the four sides of the fabric contributes to locate the stress and strain distributions around the impact area where the fabric generates a pyramid shape until failure.

Figure 7 shows the evolution of the ballistic curves in terms of variation of the residual velocity with impact velocity for different projectile geometries. It can be observed that the flat nose projectile presents the highest ballistic limit, V_{BL} = 49 m/s (velocity at which all impact energy is absorbed by the fabric and no penetration takes place). The round and truncated conical nose projectiles presented the same ballistic limit valued at 44 m/s, and the conical projectile at 46 m/s. The velocity value is reduced by 11% and 8% respectively, compared with the flat nose. This result implies that any sharper projectile presents a lower ballistic limit compared to a flat projectile [4]. This can be explained because the sharp projectile favors the separation of the transversal yarns generating a hole through which the projectile can penetrate the fabric. At the same time, the contact area was reduced to a smaller surface, causing greater stress. Under this situation, the fibers are more easily ruptured.

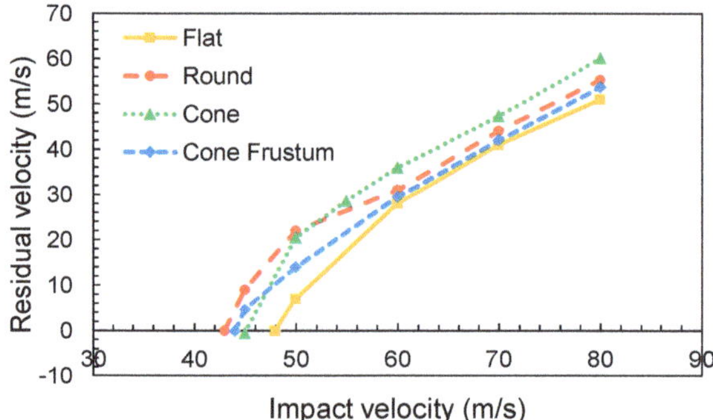

Figure 7. Ballistic curves for different projectile noses.

Figure 8 represents the projectile velocity evolution from the instant of impact moment. The initial velocity of the projectile fixed at 60 m/s is above the calculated ballistic limit. A few observations were derived from the relationship depicted. The cone geometry offers the lowest rate of deceleration while the rest of the geometries have relatively higher or similar rates. The projectile velocity histories of the round nose and truncated conical nose were quite similar, because the second geometry can be considered as a rough approximation to the round case. The energy dissipated by friction between the projectile and woven fabric increases with the frontal contact area, which was higher for the conical frustum projectile. Finally, the cone nose projectile has a higher residual velocity due to the sharp nose of the impact. This geometry favors the penetration of the fabric, as can be seen in [27].

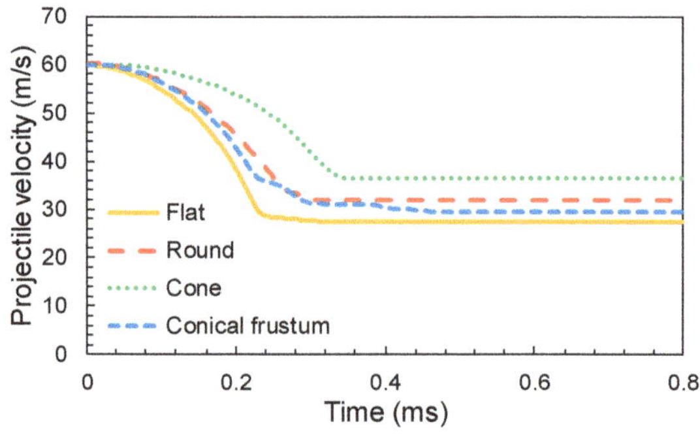

Figure 8. Typical velocity time history for different projectile geometries (V_i = 60 m/s).

Based on the principle of energy conservation, the energy dissipated by the fabric (Figure 9) is equal to the loss of the projectile kinetic energy, given by Equation (1), where m_p, represents the projectile mass, V_i is the impact velocity and V_r is the residual velocity.

$$\Delta E = \frac{1}{2}m_p\left(V_i^2 - V_r^2\right). \tag{1}$$

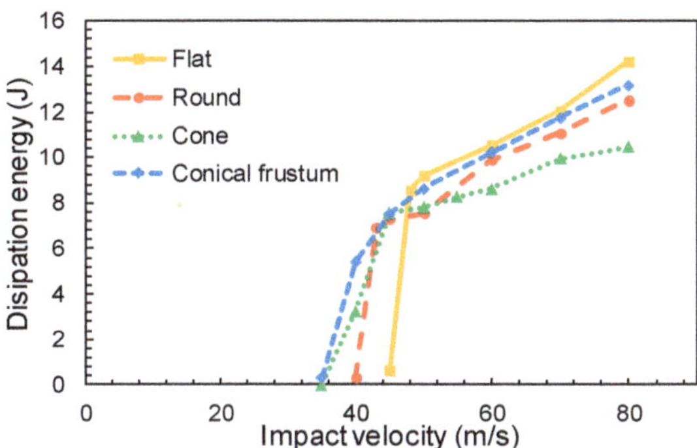

Figure 9. Variation of dissipated energy for different projectile geometries.

It can be observed in Figure 9 that the front face of the projectile plays an important role in the yarn breakage. While a sharp front face favors the separation of the yarns (cone nose projectile), the flat nose projectile penetration is entirely governed by yarn breakage which involves a higher load in the fibers and reduces the residual velocity [27,36].

5.2. Influence of the Impact Angle

The influence of the impact angle, defined as the angle between the projectile and the normal direction to the target (represented as α in Figure 2), was carried out with the round nose projectile. This projectile was chosen among the others because it is an intermediate geometry between flat and conic nose projectiles. It also has a close resemblance to common ammunition, like 9 mm "parabellun". The velocity magnitude, V, was split into directions Y and Z as follows:

$$\begin{aligned} V_Y &= V \cdot \cos(\alpha) \\ V_Z &= V \cdot \sin(\alpha) \end{aligned} \tag{2}$$

For a particular oblique angle, $\alpha = 45°$, three cases could be distinguished as function of the impact velocity (sketched in Figure 10 and detailed in Figure 11):

1. Below the ballistic limit, 30 m/s < V < 45 m/s: The projectile reduces its velocity components in both directions (V_Y and V_Z) during impact. The projectile changes the direction along Y axis because of the rebound (Figure 10a).
2. Close to the ballistic limit, 50 m/s < V < 55 m/s: The projectile suffers first a small rebound and then a small push out due to the wave generated in the fabric during the impact (Figure 10b).
3. Above the ballistic limit, 35 m/s < V < 60 m/s: The projectile continues its trajectory through the fabric with decreasing velocity in both the directions (Figure 10c).

Figure 11 illustrates the previous example. The left side plots correspond to impact velocities below the ballistic limit: the magnitude V (Figure 11a), Y-component V_Y (Figure 11e) and Z-component V_Z (Figure 11c). The right side plots correspond to impact velocities above the ballistic limit: the module V (Figure 11b), Y-component V_Y (Figure 11f) and Z-component V_Z (Figure 11d).

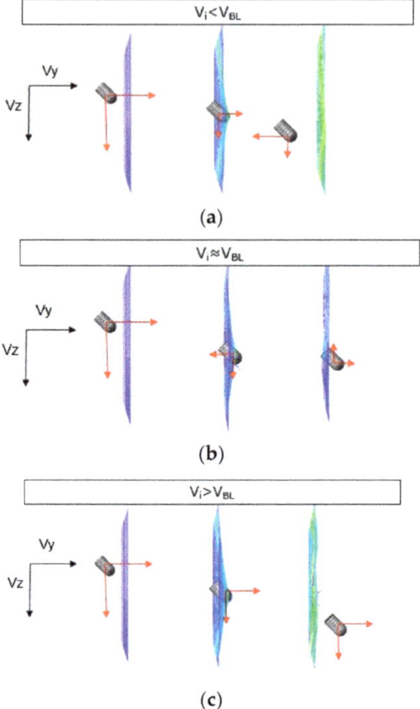

Figure 10. Schematic representation of V_Y and V_Z components as a function of the initial velocity V_i. (**a**) Impact velocity below the ballistic limit, (**b**) impact velocity close to the ballistic limit, and (**c**) impact velocity above the ballistic limit.

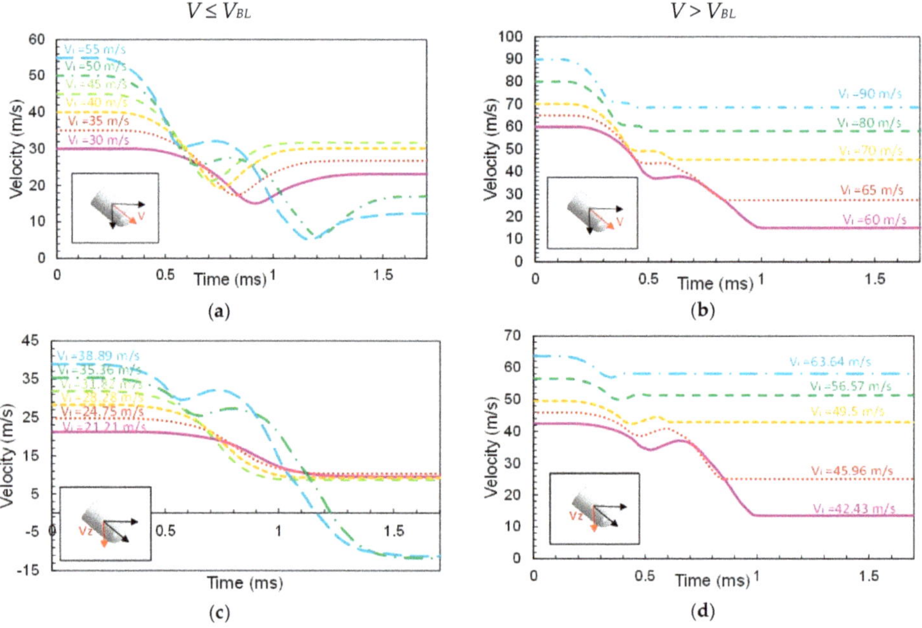

Figure 11. *Cont.*

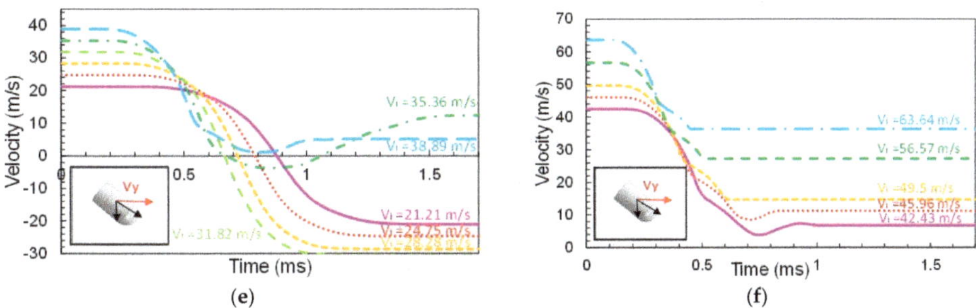

Figure 11. Velocity time histories for different initial velocities (impact angle = 45°): (**a,b**) magnitude of the velocity V, (**c,d**) velocity in the direction Z and (**e,f**) velocity in the direction Y.

The evolution of the residual velocity as a function of the impact angle is represented in Figure 12. Increasing the impact angle involves increasing the residual velocity in Z direction, due to the increase of V_Z with the angle even in rebound cases (Figure 12a). Below the ballistic limit, rebound occurs for any angle (Figure 12b, $V_i = 30$ m/s). Results show that the trend for V_Y changes above the ballistic limit. It increases with the oblique angle, but at high angles the rebound may appear ($V_i = 60$ m/s, $\alpha > 50°$).

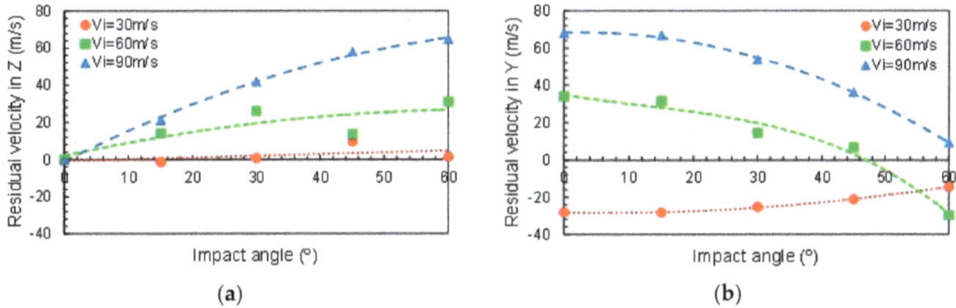

Figure 12. Residual velocity as a function of the impact angle (**a**) in Z direction (V_Z) and (**b**) in Y direction (V_Z).

The evolution of the absorbed energy during impact for different angles is represented in Figure 13. Around the ballistic limit, the fabric absorbed almost all of the impact energy to decelerate the projectile. It is observed that the ballistic limit increases with the angle. Keeping the impact velocity constant (V), the kinetic energy component due to velocity in the Y direction decreased with the angle (Figure 11). In practice, the impact component in this direction is what produces the rupture of the yarns. Consequently, the projectile would need a higher initial velocity to break the yarns. It was found that increasing the angle to 66% leads to an increment of the ballistic limit of 58%.

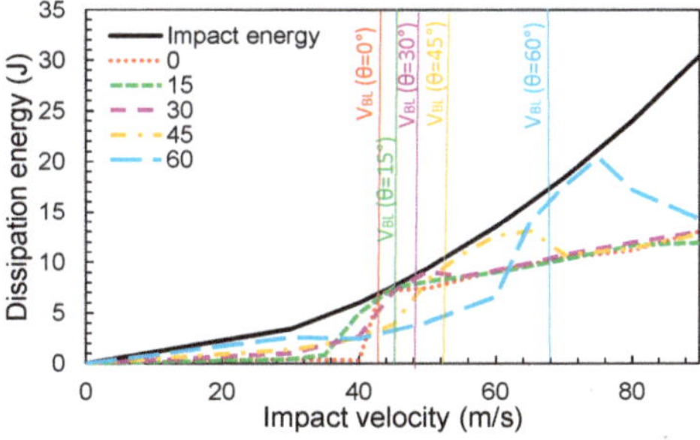

Figure 13. Variation of the dissipation energy with the impact velocity for different impact angles.

For the hemispherical projectile, the relationship to estimate the dissipation energy and the ballistic limit as a function of the impact angle were calculated (Figure 14). For both cases, the R^2 value was greater than 0.9, which supports the validity of the equations.

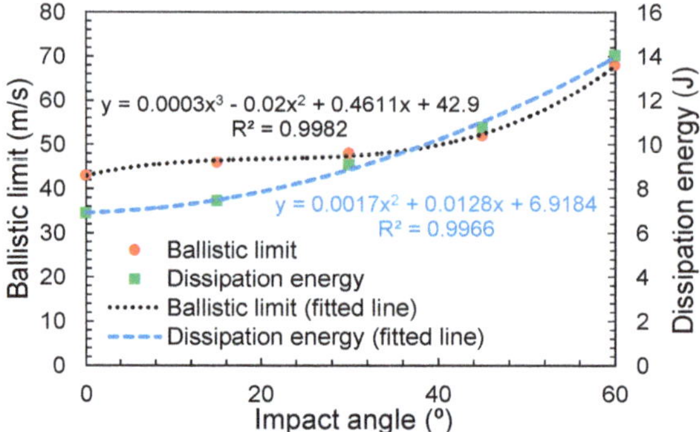

Figure 14. Variation of the ballistic limit and the dissipation energy with the impact angle.

5.3. Influence of the Number of Layers

The combination of the impact angle and the number of layers provides a multiple-choice problem. An initial study was carried out with the numerical model implemented and based on the round nose projectile.

A fitted equation calculated with the results obtained from the numerical model is presented in Equation (3), where n represents the number of layers and θ is the impact angle. The R^2 value was 0.92, highlighting the validity of the fitted model. The mechanistic model shows a good correlation between the ballistic limit obtained with the numerical model and the ballistic limit calculated from Equation (3) as can be seen in Figure 15b.

$$V_{BL} = 22.250 + 0.458 \cdot \theta + 22.425 \cdot n + 0.335 \cdot \theta \cdot n \qquad R^2 = 0.927 \qquad (3)$$

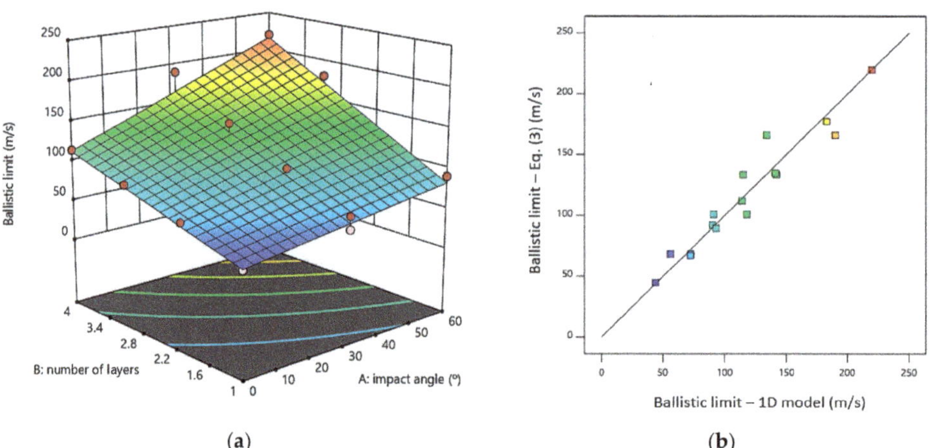

Figure 15. (**a**) Variation of the ballistic limit as a function of the number of layers and the impact angle; (**b**) correlation of the ballistic limit between the estimated values using the 1D model versus values estimated from Equation (3).

Figure 15a shows a response surface based on Equation (3). This Figure shows how the ballistic limit is enhanced by increasing the number of layers in the material. The ballistic limit value increases by 48%, 109% and 159% for two, three and four layers respectively. On the other hand, increasing the impact angle, raises the ballistic limit by nearly 27%, 63% and 95% for angles of 30°, 45° and 60° respectively. The evolution of the ballistic limit can be estimated rapidly using the fitted equation.

6. Conclusions

Based on the work presented in this paper, the following main conclusions can be drawn:

- A simplified model to study the impact in aramid fabrics at low velocities is been developed and parameterized. To validate the model, the results obtained were compared with the experimental tests reported in literature, obtaining a good agreement between the predicted values and the experimental results.

- The comparison of the 1D element based model with a 3D element based model demonstrated that the simplified models can reduce the computation time by 90%. This modelling methodology could be considered when designing personal protections with different woven structures and for various projectile geometries. The implementation of the numerical models in the industry, to help during the design process, requires simple and fast simulation tools.

- The computational analysis was also able to delineate the influence of different factors such as projectile geometry, number of layers and impact angle. Sharper projectiles lead to a higher residual velocity and a lower energy absorption, because the specific geometric feature of the projectile causes a higher deformation of the fibers allowing an improved slip through the fabric and facilitating rupture of the fibers. An increase in the impact angle and the number of layers lead to an increment of the ballistic limit.

- A mechanistic model developed for rapid estimation of the ballistic limit has been presented and validated with a very good confidence level. The expressions and surface diagrams obtained in this paper allowed to predict the critical velocity of impact once the number of layers and impact angle are known. This complementary analysis has elevated potential to be used in industry because of its simplicity. However, it is worth noting the necessity to carry out some previous work, both experimental and numerical, required to develop these types of mechanistic models with applicability in industrial environment.

Author Contributions: Conceptualization, N.F.; data curation, R.D.; formal analysis, N.F. and A.M.-S.; funding acquisition, N.F.; investigation, N.F., J.A.L. and A.M.-S.; methodology, N.F. and R.D.; project administration, R.D.; software, R.D.; supervision, J.A.L. and R.D.; validation, N.F., J.A.L. and R.D.; Writing—Original Draft, N.F.; Writing—Review and Editing, J.A.L., A.M.-S. and R.D.

Funding: This research was funded by the Ministry of Economy and Competitiveness from Spain, grant number BES-2012-055162 and the international collaborations subprogram under the reference EEBB-I-2016-11586.

Acknowledgments: We would like to thank to the Centre for Advanced Composite Materials for supplies the means to carry out this study.

Conflicts of Interest: The authors declare no conflict of interest. The funders had no role in the design of the study; in the collection, analyses, or interpretation of data; in the writing of the manuscript, or in the decision to publish the results.

References

1. Tabiei, A.; Nilakantan, G. Ballistic Impact of Dry Woven Fabric Composites: A Review. *Appl. Mech. Rev.* **2008**, *61*, 10801–10813. [CrossRef]
2. Grujicic, M. Multiscale modeling of polymeric composite materials for ballistic protection. In *Advanced Fibrous Composite Materials for Ballistic Protection*; Woodhead Publishing: Cambridge, UK, 2016; pp. 323–361.
3. Lim, C.; Tan, V.B.; Cheong, C. Perforation of high-strength double-ply fabric system by varying shaped projectiles. *Int. J. Impact Eng.* **2002**, *27*, 577–591. [CrossRef]

4. Tan, V.B.; Lim, C.; Cheong, C. Perforation of high-strength fabric by projectiles of different geometry. *Int. J. Impact Eng.* **2003**, *28*, 207–222. [CrossRef]

5. Shim, V.P.W.; Tan, V.B.C.; Tay, T.E. Modelling deformation and damage characteristics of woven fabric under small projectile impact. *Int. J. Impact Eng.* **1995**, *16*, 585–605. [CrossRef]

6. Park, Y.; Kim, Y.; Baluch, A.H.; Kim, C.-G. Empirical study of the high velocity impact energy absorption characteristics of shear thickening fluid (STF) impregnated Kevlar fabric. *Int. J. Impact Eng.* **2014**, *72*, 67–74. [CrossRef]

7. Taraghi, I.; Fereidoon, A.; Taheri-Behrooz, F. Low-velocity impact response of woven Kevlar/epoxy laminated composites reinforced with multi-walled carbon nanotubes at ambient and low temperatures. *Mater. Des.* **2014**, *53*, 152–158. [CrossRef]

8. Nilakantan, G.; Merrill, R.L.; Keefe, M.; Gillespie, J.W.; Wetzel, E.D. Experimental investigation of the role of frictional yarn pull-out and windowing on the probabilistic impact response of kevlar fabrics. *Compos. Part B Eng.* **2015**, *68*, 215–229. [CrossRef]

9. López-Gálvez, H.; Rodriguez-Millán, M.; Feito, N.; Miguelez, H. A method for inter-yarn friction coefficient calculation for plain wave of aramid fibers. *Mech. Res. Commun.* **2016**, *74*, 52–56. [CrossRef]

10. Duan, Y.; Keefe, M.; Bogetti, T.A.; Cheeseman, B.A.; Powers, B. A numerical investigation of the influence of friction on energy absorption by a high-strength fabric subjected to ballistic impact. *Int. J. Impact Eng.* **2006**, *32*, 1299–1312. [CrossRef]

11. Cunniff, P.M. An Analysis of the System Effects in Woven Fabrics under Ballistic Impact. *Text. Res. J.* **1992**, *62*, 495–509. [CrossRef]

12. Pan, N.; Lin, Y.; Wang, X.; Postle, R. An Oblique Fiber Bundle Test and Analysis. *Text. Res. J.* **2000**, *70*, 671–674. [CrossRef]

13. Ha-Minh, C.; Imad, A.; Boussu, F.; Kanit, T. Experimental and numerical investigation of a 3D woven fabric subjected to a ballistic impact. *Int. J. Impact Eng.* **2016**, *88*, 91–101. [CrossRef]

14. Chocron Benloulo, I.S.; Rodríguez, J.; Martínez, M.A.; Sánchez Gálvez, V. Dynamic tensile testing of aramid and polyethylene fiber composites. *Int. J. Impact Eng.* **1997**, *19*, 135–146. [CrossRef]

15. Cheeseman, B.A.; Bogetti, T.A. Ballistic impact into fabric and compliant composite laminates. *Compos. Struct.* **2003**, *61*, 161–173. [CrossRef]

16. Rodriguez, J.; Chocron, I.S.; Martinez, M.A.; Sánchez-Gálvez, V. High strain rate properties of aramid and polyethylene woven fabric composites. *Compos. Part B Eng.* **1996**, *27*, 147–154. [CrossRef]

17. Garcia, C.; Trendafilova, I.; Zucchelli, A. The Effect of Polycaprolactone Nanofibers on the Dynamic and Impact Behavior of Glass Fibre Reinforced Polymer Composites. *J. Compos. Sci.* **2018**, *2*, 43. [CrossRef]

18. Garcia, C.; Trendafilova, I. Triboelectric sensor as a dual system for impact monitoring and prediction of the damage in composite structures. *Nano Energy* **2019**, *60*, 527–535. [CrossRef]

19. Aruniit, A.; Kers, J.; Goljandin, D.; Saarna, M.; Tall, K.; Majak, J.; Herranen, H. Particulate filled composite plastic materials from recycled glass fibre reinforced plastics. *Mater. Sci.* **2011**, *17*, 276–281. [CrossRef]

20. Ramaiah, G.B.; Chennaiah, R.Y.; Satyanarayanarao, G.K. Investigation and modeling on protective textiles using artificial neural networks for defense applications. *Mater. Sci. Eng. B* **2010**, *168*, 100–105. [CrossRef]

21. Lopes, C.S.; Seresta, O.; Coquet, Y.; Gürdal, Z.; Camanho, P.P.; Thuis, B. Low-velocity impact damage on dispersed stacking sequence laminates. Part I: Experiments. *Compos. Sci. Technol.* **2009**, *69*, 926–936. [CrossRef]

22. Duan, Y.; Keefe, M.; Bogetti, T.A.; Cheeseman, B.A. Modeling the role of friction during ballistic impact of a high-strength plain-weave fabric. *Compos. Struct.* **2005**, *68*, 331–337. [CrossRef]

23. Rao, M.P.; Duan, Y.; Keefe, M.; Powers, B.M.; Bogetti, T.A. Modeling the effects of yarn material properties and friction on the ballistic impact of a plain-weave fabric. *Compos. Struct.* **2009**, *89*, 556–566. [CrossRef]

24. Nilakantan, G.; Keefe, M.; Wetzel, E.D.; Bogetti, T.A.; Gillespie, J.W. Computational modeling of the probabilistic impact response of flexible fabrics. *Compos. Struct.* **2011**, *93*, 3163–3174. [CrossRef]

25. Nilakantan, G.; Gillespie, J.W. Ballistic impact modeling of woven fabrics considering yarn strength, friction, projectile impact location, and fabric boundary condition effects. *Compos. Struct.* **2012**, *94*, 3624–3634. [CrossRef]

26. Nilakantan, G.; Wetzel, E.D.; Bogetti, T.A.; Gillespie, J.W. Finite element analysis of projectile size and shape effects on the probabilistic penetration response of high strength fabrics. *Compos. Struct.* **2012**, *94*, 1846–1854. [CrossRef]

27. Nilakantan, G.; Wetzel, E.D.; Bogetti, T.A.; Gillespie, J.W. A deterministic finite element analysis of the effects of projectile characteristics on the impact response of fully clamped flexible woven fabrics. *Compos. Struct.* **2013**, *95*, 191–201. [CrossRef]

28. Nilakantan, G.; Nutt, S. Effects of fabric target shape and size on the V_{50} ballistic impact response of soft body armor. *Compos. Struct.* **2014**, *116*, 661–669. [CrossRef]

29. Grujicic, M.; Bell, W.C.; Arakere, G.; He, T.; Cheeseman, B.A. A meso-scale unit-cell based material model for the single-ply flexible-fabric armor. *Mater. Des.* **2009**, *30*, 3690–3704. [CrossRef]

30. Grujicic, M.; Arakere, G.; He, T.; Bell, W.C.; Glomski, P.S.; Cheeseman, B.A. Multi-scale ballistic material modeling of cross-plied compliant composites. *Compos. Part B Eng.* **2009**, *40*, 468–482. [CrossRef]

31. Barauskas, R.; Abraitienė, A. Computational analysis of impact of a projectile against the multilayer fabrics in LS-DYNA. *Int. J. Impact Eng.* **2007**, *34*, 1286–1305. [CrossRef]

32. Ha-Minh, C.; Boussu, F.; Kanit, T.; Crépin, D.; Imad, A. Analysis on failure mechanisms of an interlock woven fabric under ballistic impact. *Eng. Fail. Anal.* **2011**, *18*, 2179–2187. [CrossRef]

33. Ha-Minh, C.; Imad, A.; Kanit, T.; Boussu, F. Numerical analysis of a ballistic impact on textile fabric. *Int. J. Mech. Sci.* **2013**, *69*, 32–39. [CrossRef]

34. Park, Y.; Kim, Y.; Baluch, A.H.; Kim, C.-G. Numerical simulation and empirical comparison of the high velocity impact of STF impregnated Kevlar fabric using friction effects. *Compos. Struct.* **2015**, *125*, 520–529. [CrossRef]

35. Chu, T.-L.; Ha-Minh, C.; Imad, A. A numerical investigation of the influence of yarn mechanical and physical properties on the ballistic impact behavior of a Kevlar KM2® woven fabric. *Compos. Part B Eng.* **2016**, *95*, 144–154. [CrossRef]

36. Das, S.; Jagan, S.; Shaw, A.; Pal, A. Determination of inter-yarn friction and its effect on ballistic response of para-aramid woven fabric under low velocity impact. *Compos. Struct.* **2015**, *120*, 129–140. [CrossRef]

37. Nilakantan, G.; Gillespie, J.W. Yarn pull-out behavior of plain woven Kevlar fabrics: Effect of yarn sizing, pullout rate, and fabric pre-tension. *Compos. Struct.* **2013**, *101*, 215–224. [CrossRef]

38. Nilakantan, G.; Nutt, S. Effects of clamping design on the ballistic impact response of soft body armor. *Compos. Struct.* **2014**, *108*, 137–150. [CrossRef]

39. Rao, M.P.; Nilakantan, G.; Keefe, M.; Powers, B.M.; Bogetti, T.A. Global/Local Modeling of Ballistic Impact onto Woven Fabrics. *J. Compos. Mater.* **2009**, *43*, 445–467. [CrossRef]

40. Nilakantan, G.; Keefe, M.; Bogetti, T.A.; Adkinson, R.; Gillespie, J.W. On the finite element analysis of woven fabric impact using multiscale modeling techniques. *Int. J. Solids Struct.* **2010**, *47*, 2300–2315. [CrossRef]

41. Nilakantan, G.; Keefe, M.; Bogetti, T.A.; Gillespie, J.W. Multiscale modeling of the impact of textile fabrics based on hybrid element analysis. *Int. J. Impact Eng.* **2010**, *37*, 1056–1071. [CrossRef]

42. Ha-Minh, C.; Kanit, T.; Boussu, F.; Imad, A. Numerical multi-scale modeling for textile woven fabric against ballistic impact. *Comput. Mater. Sci.* **2011**, *50*, 2172–2184. [CrossRef]

43. Lozano-Mínguez, E.; Palomar, M.; Infante-García, D.; Rupérez, M.J.; Giner, E. Assessment of mechanical properties of human head tissues for trauma modelling. *Int. J. Numer. Method. Biomed. Eng.* **2018**, *34*, 1–17. [CrossRef] [PubMed]

44. Palomar, M.; Lozano-Mínguez, E.; Rodríguez-Millán, M.; Miguélez, M.H.; Giner, E. Relevant factors in the design of composite ballistic helmets. *Compos. Struct.* **2018**, *201*, 49–61. [CrossRef]

45. Moure, M.M.; Feito, N.; Aranda-Ruiz, J.; Loya, J.A.; Rodriguez-Millan, M. On the characterization and modelling of high-performance para-aramid fabrics. *Compos. Struct.* **2019**, *212*, 326–337. [CrossRef]

46. Abtew, M.A.; Boussu, F.; Bruniaux, P.; Loghin, C.; Cristian, I. Ballistic impact mechanisms—A review on textiles and fibre-reinforced composites impact responses. *Compos. Struct.* **2019**, *223*, 110966. [CrossRef]

47. Yu, J.H.; Dehmer, P.G.; Yen, C. *High-Speed Photogrammetric Analysis on the Ballistic Behavior of Kevlar Fabrics Impacted by Various Projectiles*; Report No. ARL-TR-5333; Army Research Laboratory: Adelphi, MD, USA, 2010.

48. Smith, M. *ABAQUS/Standard User's Manual, Version 6.9*; Simulia: Providence, RI, USA, 2009.

Article

A Semi-Empirical Deflection-Based Method for Crack Width Prediction in Accelerated Construction of Steel Fibrous High-Performance Composite Small Box Girder

Bishnu Gupt Gautam [ID]**, Yi-Qiang Xiang *** [ID]**, Zheng Qiu and Shu-Hai Guo**

College of Civil Engineering and Architecture, Zhejiang University, Hangzhou 310058, China;
bishnu@zju.edu.cn (B.G.G.); qiuzheng_zju@126.com (Z.Q.); 15158016702@163.com (S.-H.G.)
* Correspondence: xiangyiq@zju.edu.cn; Tel.: +86-571-8820-8700

Received: 23 February 2019; Accepted: 19 March 2019; Published: 22 March 2019

Abstract: Accelerated construction in the form of steel–concrete composite beams is among the most efficient methods to construct highway bridges. One of the main problems with this type of composite structures, which has not yet been fully clarified in the case of continuous beam, is the crack zone initiation that gradually expands through the beam width. In the current study, a semi-empirical model was proposed to predict the size of cracks in terms of small box girder deflection and intensity of load applied on the structure. To this end, a set of steel–concrete composite small box girders were constructed by the use of steel fibrous concrete and experimentally tested under different caseloads. The results were then used to create a dataset of the box girder response in terms of beam deflection and crack width. The obtained dataset was then utilized to develop a simplified formula providing the maximum width of cracks. The results showed that the cracks initiated in the hogging moment region when the load exceeded 80 kN. Additionally, it was observed that the maximum cracked zone occurred in the center of the beam due to the maximum negative moment. Moreover, the crack width of the box girder at different loading cases was compared with the test results obtained from the literature. A good agreement has been found between the proposed model and experiment results.

Keywords: static behavior; studs; deflection; finite element model; composite beam; crack width

---

1. Introduction

Advancement in new construction techniques has received much attention in the last decades. The conventional bridge constructions and rehabilitation are usually accompanied by several operational difficulties. In order to overcome the traditional bridge construction problems, quite a few construction techniques were established such as accelerated, rapid, modular, mechanized, or precast constructions. Among them, the accelerated construction is an effective technique to reduce the impact of the bridge construction period on the surrounding traffic flow [1]. Moreover, it can properly help the engineering projects to save the implementation cost, improve the construction safety, reduce the adverse effects on the surrounding environment, ensure the higher construction quality, etc. [2].

Accelerated construction of steel fibrous high-performance continuous composite box girder bridge (ACHPCBG-bridge) is classified into two types—steel–concrete composite box girder bridge with single or double cell and small box girder bridge with multiple cells [3]. If the depth of the box girder exceeds 1/6 or 1/5 of the bridge width, single-cell box girder is recommended to be used in practice; however, if the depth is smaller than 1/6 of the width of bridge, double-cell or multiple-cell box girders can be considered [4]. The boxes used in the ACHPCBG-bridge are mostly rectangular, trapezoidal, or triangular in shape [5,6] as shown in Figure 1. It is reported that triangular boxes with apex down suffer from some disadvantages [7]. They usually have to be deeper than rectangular or trapezoidal boxes. Additionally, because of smaller area, triangular boxes have less torsional resistance. Furthermore, their bottom flange often has to be a heavy built-up section combined with bent plates for connection to the webs. Bending moment and shear force on rectangular box girders are greater than that of trapezoidal cross section [8]. Therefore, trapezoidal box girders may be a better choice for accelerated construction in comparison to other box girders. Characteristics of concrete material is also an important part in steel–concrete composite box girders, especially for cases where a large strength-to-weight ratio is required [9]. A box girder is either simply supported or continuous. Many researchers simulated the composite structures in bridge application to study various aspects of their structural behaviors [10–17].

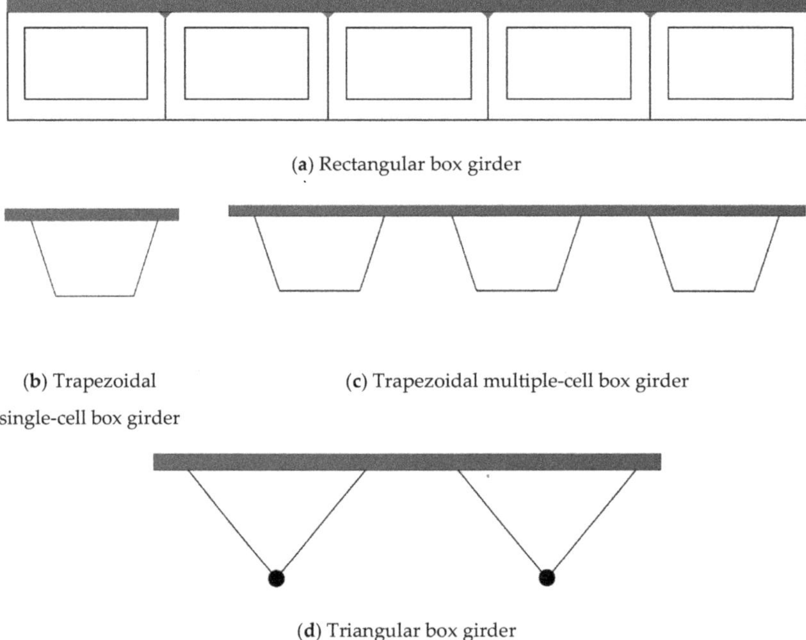

(**a**) Rectangular box girder

(**b**) Trapezoidal
single-cell box girder

(**c**) Trapezoidal multiple-cell box girder

(**d**) Triangular box girder

Figure 1. Different cross-sections of box girders: (**a**) Rectangular box girder; (**b**) trapezoidal single-cell box girder; (**c**) trapezoidal multiple-cell box girder; (**d**) triangular box girder.

The ultimate loading capacity of the ACHPCBG-bridge is usually determined by either flexural or shear bearing capacity and is governed by the compressive strength of the concrete or tensile strength of the steel girder. Moreover, in some areas with a continuous composite beam subjected to hogging moment, high strength is required to resist the negative bending. In the support region where the negative bending moment acts, a relatively high tensile stress is generated in a concrete slab and compressive stresses are induced in the lower steel region. As such, the mechanical behavior of these girders is strongly nonlinear even for low-level stress at that region, resulting in crack initiation in the slab, which is generally considered as a shortcoming for the durability and service life of a structure. [18–23].

In any construction, the characteristics of concrete play an essential role. Concrete is brittle and has limited ductile behavior. Therefore, a form of reinforcement is required to enhance structural stability. Steel bars are used as reinforcement in concrete structures; however, there is still a possibility of crack formation internally or externally, which may lead to a major problem in overall stability of structure [24,25].

The cracks developed in reinforced concrete members extend freely until encountering a reinforcing bar. We need to arrest the cracks to lengthen the lifespan of structures. Hence, a multi-directional and closely spaced reinforcement is required to be used in concrete. The fibers are, thus, an excellent choice to overcome this type of problem in practice. Fiber reinforced concrete (FRC) is composed of short discrete fibers that are uniformly distributed and randomly oriented within the concrete matrix. The fibers are mainly classified into four types—steel, glass, natural, and synthetic—and each of the different fibers have different properties [26–30].

Fibers in a concrete element increase the structural integrity, provide high tensile strength to plain concrete, reduce the permeability of concrete, and increase the resistance to impact load. Fibers can reduce the number of rebars without loss of strength. They can also eliminate cracks propagation by bridging action. The flexural behavior, bond strength, and especially toughness of SFRC increase as the fiber content increases. Carbon or steel fibers can be added to a cement matrix at a high volume fraction (0.5–3%) to increase the conductivity of the composite. The properties of fiber concrete depend upon the volume of fibers used [31–36].

Various types of problems occur in composite materials, mostly due to cracks and delamination. The crack in a composite structure may reduce the structural stiffness and strength and redistribute the load, which may either delay the structural failure or accelerate the structural collapse. The crack is not the main cause of structural failure, but rather is the part of the failure process that may lead to the loss of structural integrity. Therefore, it plays an important role in the failure mechanism of steel–concrete composite structures [37–43].

The composite beam usually consists of a steel section jointly acting with one (or two) flange(s) made of reinforced concrete that is mainly subjected to bending [44]. These two materials are interconnected by means of mechanical shear connectors. Therefore, a composite beam, even with small steel sections, has high stiffness and carries heavy loads on long spans. However, by increasing the load intensity, additional issues such as slip and deflection occur along the beam [45]. In the current study, different deformation gauges were installed at the bottom of flanges of the beam to measure the maximum deflection induced along the beam.

The current study mainly aimed at investigating the static behavior of the ACHPCBG-bridge under vertical loading and proposing a simplified methodology to estimate the width of cracks. For this purpose, an ACHPCBG-bridge was experimentally tested under several caseloads to get deflection response. Additionally, the cracking process at hogging moment region was measured at each load step. The details of experimental models are discussed in the following sections.

2. Design and Fabrication of Experimental ACHPCBG-bridge

Accelerated construction steel–concrete composite small box girder is a construction technique consisting of an open steel box girder, diaphragm, welding stud, and concrete slab. It is difficult to make corresponding scale models, especially of the concrete bridge deck and the thickness of the steel plate. Therefore, based on the 25 m span prototype bridge, a stereotype model of the steel–concrete composite small box girder was prepared with the members-to-actual length ratio of 1:4. The basic parameters of box girder are shown in Table 1.

In the accelerated construction of steel–concrete composite small box girder, concrete material, stud group arrangement, and stud group spacing are three main parameters. At present, ordinary concrete by a grade size below C50 is mainly used in steel–concrete composite structures. However, because of the high strength and low weight, high-performance concrete not only reduces the weight of superstructure and decrease the costs, but also provides higher durability to the main structure. Hence, it is gradually getting more applications in bridge engineering. As to the knowledge of the authors, there is no specific literature or standard available to give unified understanding of the arrangement and spacing of studs in group. From the design point of view, in order to ensure the mechanical performance of the bridge structure, the degree of shear connection should not be too small; however, from the viewpoint of construction process, it is desirable to increase the stud spacing as much as possible to adapt the accelerated construction. The current code for design of steel–concrete composite structures in China [46] requires that the number of welding studs in each shear span area should be greater than the ratio of longitudinal shear force at the interface between steel beam and concrete slab and the shear capacity of single welding stud, i.e., complete shear design is required. In the steel structure design code [47], when the strength and deformation are satisfied, the longitudinal and horizontal shear capacity of the shear connectors at the interface of composite beams can guarantee the full flexural capacity of the maximum moment section; then, they can be designed according to the partial shear connection. However, partial shear connections are limited to composite beams with equal cross-section spans not exceeding 20 m. The AASHTO [48] bridge design code stipulates that the spacing of reserved hole should not be more than 610 mm. EC4 [49] requires that the maximum spacing of uniform welding studs should not exceed the minimum of 4 times the thickness of concrete slab and 800 mm; however, there is no specific provision for group stud shear connectors.

Table 1. Material and geometrical details of specimen girder.

Material	Parameter	Value	Parameter	Value
Concrete	Density, (kg/m^3)	2400	Combined beam width, (mm)	700
	Elastic modulus, (MPa)	36,400	Beam length (calculated span), (cm)	630(600)
	Poisson's ratio	0.2	Beam height, (mm)	327
	Yield strength, (MPa)	76.24	Number of welding studs	132
Steel	Density, (kg/m^3)	7850	Design strength of bridge deck concrete, (MPa)	C60
	Elastic modulus, (GPa)	210	Thickness of concrete deck, (mm)	70
	Poisson's ratio	0.3	Reserved hole concrete design strength, (MPa)	C80
	Yield strength, (MPa)	421	Spacing center distance, (mm)	600
Reinforcement	Density, (kg/m^3)	7800	-	-
	Elastic modulus, (GPa)	206	-	-
	Poisson's ratio	0.3	-	-
	Yield strength, (MPa)	445	-	-
Stud	Density, (kg/m^3)	7800	-	-
	Elastic modulus, (GPa)	210	-	-
	Poisson's ratio	0.3	-	-
	Yield strength, (MPa)	360	-	-
	Size, (mm)	$\Phi13 \times 50$	-	-

The structure and main dimension of the composite test beam are shown in Figure 2, which is made up of an open steel box girder and precast concrete slab. The test beam was 6.3 m in total length. It had two spans, with each span of 3 m length. The total height of the composite box girder was 327 mm with 70 mm slab thickness and 257 mm height of steel box girder. The upper flange was 6 mm thick and 135 mm wide. The web and bottom plates were 6 mm and 8 mm thick, respectively. A solid web diaphragm was set at every 600 mm from the support position. The diaphragm was 6 mm thick and 220 mm high. The test beam was made of Q345qc steel. The upper flange of the steel beam was arranged with shear group studs. The center distance of the group stud was 600 mm. The lateral distance of the group stud was 50 mm and the longitudinal distance was 65 mm. The welding stud was $\Phi13 \times 50$ made of ML15AL. The configuration of the steel–concrete composite beam is given in Figure 2. Construction of prefabricated concrete slabs, welding of steel beams, and welding of studs are strictly in accordance with the requirements of design, drawings, and construction technique specifications. The main fabrication process is shown in Figure 3. Copper-plated steel fiber with a diameter 0.2 mm and length of 13 mm, tensile strength of 2000 MPa, and volume fraction of 1.5% was used in a concrete element. The physical properties of steel fiber used are shown in Appendix A, Table A3. The information about different properties and materials used, such as cement, fine aggregates, coarse aggregate, water, and chemicals have been described in details in Appendix A.

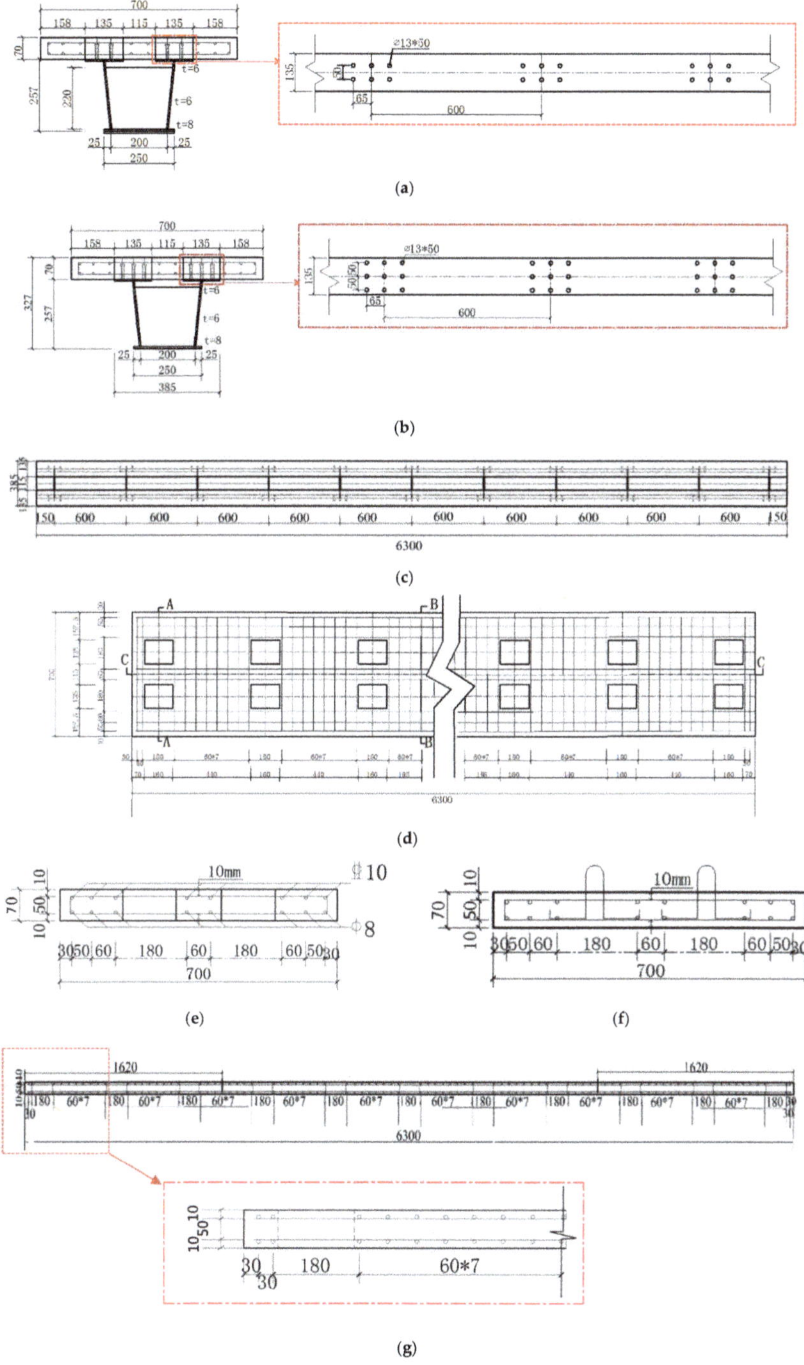

Figure 2. Details of experimental accelerated construction of steel fibrous high-performance continuous composite box girder bridge (ACHPCBG-bridge) (mm). (**a**) Composite beam cross-section and stud arrangement of specimen 1; (**b**) composite beam cross-section and stud arrangement of specimen 2; (**c**) plan of steel; (**d**) plan of concrete; (**e**) cross-section at A-A; (**f**) cross-section at B-B; (**g**) cross-section at C-C.

(a)   (b)   (c)   (d)

(e)   (f)   (g)   (h)

Figure 3. Fabrication process of specimen: (**a**) Formwork erection and reinforcing cage binding; (**b**) casting and maintenance of concrete slab; (**c**) steel beam welding; (**d**) stud welding of test specimen 1; (**e**) stud welding of test specimen 2; (**f**) completion of concrete slab and open steel box girder; (**g**) placing of concrete slab on steel box girder to make it composite; (**h**) reserve hole filling.

3. Experimental Test

3.1. Test Method and Instrumentation

3.1.1. Deflection

Deflection was measured by means of digital displacement transducer deformation gauges at key sections as shown in Figure 4. At the key sections, two sensors were used on each side of the steel bottom to get the accurate measurement of displacement. Due to limitation of instrument available in the lab, a set of two actuators with a total imposing force of 900 kN was applied to the model. The numerical model was established as a first approximation in order to find the experimental result properly. The deflection at 900 kN was experimentally found as 13.211 mm, which is very close to the analytical value of 12.8 2 mm. The details about obtained results from the experimental test and numerical analysis is presented in the establishment of numerical modeling section.

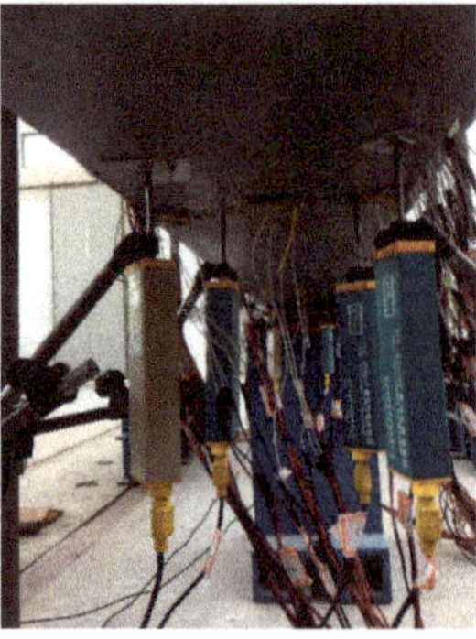

Figure 4. The arrangement of the displacement sensor.

3.1.2. Crack Measure

The digital concrete crack gauges were used to measure the crack width at a key section (i.e., the negative bending moment region) of the concrete slab. This crack gauge was used for quantitative detection of the crack width on the concrete surface, known as the common concrete non-destructive testing equipment. The gauge could automatically interpret the measurement and directly display the crack width value on the screen. It had a measuring range of 0.01 to 3 mm, the magnification of 60×, and the estimated accuracy of 0.005. At different loading cases, crack length and crack width were measured. The observed maximum crack width was considered for formulation and verification. In the experimental test of specimen 1, crack number 3 had the maximum crack width of 0.456 mm at the 900 kN load; for specimen 2, however, crack number 2 had the maximum crack width of 0.349 mm. The detail of crack width, crack length, and loading cases of specimens 1 and 2 are presented in Tables 2 and 3, respectively. More details about crack formation is discussed in Section 6, simplified model.

3.2. Loading Procedure

The experiment was performed in the structural lab of the Quzhou University. A set of two actuators with total 900 kN loading capacity equipment was used in the experiment to apply the load at the mid-span of the steel–concrete composite beam. The test specimen was supported by a roller system at both ends and hinge support at the center of the two-span continuous beam. The change in behavior of composite girders was carefully observed through the static test. The test girder was loaded up to 100 kN in 20 kN intervals, followed by 50 kN intervals up to 900 kN. The maximum capacity of the instrument was 900 kN. A loading plate (110 cm × 30 cm × 20 cm) and a supporting plate (70 cm × 15 cm × 8 mm) have been used in the experiment. The test setup for the steel–concrete composite beam is illustrated in Figure 5

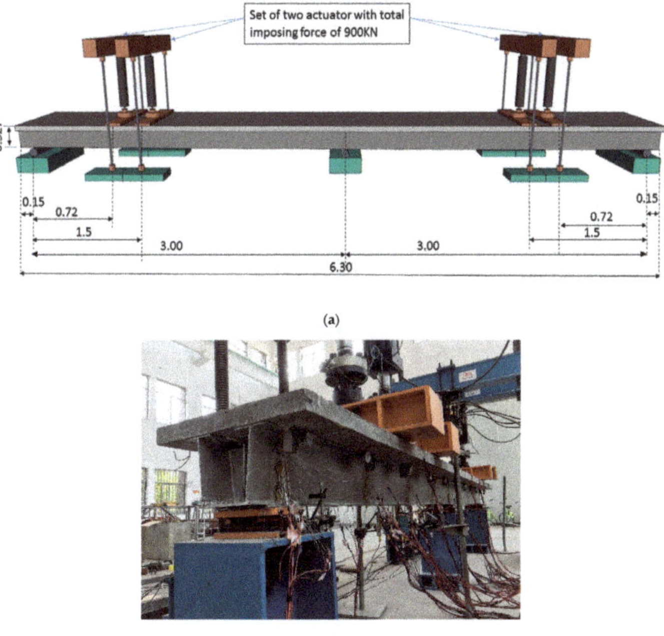

Figure 5. The typical geometry and dimensions of the test specimen: (**a**) Schematic front view (m); (**b**) side view.

Table 2. Observed experimental results of specimen 1 (Unit: Crack length, m; crack width, mm).

Set	80 kN		150 kN		300 kN		450 kN		600 kN		800 kN		900 kN	
Crack Number	Crack Length	Crack Width	Crack Length	Crack Width	Crack Length	Crack Width	Crack Length	Crack Width	Crack Length	Crack Width	Crack Length	Crack Width	Crack Length	Crack Width
1	0.052	0.041	0.7	0.063	0.7	0.134	0.7	0.174	0.7	0.221	0.7	0.309	0.7	0.309
2	0.049	0.04	0.7	0.092	0.7	0.148	0.7	0.215	0.7	0.228	0.7	0.228	0.7	0.228
3	0.058	0.045	0.7	0.08	0.7	0.215	0.7	0.336	0.7	0.389	0.7	0.403	0.7	0.456
4	-	-	0.7	0.064	0.7	0.121	0.7	0.161	0.7	0.186	0.7	0.228	0.7	0.228
5	-	-	-	-	0.42	0.107	0.42	0.107	0.7	0.134	0.7	0.161	0.7	0.161
6	-	-	-	-	0.45	0.127	0.45	0.161	0.53	0.174	0.53	0.188	0.53	0.188
7	-	-	-	-	0.11	0.181	0.11	0.181	0.11	0.181	0.11	0.181	0.11	0.181

Table 3. Observed experimental results of specimen 2 (Unit: crack length, m; crack width, mm).

Set	80 kN		150 kN		300 kN		450 kN		600 kN		800 kN		900 kN	
Crack Number	Crack Length	Crack Width	Crack Length	Crack Width	Crack Length	Crack Width	Crack Length	Crack Width	Crack Length	Crack Width	Crack Length	Crack Width	Crack Length	Crack Width
1	0.7	0.054	0.7	0.107	0.7	0.174	0.7	0.161	0.7	0.255	0.7	0.295	0.7	0.322
2	0.58	0.054	0.7	0.094	0.7	0.188	0.7	0.223	0.7	0.265	0.7	0.336	0.7	0.349
3	-	-	-	-	0.13	0.054	0.17	0.067	0.44	0.121	0.7	0.188	0.7	0.215
4	-	-	-	-	0.12	0.094	0.11	0.107	0.11	0.134	0.12	0.188	0.12	0.188
5	-	-	-	-	0.21	0.054	0.21	0.054	0.21	0.067	0.21	0.107	0.21	0.094
6	-	-	-	-	0.57	0.094	0.7	0.134	0.7	0.188	0.7	0.215	0.7	0.242
7	-	-	-	-	0.18	0.054	0.21	0.094	0.22	0.107	0.22	0.134	0.7	0.134

4. Establishment of Numerical Modeling

4.1. Modeling

The specimen had two spans of 3 m and the total length of the girder was 6.3 m. A 3D numerical model of the steel–concrete composite girder beam with a clear span of 3 m was simulated in ABAQUS version 6.14 [50]. A detailed view of the numerical model is shown in Figure 6. Loading was applied to the top surface of the concrete slab and distributed over the full width of the girder. The load increased with 20 kN intervals up to 100 kN and, then, the load interval was set to 50 kN up to the total imposing force of the actuator. Due to the symmetry in geometry, loading, and boundary conditions, only half of the beam was modeled. The coordinate system represented axes X, Y, and Z as axes 1, 2, and 3 in the model, respectively. The symmetry boundary conditions are shown in Figure 6e with a restrained degree of freedom [51].

ABAQUS/standard solver with a linear geometric order was used in the study. Eight-node brick elements with reduced integration (C3D8R) were used to model the concrete and stud. The first-order interpolation, three-dimensional beam element (B31), and four-node shell element with reduced integration (S4R) were used to model reinforcement bars and steel, respectively. The final mesh included 55,496 elements and 67,327 nodes. Steel reinforcement bars were modeled using embedded rebar element. Surface-to-surface contact was used for stud-to-concrete and steel-to-concrete interactions. The material nonlinearities of concrete and steel were modeled using the concrete damage plasticity model and the elastic-plastic bilinear model, respectively. The accuracy of the results basically depends on the size of the mesh, the constitutive model, and boundary conditions [52]. Therefore, these aspects should be carefully incorporated into the proposed finite element model. Adequate attention was paid to the development of hexahedral mesh and the assigning interaction between various surfaces. Various components, namely, concrete slab, steel beam, stud connectors, reinforcement bars, and stiffeners, were meshed part by part instead of using global or sweep features. Thus, a regular structured hexahedral mesh was generated. To get an acceptable level of accuracy, the approximate global mesh size of 0.025 was used for reinforcement bars and concrete; whereas an approximate global mesh size of 0.015 was used for studs and steels. The modeling of different parts of the beam is shown in Figure 6. Material and geometrical details of the specimen girder are given in Table 1.

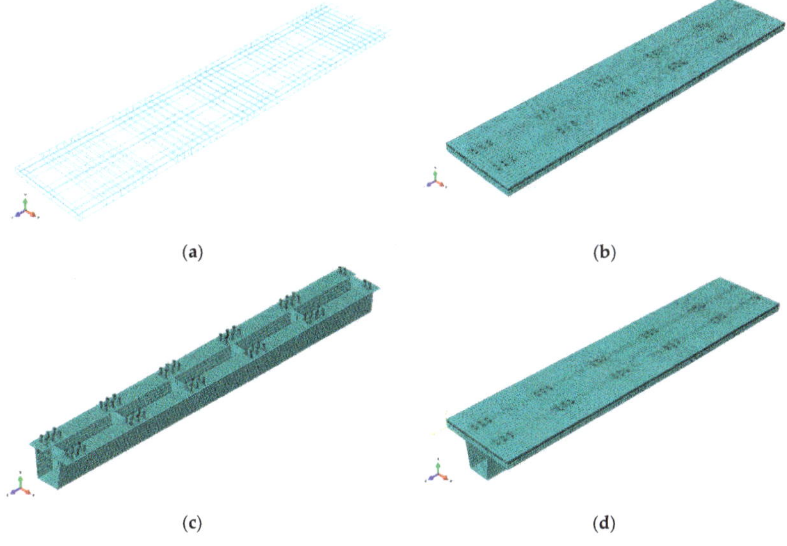

(a)   (b)

(c)   (d)

Figure 6. *Cont.*

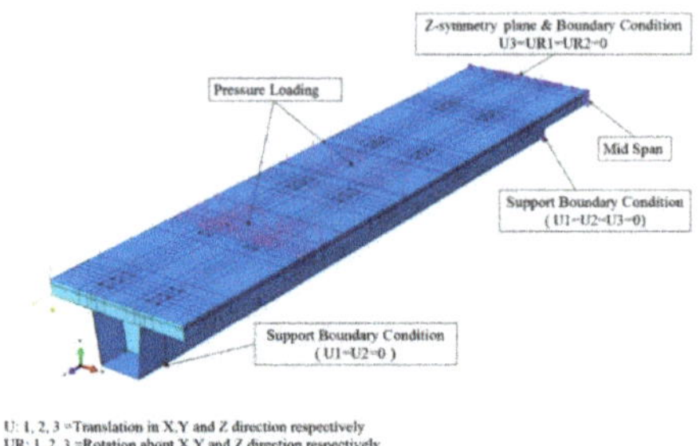

(e)

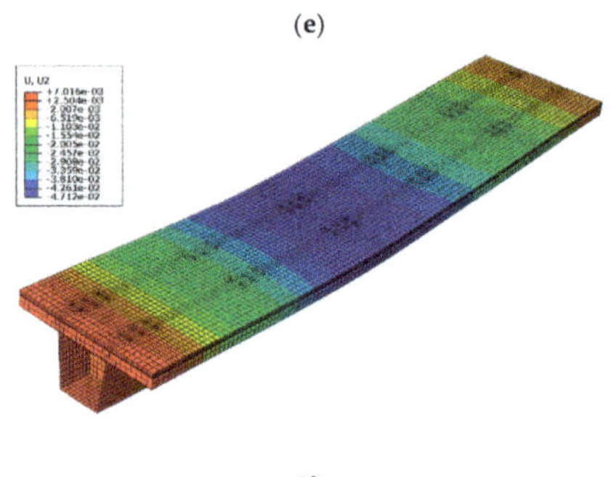

(f)

Figure 6. The finite element model: (**a**) Reinforced-steel meshing; (**b**) concrete-slab meshing; (**c**) steel beam, diaphragm, and stud meshing; (**d**) composite box girder meshing; (**e**) boundary conditions; (**f**) load-deflection behavior at 1000 kN load.

4.2. Material Description

4.2.1. Concrete

In order to simulate the mechanical behavior of concrete, a damage plastic model of concrete was utilized in the current study. The constitutive relationship of concrete is adopted from the uniaxial tension and compressive stress–strain curve of concrete given in the "code for the design of concrete structure" GB 50010-2010 [53]. On the day of the experiment, the measured average compressive strength of C60 concrete cube was 76.24 MPa, and the design strength of reserved-hole C80 concrete was 87.4 MPa. Additionally, the average splitting strength and the average elastic modulus were 5.09 MPa and 36.4 GPa, respectively. The uniaxial tension and the uniaxial compression parameters are listed in Table 4.

Table 4. Concrete uniaxial tension and compression parameters.

Parameter	α_t	$f_{t,r}$ (Mpa)	$\varepsilon_{t,r}$	α_c	$f_{c,r}$ (Mpa)	$\varepsilon_{c,r}$
C60	3.82	3.5	1.28×10^{-4}	3.81	76.24	2.19×10^{-3}

The stress–strain relationship of concrete under uniaxial tension is expressed as follows:

$$\sigma = (1 - d_t)E_c\varepsilon \tag{1}$$

$$d_t = 1 - \rho_t\left[1.2 - 0.2x^5\right], x \leq 1 \tag{2}$$

$$d_t = 1 - \frac{\rho_t}{\alpha_t(x-1)^{1.7} + x}, x > 1 \tag{3}$$

$$x = \frac{\varepsilon}{\varepsilon_{t,r}} \tag{4}$$

$$\rho_t = \frac{f_{t,r}}{E_c \varepsilon_{t,r}} \tag{5}$$

where α_t is the parameter of concrete uniaxial tension stress–strain curve in the decline period, $f_{t,r}$ is the representation of concrete uniaxial tensile strength, $\varepsilon_{t,r}$ is the peak tensile strain corresponding to $f_{t,r}$, and d_t is the evolution parameter of concrete under uniaxial tension.

Compression stress–strain relationships are assumed as follows:

$$\sigma = (1 - d_c) E_c \varepsilon \tag{6}$$

$$d_c = 1 - \frac{\rho_c n}{n - 1 + x^n}, x \leq 1 \tag{7}$$

$$d_c = 1 - \frac{\rho_c}{\alpha_c(x-1)^2 + x}, x > 1 \tag{8}$$

$$\rho_c = \frac{f_{c,r}}{E_c \varepsilon_{c,r}} \tag{9}$$

$$n = \frac{E_c \varepsilon_{c,r}}{E_c \varepsilon_{c,r} - f_{c,r}} \tag{10}$$

$$x = \frac{\varepsilon}{\varepsilon_{c,r}} \tag{11}$$

where α_c is the parameter of concrete uniaxial compression stress–strain curve in the decline period, $f_{c,r}$ is the representation of concrete uniaxial compressive strength, $\varepsilon_{t,r}$ is the peak compressive strain corresponding to $f_{c,r}$, and d_c is the evolution parameter of concrete under uniaxial compression.

The stress–strain relationship and damage model of C60 concrete are shown in Figure 7.

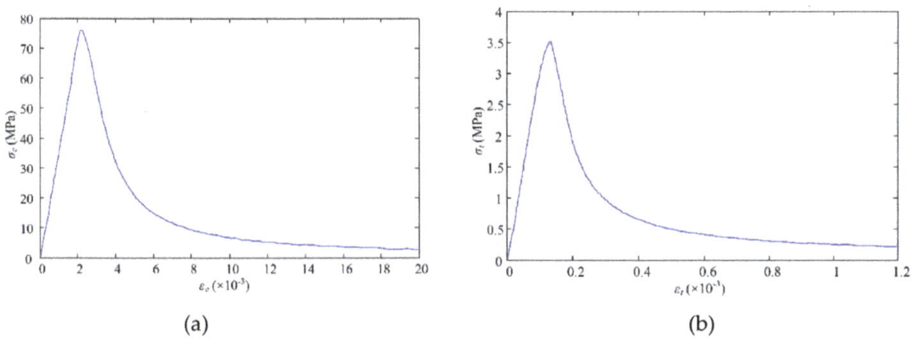

Figure 7. The stress–strain curve of C60 Concrete. (a) Compression; (b) tension.

4.2.2. Steel Beam, Stiffener, and Stud

The steel and stiffener were modeled considering the nonlinear behavior of the materials. The elastic-plastic material model was employed based on the nominal stress–strain behavior of steel. In the stud connection, an elastic-plastic bilinear model was utilized. However, the material of steel plates, studs, and steel bars was defined by the ideal elastoplastic model. That is, when the steel yields, the bearing capacity does not increase, but the deformation continues to increase. The stress–strain relationships of steel plates, studs, and bars are shown in Figure 8.

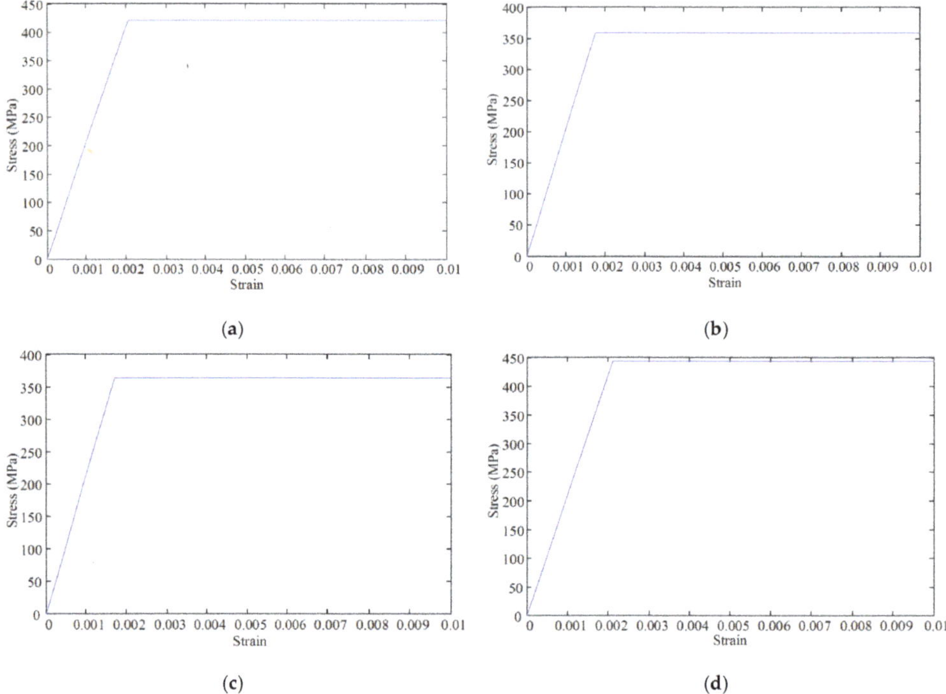

Figure 8. Stress–strain relationships: (**a**) The stress–strain curve of steel; (**b**) the stress–strain curve of stud; (**c**) the stress–strain curve of φ8 rebar; (**d**) the stress–strain curve of φ10 rebar.

4.3. Bonding

The bonding between the materials was done by the use of interaction in Abaqus. The stud and concrete was modeled by the penalty method considering a friction coefficient of 0.4 in the tangential direction and hard contact in the normal direction to avoid penetration between the two contact surfaces [51]. On account of the interaction between the flange of steel and concrete slabs, the steel was determined as the "slave surface" and the concrete as the "master surface". The finite sliding method was employed for the interaction between studs and concrete. To simulate the steel bar–concrete interaction, the reinforcement bar was selected as the embedded region and concrete was set to be the host region.

4.4. Comparison Between Numerical Analysis Values and Experimental Results

The validation of experimental results was performed using the numerical analysis data as described earlier in Section 4. There was a good agreement between numerical and experimental results. Due to the maximum capacity of instrument available in the lab, a set of two actuators with a total imposing load of 900 kN was applied to the model. However, in the numerical model, the final step of the model was set to 1000 kN with the full shear interaction. The load increment was considered 10% in each load step. The numerical results showed a relatively stiffer behavior compared to the experimental model. It could be due to the stabilization process that occurred during the loading stage because of the change in loading system from one set of actuators to another set of actuators. However, similar results were obtained at the 900 kN load. Afterward, by a small increment in load, a significant change was observed in the deflection that resulted from the numerical model. Due to the symmetry in the geometric shape of the beam, the experimental measurement was performed in just one span.

The deflection at 900 kN was analytically found as 12.82 mm, which is very close to the experimental value of 13.211 mm. The deflection obtained from the numerical analysis is presented in Table 5.

Table 5. Load-deflection comparison of beam.

Load, kN	80	150	200	300	400	450	500	600	700	800	900	950	1000
Deflection, FEM (mm)	1.048	1.932	2.563	3.825	5.102	5.745	6.169	7.053	8.041	9.451	12.82	20.41	47.12
Deflection Specimen 1 (mm)	1.6	2.456	3.739	4.535	6.1465	7.1025	7.4925	8.0125	9.977	11.71	13.211	-	-

5. Simplified Model

The first crack observed when the load exceeded 80 kN; then, by increasing the load intensity, further crack formations were observed in the specimen. The cracks formed on the box girder were marked by a set of numbers demonstrating the occurrence order. Crack number 3 had the maximum crack width and was considered in the formulation process. The major cracks observed in the specimen are presented in Figures 9 and 10. Additionally, the related data are listed in Tables 2 and 3. According to EC4 [49], the effective length of the beam was considered 1500 mm. To observe the crack width and length and measure them accurately, a 5 × 5 cm grid size was made on concrete slab as demonstrated in Figures 9 and 10. The relationship between crack width and maximum central deflection of the ACHPCBG-bridge is addressed in this section. This relationship relies on the evaluation of load-deflection behaviors and load-crack width behaviors of experimental model outcomes.

There are many situations where an individual wants to use a simplified model and finds a formula that best fits a given set of data. Simplifying the model is the best solution and is the process of constructing a mathematical function with the best fit to a series of data points, giving a mathematical ideal solution.

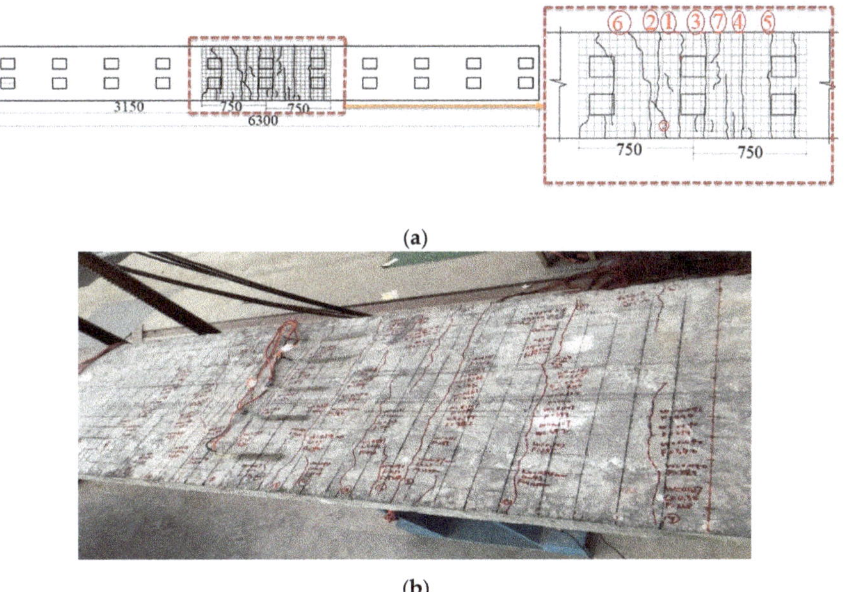

(a)

(b)

Figure 9. Formation and distribution of cracks in specimen 1: (**a**) Schematic view of cracks; (**b**) actual view of cracks.

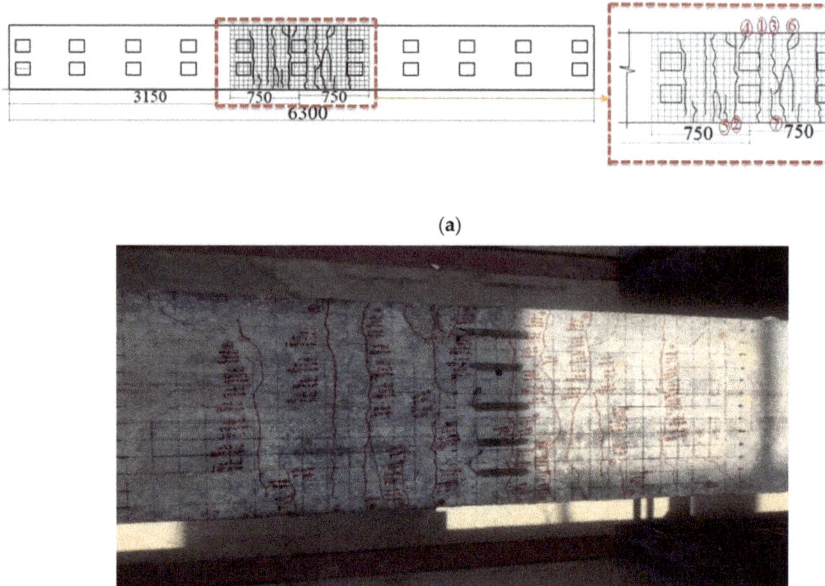

(a)

(b)

Figure 10. Formation and distribution of cracks in specimen 2: (**a**) Schematic view of cracks; (**b**) actual view of cracks.

As the first step, with experimental results a simplified formula was developed between the deflection and the load applied on the beam. It was done by fitting a first-order polynomial function to the data presented in Figure 11. The result was as follows:

$$L = 72.183d - 36.781$$

(12)

where d (m) is the deflection measured at upper yield point of elastic stage in the center of the beam and L (kN) is the intensity of the load excreted on the beam.

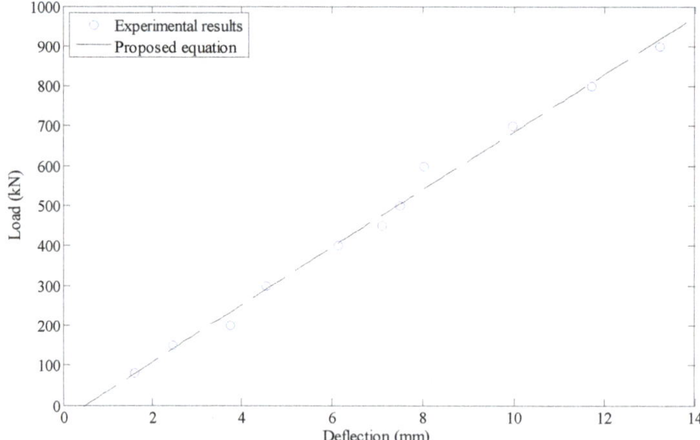

Figure 11. Fitting the load–deflection curve.

Next, using the data obtained from the experimental tests, a third-order polynomial regression model was developed between the load and crack width. It was done by fitting a third-order polynomial function to the data presented in Figure 12. The result was as follows:

$$C = 4 * 10^{-10}L^3 - 9 * 10^{-7}L^2 + 0.001L - 0.0317 \tag{13}$$

where C (mm) is the maximum width of crack observed before failure. Substituting Equation (12) in Equation (13) gives a new expression, as follows:

$$C = 1.5 * 10^{-4}d^3 - 4.9193 * 10^{-3}d^2 + 0.07707d - 0.06971 \tag{14}$$

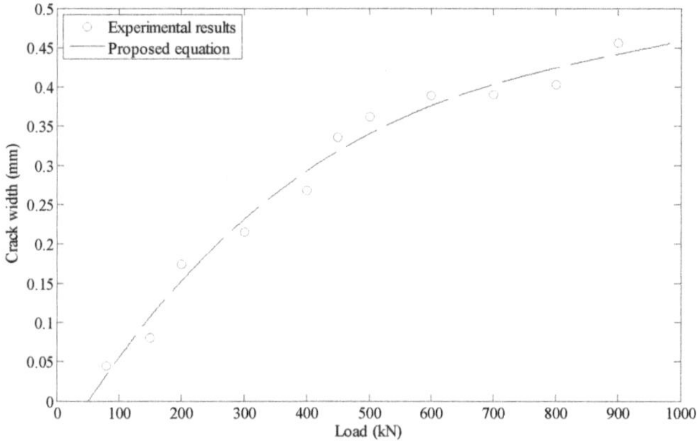

Figure 12. Fitting the load–crack width curve.

To adopt the proposed model for the prediction of crack size in steel–concrete composite small box girder, one can simply calculate the maximum deflection under the static load case. The equivalent intensity of the load can then be calculated using Equation (12). Next, Equation (13) can be utilized to approximate the maximum width of the crack generated in box girder beam. However, the limitation of the proposed model regarding the dimension and construction size should also be considered while utilizing the proposed model. Having Equation (14), the crack width is available at any desired beam deflection. The behavior of central deflection and crack width at negative bending moment region is shown in Figure 13.

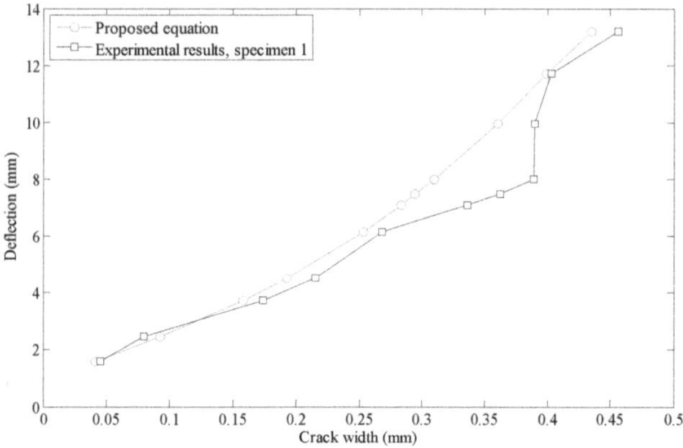

Figure 13. Crack width-deflection comparison of beam.

6. Discussion

It is noteworthy that the provided relationship is limited to the upper yield points of the elastic stage. In order to validate the results and evaluate the accuracy of the proposed formula, a laboratory model was created. The experimental model was performed under different loads, and the maximum deflection was calculated in each case. Additionally, the crack width was measured from the experimental test. The results are reported in Tables 6 and 7. The load-deflation comparisons of tested specimen are shown in Figure 14. Furthermore, the crack width was approximated by the proposed model for each load intensity and compared with the experimental results of specimen 1 and 2.

Table 6. Validation of the proposed relationship for crack width prediction of specimen 1.

Load, KN	Deflection, mm Experiment	Crack Width, mm (Experimental model)	Crack width, mm (Proposed formula)
80	1.6	0.045	0.041
150	2.456	0.08	0.092
300	4.535	0.215	0.192
450	7.102	0.336	0.283
600	8.012	0.389	0.31
800	11.71	0.403	0.399
900	13.211	0.456	0.435

Table 7. Validation of the proposed relationship for crack width prediction of specimen 2.

Load, KN	Deflection, mm Experiment	Crack width, mm (Experimental model)	Crack width, mm (Proposed formula)
80	1.2975	0.054	0.0223
150	2.456	0.094	0.0921
300	4.741	0.188	0.201
450	7.177	0.223	0.285
600	9.1455	0.265	0.338
800	11.8885	0.336	0.403
900	13.0965	0.349	0.432

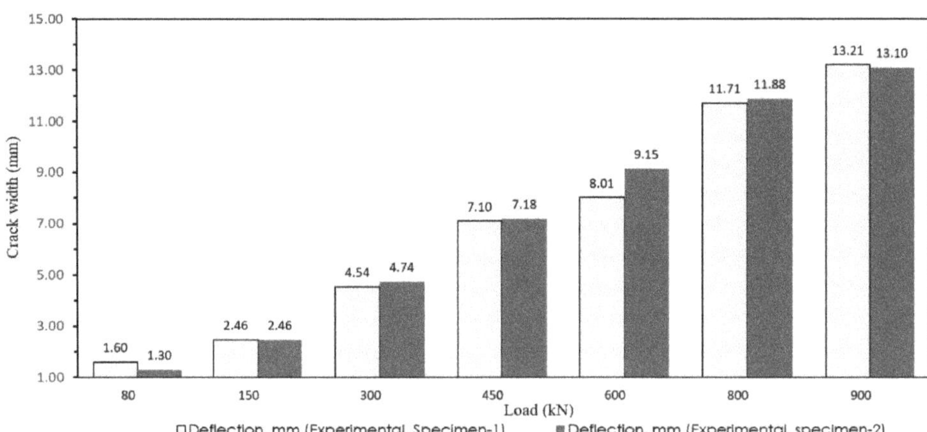

Figure 14. Load deflection comparison of tested specimens.

As observed, the crack widths obtained from the simplified model are appropriately close to the results of the experiments. To better observe the accuracy, the results of the proposed and experimental models are depicted in Figures 15 and 16. The white bars show the crack width that resulted from Equation (14) and the gray bars depict the experimental results.

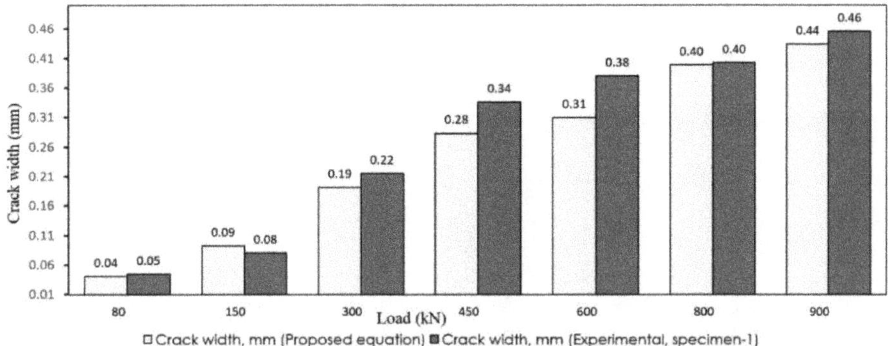

Figure 15. Comparison of crack width (specimen 1).

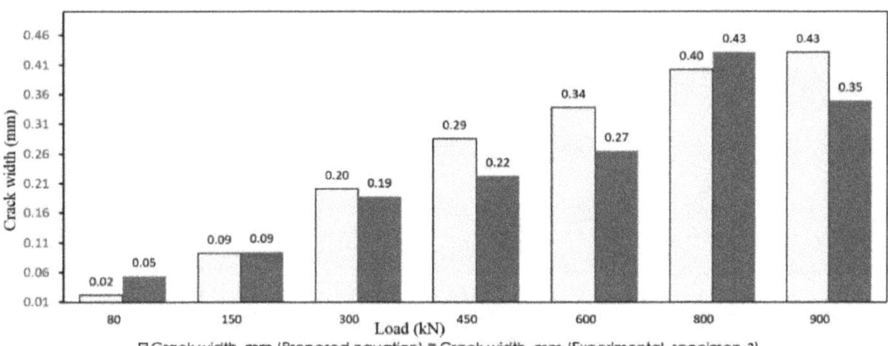

Figure 16. Comparison of crack width (specimen 2).

To further evaluate the model, a few studies were selected from the literature to compare with the proposed formula. Su et al. [54] experimentally analyzed two different types of continuous composite box girders. One specimen was a conventional composite box girder with cast in situ (specimen NCN-1) slab whereas the other was a composite box girder with a prefabricated prestressed concrete slab (specimen NCN-2). Due to using prestressed concrete, the proposed formula yielded slightly higher results than the experimental results regarding NCN-2; however, in the case of NCN-1, the results of the proposed formula were in good agreement with the experimental results as shown in Figure 17. According to their research, no great increase of the crack width was found when load increased from 312 to 700 kN; but, in this stage, there was substantial increase in the amount of cracking. This stage is the stabilization process of the crack, which means the crack distribution experiences a transition from its randomly distributed state to a quasi-uniformly distributed state. Similarly, Xu et al. [12] manufactured and tested a continuous double composite girder (DCG) to study the mechanical behavior in negative flexural region. Their comparison results are shown in Figure 18.

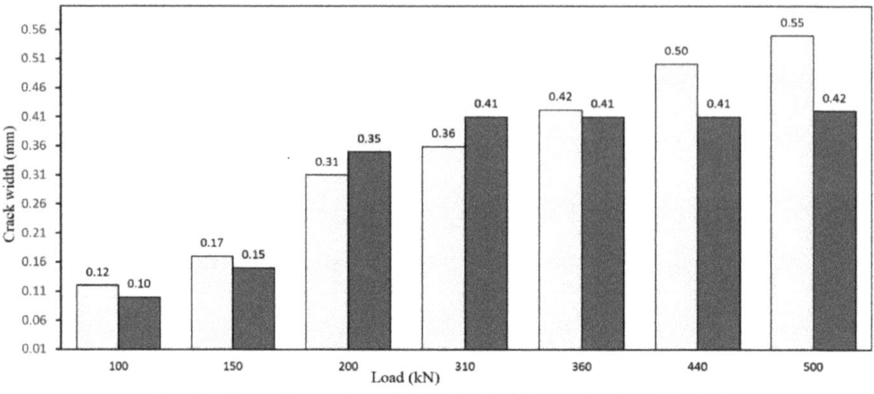

Figure 17. Comparison of crack width resulted from the proposed formula and experiments for the NCN-1 sample in Su et al.'s study [54].

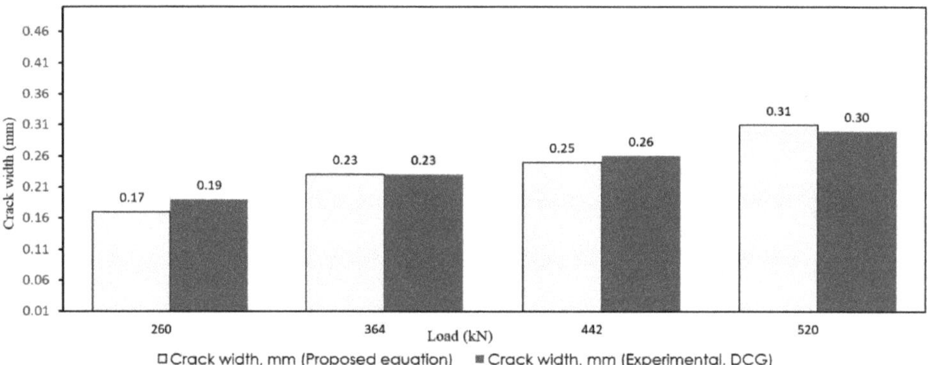

Figure 18. Comparison of crack width resulted from the proposed formula and experiments for the DCG sample in Xu et al.'s study [12].

7. Conclusions

The current study aimed at investigating the behavior of ACHPCBG-bridge utilizing experimental models. Therefore, a vertical loading was gradually applied to the beam, and the maximum deflection along the beam was observed at certain points. Additionally, the cracking mechanism was investigated by the experimental model and the maximum width of the cracks were measured by a digital crack gauge on the beam surfaces. Finally, a simplified formula was developed approximating the crack width as a function of deflection

The main conclusions drawn from the current study are as follows:

- A semi-empirical formula was developed based on experimental studies to approximate the width of crack in ACHPCBG-bridge.
- It was observed that the crack zone initiated when the load exceeded 80 kN with a crack width of 0.045 mm. The appeared crack propagated in full width, i.e., 0.7 m of the beam, when the load reached 150 kN. In this condition, the crack width was recorded as 0.08 mm.
- By the use of steel fiber in a concrete, the integrity and load resistance capacity are consequently increased. In this case, the maximum width of the crack was limited to 0.456 mm at 900 kN load.
- If construction can be done properly in a systematic manner, crack propagation would be effectively limited due to the bridging action of fibers.

Since the proposed formula is presented as an explicit function, it can be practically used to predict crack width. The proposed formula can also be used as a limit state function in reliability analysis to calculate the probability of failure for ACHPCBG-bridge. As the experiment models were designed by a 1:4 ratio based on 25 m prototype model, the application of the proposed method is limited to the models within a similar range of parameters. This issue can be investigated by the authors in future research.

Author Contributions: Conceptualization, B.G.G. and Y.-Q.X.; formal analysis, B.G.G.; data curation, B.G.G., Z.Q., and S.-H.G.; investigation, B.G.G., Y.-Q.X. and S.-H.G.; methodology, B.G.G.; project administration, Y.-Q.X. and B.G.G.; resources, Y.-Q.X. and B.G.G.; software, B.G.G.; validation, B.G.G., Y.-Q.X., S.-H.G. and Z.Q.; visualization, B.G.G.; writing—original draft, B.G.G.; writing—review and editing, B.G.G.; funding acquisition, Y.-Q.X.; supervision, Y.-Q.X.

Acknowledgments: The authors appreciate the financial support from Zhejiang University, the Cyrus Tang Foundation in China, and National Natural Science Foundation of China (NSFC, No. 51541810). Additionally, the cooperation of staff members of the structural laboratory of Quzhou University and the help from Dr. Ying Yang in the experimental research are highly appreciated.

Conflicts of Interest: The authors declare no conflict of interest. The funding sponsors had no role in the design of the study; in the collection, analyses, or interpretation of data; in the writing of the manuscript; or in the decision to publish the results.

Notations

α_t	parameter of concrete uniaxial tension stress–strain curve in the decline period
$f_{t,r}$	representation of concrete uniaxial tensile strength
$\varepsilon_{t,r}$	peak tensile strain corresponding to $f_{t,r}$
d_t	evolution parameter of concrete under uniaxial tension
α_c	parameter of concrete uniaxial compression stress–strain curve in the decline period
$f_{c,r}$	representation of concrete uniaxial compressive strength
d_c	evolution parameter of concrete under uniaxial compression
$\varepsilon_{c,r}$	peak compressive strain corresponding to $f_{c,r}$
C	crack width
d	deflection
L	applied load
Φ	diameter of rebar and stud
t	thickness of web and flange

Appendix A

The cement was made of local brand of 52.5 grade OPC, and Ganjiang river medium sand was used for fine aggregate with a fineness modulus of 2.8; well-graded basalt was used for coarse aggregate with a maximum particle size less than 16mm. The main chemical composition for bonding properties of the constituent materials are shown in Table A1. The average fineness of the slag powder, density, water reduction rate, activity index on 28d, and activity index on 56d were 3 microns, 2.40 g/cm^3, 10–15%, 105–110% on 28d, and 110–125%, respectively.

Table A1. Chemical constituents of Diatomite, SiO_2 and mineral powder (%).

Component	SiO_2	Al_2O_3	MgO	CaO	f-CaO	SO_3	MnO	Density	Loss
Diatomite	76.11	11.21	3.5	3.8	-	-	-	2.6	<1
Sub-nano SiO_2	80.11	1.49	2.69	4.64	1.98	0.65	-	-	<1.5
Micron grade Mineral powder	33.84	11.68	10.61	38.13	-	-	0.34	-	<3

Silicon Powder was made by local Building Materials Co., Ltd. according to GB/T176-1996 [55]. The test results of Silicon powder measured by GB/T18736/2002 [56] is reported in Table A2.

Table A2. Silicon powder performance index.

Item	SiO$_2$ (%)	Moisture Content (%)	Ignition Loss (%)	Water Demand Ratio (%)	Fineness (45 um) (%)	28d Activity Index
Control index	$\geq$85	$\leq$3.0	$\leq$6	$\leq$125	-	$\geq$85
Test Result	95.3	0.90	2.1	119.5	0.9	90

Steel fibers: Straight copper-plated steel fibers with a diameter of 0.2 mm, length of 13 mm, tensile strength of 2000 MPa, and volume fraction of 1.5% were used as steel fibers. Figure A1 shows the steel fibers mixed with C80 high-performance concrete. Table A3 shows the physical properties of the steel fibers utilized in the experimental test.

Table A3. Physical properties of steel fibers.

Fiber type	Length (mm)	Diameter (mm)	Aspect Ratio	Tensile Strength (MPa)	Modulus of Elasticity (GPa)	Density (Kg/m^3)
Steel fiber	13	0.2	65	2000	210	7800

Figure A1. Steel fiber in concrete.

Water reducer: Superplasticizer of a local chemical company was used. The index parameters are listed in Table A4.

Table A4. Performance index of water reducer.

Item	PH Value	Density (g/ml)	Solid Content (%)	Chloride Ion Content (%)	Alkali Content (%)	1 h Loss	Water Reduction Rate.
Control Index	6.5 ± 0.02	1.04 ± 0.02	21 ± 1	$\leq$0.2	$\leq$3.0	40	$\geq$25
Test result	6.5	1.046	21.5	0.05	0.12	23	30.4

Water: Ordinary tap water was used in the experiment.

C60 concrete and C80 concrete were tested, and each concrete was designed with three mixing ratios. Three sets of test pieces were made for each combination, and each group had three cube test pieces. After conducting the strength test, the design mix ratio was optimized. The concrete mix of C60 and C80 is shown in Table A5.

When the C60 high-performance concrete bridge deck of the test beam was prefabricated, eight groups of 150 mm cubic test blocks and two groups of 150 × 150 × 300 mm prism test blocks were made. Four groups of cube blocks with the same curing conditions of the bridge deck were tested. For two groups of cubes, compressive strength was measured when the cumulative temperature reached 600 °C. One group was tested for cube compressive strength on the test day, and the other was preserved. The remaining four groups were them used for standardization; among them, two groups were used for measuring 7-day compressive strength, one group was used for measuring 28-day compressive strength of concrete cube, and one group was used for testing splitting strength. The prismatic test block is standard for testing the 28-day axial compressive strength of concrete and elastic modulus. Three groups of 150 mm cubic test blocks and two groups of 150 × 150 × 300 mm prismatic test blocks were made when C80 steel fiber high-performance concrete was poured into the reserved holes. Two groups of cubic specimens were used to obtain the 28-day cubic strength and splitting strength of C80 steel fiber high-performance concrete and the remaining one group was preserved. At the same time, two groups of prism specimens were used to test the 28-day axial compressive strength and elastic modulus of concrete. Both cube and prism specimens were loaded by universal testing machine. The test results and elastic modulus of C60 high-performance concrete cube under standard curing and actual curing condition (i.e., the same curing condition with bridge deck) are shown in Tables A6–A8. The test results of C80 steel fiber high-performance concrete cube and elastic modulus are shown in Tables A9 and A10.

Table A5. Benchmark mixture proportion of C60 and C80 high-performance concrete.

Concrete	Water Binder Ratio	Sand Rate (%)	The Amount of Raw Mterial Used Oer Concrete							
			Cement	Slag Powder	Fine Aggregate	Coarse Aggregate	Water	Water Reducer	Silica Fume	Steel Fiber
C60 *	0.3	40	388	129	699	1049	155	6.2	-	-
C80 **	0.25	38	435	87	652	1063	157	10.44	58	87

* Cement = 0.75, slag powder = 0.25, fine aggregate = 1.35, coarse aggregate = 2.03, water = 0.3, Admixture = 0.012.
** Cement = 0.75, slag powder = 0.15, Silica powder = 0.1, fine aggregate = 1.124, coarse aggregate = 1.83, steel fiber = 0.15, water = 0.27, and water reducer = 0.018.

Table A6. Concrete strength obtained from the experimental test on standard C60 HPC cubic specimen.

Test Block Number	Standardized 7-day Cube Strength (MPa)		Standardize 28-day Cube Strength (MPa)	Splitting Strength (MPa)
Test block 1	63.4	55.9	77.8	5.63
Test block 2	55.2	54.1	69.9	5.08
Test block 3	48.9	57.9	67.2	4.56
Average value		55.9	71.6	5.09

Table A7. Concrete strength obtained from experimental test on C60 HPC cubic specimen at the same condition with bridge deck.

Test Block Number	Compressive Strength (MPa) *		Strength Measured at the Test Day (MPa)
Test block 1	71.3	64.9	77.87
Test block 2	71.4	61.3	75.37
Test block 3	64.7	69.3	75.47
Average value		67.15	76.24

* Strength measured at the same curing condition with bridge deck.

Table A8. Elastic modulus obtained from the experimental test on C60 HPC.

Item	First Group	Second Group	Third Group	Average Value
Modulus of Elasticity (GPa)	36.0	36.7	36.5	36.4

Table A9. Concrete strength obtained from the experimental test on standard C80 HPC cubic specimen.

Test Block Number	Standard Curing 28-day Strength (MPa)	Splitting Strength (MPa)
Test block 1	90.4	8.27
Test block 2	87.5	9.91
Test block 3	84.4	8.83
Average value	87.4	9

Table A10. Elastic modulus and Poisson's ratio obtained from the experimental test on C80 HPC.

Item	First Group	Second Group	Third Group	Average Value
Modulus of elasticity (GPa)	37.1	36.1	36.3	36.5
Poisson ratio	0.183	0.181	0.195	0.186

References

1. Khan, M.A. *Accelerated Bridge Construction: Best Practices and Techniques*; Elsevier/BH: Amsterdam, the Netherlands; Boston, MA, USA, 2015; pp. 3–4. ISBN 978-0-12-407224-4.
2. Ofili, M. *State of Accelerated Bridge Construction (ABC) in the United States*; Georgia Southern University, University Honors Program Thesis: Stateboro, GA, USA, 2015.
3. Xiang, Y.Q.; Zhu, S.; Zhao, Y. Research and development on Accelerated Bridge Construction Technology. *Chin. J. Highway Transport* **2018**, *31*, 1–27.
4. Jorg, S.; Hartmut, C.S. *Concrete Box-Girder Bridges*; IABSE-AIPC-IVBH, ETH-Honggerberg: Zurich, Switzerland, 1982; pp. 16–18. ISBN 3857480319.
5. Chen, Y.Y.; Dong, J.C.; Xu, T.H. Composite box girder with corrugated steel webs and trusses- A new type of bridge structure. *Eng. Struct.* **2018**, *166*, 354–362. [CrossRef]
6. Feng, M.R. Modern Bridge in China. *Struct. Infrastruct. E* **2014**, *10*, 429–442. [CrossRef]
7. Composite Box-Girder Bridges Page. Available online: http://www.civilengineeringx.com/structural-analysis/structural-steel/composite-box-girder-bridges/ (accessed on 11 January 2019).
8. Satwik, M.B.; Ashwin, K.N.; Dattatreya, J.K. Comparative study of rectangular and trapezoidal concrete box girder using finite element method. *J. Civ. Eng. Environ. Tech.* **2016**, *3*, 489–495. [CrossRef]
9. Transpiration Research Board. *Innovative Bridge Designs for Rapid Renewal: ABC Toolkit*; S2-R04-RR-2; Transpiration Research Board: Washington, DC, USA, 2013; pp. 3–5.
10. Liang, Q.Q.; Uy, B.; Bradford, M.A.; Ronagh, H.R. Strength Analysis of Steel–Concrete Composite Beams in Combined Bending and Shear. *J. Struct. Eng.* **2005**, *131*, 1593–1600. [CrossRef]
11. Sun, Q.L.; Yang, Y.; Fan, J.S.; Zhang, Y.L.; Bai, Y. Effect of longitudinal reinforcement and prestressing on stiffness of composite beams under hogging moments. *J. Construct. Steel Res.* **2014**, *100*, 1–11. [CrossRef]
12. Xu, C.; Su, Q.T.; Wu, C.; Sugiura, K. Experimental study on double composite action in the negative flexural region of two-span continuous composite box girder. *J. Construct. Steel Res.* **2011**, *67*, 1636–1648. [CrossRef]
13. Xiang, Y.Q.; He, X.Y. Short- and long-term analyses of shear lag in RC box girders considering axial equilibrium. *Struct. Eng. Mech.* **2017**, *62*, 725–737.
14. Xiang, Y.Q.; Guo, S.H.; Qiu, Z. Influence of group studs layout style on static behavior of the steel-concrete composite small box girder models. *J. Build. Struct.* **2017**, *38*, 376–383.
15. Guo, S.H. Static behavior Analysis and Experimental Investigation of Accelerated Construction High Performance Steel-Concrete Composite Small multi-box Girder Bridges. Master's Thesis, Zhejiang University, Hangzhou, China, June 2017.
16. Qiu, Z. Dynamic Behavior Analysis and Experimental Investigation of Accelerated Construction Steel-Concrete Composite Small Box Girder Bridges. Master's Thesis, Zhejiang University, Hangzhou, China, June 2018.
17. Xiang, Y.Q.; Qiu, Z.; Gautam, B.G. Analysis on Vibration Frequency of Steel-Concrete Composite Beam With External Prestressed Tendons by Considering Slip Effect. *J. Southeast Univ.* **2017**, *47*, 107–111.
18. Lin, W.W.; Yoda, T. Experimental and Numerical Study on Mechanical Behavior of Composite Girders under Hogging Moment. *Adv. Steel Construct.* **2013**, *9*, 309–333.
19. Śledziewski, K. Experimental and numerical studies of continuous composite beams taking into consideration slab cracking. *Eksploat. I Niezawodn.-Maint. Reliab.* **2016**, *18*, 578–589. [CrossRef]
20. Lin, W.W.; Yoda, T.; Taniguchi, N.; Kasano, H.; He, J. Mechanical Performance of Steel-Concrete Composite Beams Subjected to a Hogging Moment. *J. Struct. Eng.* **2014**, *140*. [CrossRef]
21. Abas, F.M.; Gilbert, R.I.; Foster, S.J.; Bradford, M.A. Strength and serviceability of continuous composite slabs with deep trapezoidal steel decking and steel fibre reinforced concrete. *J Eng. Struct.* **2013**, *49*, 866–875. [CrossRef]

22. Gholamhoseini, A.; Khanlou, A.; MacRae, G.; Scott, A.; Hicks, S.; Leon, R. An experimental study on strength and serviceability of reinforced and steel fibre reinforced concrete (SFRC) continuous composite slabs. *Eng. Struct.* **2016**, *114*, 171–180. [CrossRef]

23. Hamoda, A.; Hossain, K.M.A.; Sennah, K.; Shoukry, M.; Mahmoud, Z. Behaviour of composite high performance concrete slab on steel I-beams subjected to static hogging moment. *Eng. Struct.* **2017**, *140*, 51–65. [CrossRef]

24. Smarzewski, P. Analysis of Failure Mechanics in Hybrid Fibre-Reinforced High-Performance Concrete Deep Beams with and without Openings. *Materials* **2019**, *12*, 101. [CrossRef]

25. Ma, K.; Qi, T.; Liu, H.; Wang, H. Shear Behavior of Hybrid Fiber Reinforced Concrete Deep Beams. *Materials* **2018**, *11*, 2023. [CrossRef] [PubMed]

26. Chalioris, C.E. Analytical approach for the evaluation of minimum fibre factor required for steel fibrous concrete beams under combined shear and flexure. *Constr. Build. Mater.* **2013**, *43*, 317–336. [CrossRef]

27. Chalioris, C.E.; Karayannis, C.G. Effectiveness of the use of steel fibres on the torsional behavior of flanged concrete beams. *Cem. Concr. Compos.* **2009**, *31*, 331–341. [CrossRef]

28. Campione, G. Analytical prediction of load deflection curves of external steel fibers R/C beam–column joints under monotonic loading. *Eng. Struct.* **2015**, *83*, 86–98. [CrossRef]

29. Guerini, V.; Conforti, A.; Plizzari, G.; Kawashima, S. Influence of Steel and Macro-Synthetic Fibers on Concrete Properties. *Fibers* **2018**, *6*, 47. [CrossRef]

30. Chalioris, C.E.; Sfiri, E.F. Shear Performance of Steel Fibrous Concrete Beams. *Procedia Eng.* **2011**, *14*, 2064–2068. [CrossRef]

31. Chalioris, C.E. Steel fibrous RC beams subjected to cyclic deformations under predominant shear. *Eng. Struct.* **2013**, *49*, 104–118. [CrossRef]

32. Chalioris, C.E.; Kosmidou, P.M.K.; Papadopoulos, N.A. Investigation of a New Strengthening Technique for RC Deep Beams Using Carbon. *Fibers* **2018**, *6*, 52. [CrossRef]

33. Juaʹrez, C.; Valdez, P.; Duraʹn, A.; Sobolev, K. The diagonal tension behavior of fiber reinforced concrete beams. *Cem. Concr. Compos.* **2007**, *29*, 402–408. [CrossRef]

34. Lin, W.T.; Wu, Y.C.; Cheng, A.; Chao, S.J.; Hsu, H.M. Engineering Properties and Correlation Analysis of Fiber Cementitious Materials. *Materials* **2014**, *7*, 7423–7435. [CrossRef]

35. ACI Committee 544. *Report on the Physical Properties and Durability of Fiber-Reinforced Concrete*; ACI Manual of concrete practice (MCP), American Concrete Institute: Farmington Hills, MI, USA, 2010.

36. Mirza, F.A.; Soroushian, P. Effects of alkali-resistant glass fiber reinforcement on crack and temperature resistance of lightweight concrete. *Cem. Concr. Compos.* **2002**, *24*, 223–227. [CrossRef]

37. Ryua, H.K.; Changa, S.P.; Kimb, Y.J.; Kim, B.S. Crack control of a steel and concrete composite plate girder with prefabricated slabs under hogging moments. *Eng. Struct.* **2005**, *27*, 1613–1624. [CrossRef]

38. Muttoni, A.; Ruiz, M.F. Concrete Cracking in Tension Members and Application to Deck Slabs of Bridges. *J. Bridge Eng.* **2007**, *12*, 646–653. [CrossRef]

39. Xing, Y.; Han, Q.H.; Xu, J.; Guo, Q.; Wang, Y.H. Experimental and numerical study on static behavior of elastic concrete-steel composite beams. *J. Construct. Steel Res.* **2016**, *123*, 79–92. [CrossRef]

40. Lin, W.W.; Yoda, T.; Taniguchi, N. Application of SFRC in steel–concrete composite beams subjected to hogging moment. *Construct. Steel Res.* **2014**, *101*, 175–183. [CrossRef]

41. Ge, W.J.; Ashour, A.F.; Yu, J.; Gao, P.; Cao, D.F.; Cai, C.; Ji, X. Flexural Behavior of ECC–Concrete Hybrid Composite Beams Reinforced with FRP and Steel Bars. *J. Compos. Constr.* **2019**, *23*, 04018069. [CrossRef]

42. Shamass, R.; Cashell, K.A. Analysis of stainless steel-concrete composite beams. *J. Constr. Steel Res.* **2019**, *152*, 132–142. [CrossRef]

43. Xu, L.Y.; Nie, X.; Tao, M.X. Rational modeling for cracking behavior of RC slabs in composite beams subjected to a hogging moment. *Constr. Build. Mater.* **2018**, *192*, 357–365. [CrossRef]

44. Mahmoud, A.M. Finite element modeling of steel concrete beam considering double composite action. *Ain Shams Eng. J.* **2016**, *7*, 73–88. [CrossRef]

45. Samaaneh, M.A.; Sharif, A.M.; Baluch, M.H.; Azad, A.K. Numerical investigation of continuous composite girders strengthened with CFRP. *Steel Composite Struct.* **2016**, *21*, 1307–1325. [CrossRef]

46. *Code for Design of Steel and Concrete Composite Bridges*; GB50917-2013; China Planning Press: Beijing, China, 2014.

47. *Code for Design of Steel Structures*; GB 50017-2017; China Building Industry Press: Beijing, China, 2018.

48. *AASHTO LRFD Bridge Design Specifications*, 4th ed.; American Association of State Highway and Transportation Officials: Washington, DC, USA, 2007.
49. *Design of Composite Steel and Concrete Structures—Part 1-1: General Rules and Rules for Building*; Eurocode 4, EN, 1994-1-1; European Publication: London, UK, 2004; Available online: https://www.phd.eng.br/wp-content/uploads/2015/12/en.1994.1.1.2004.pdf (accessed on 15 March 2019).
50. ABAQUS. *Theory Manual Version 6.14*; Dassault System: Johnston, RI, USA, 2014.
51. Prakash, A.; Anandavalli, N.; Madheswaran, C.K.; Rajasankar, J.; Lakshmanan, N. Three Dimensional FE Model of Stud Connected Steel-Concrete Composite Girders Subjected to Monotonic Loading. *Int. J. Mech. Appl.* **2011**, *1*, 1–11. [CrossRef]
52. Tahmasebinia, F.; Ranzi, G.; Zona, A. Beam Tests of Composite Steel-concrete Members: A three-dimensional Finite Element Model. *Int. J. Steel Struct.* **2012**, *12*, 37–45. [CrossRef]
53. *Code for Design of Concrete Structures*; GB50010-2010; Chemical Industry Press: Beijing, China, 2015.
54. Su, Q.T.; Yang, G.T.; Bradford, M.A. Behavior of a Continuous Composite Box Girder with a Prefabricated Prestressed-Concrete Slab in Its Hogging-Moment Region. *J. Bridge Eng.* **2015**, *20*. [CrossRef]
55. *Method for Chemical Analysis of Cement*; GB/T 176-1996; National Standardization Technical Committee for Cement Products: Suzhou, China, 1996.
56. *Mineral Admixtures for High Strength and High Performance Concrete*; GB/T 18736/2002; National Standardization Technical Committee for Cement Products: Suzhou, China, 2002.

 *materials*

Article

Progressive Failure Simulation of Notched Tensile Specimen for Triaxially-Braided Composites

Zhenqiang Zhao [1,2,3], Haoyuan Dang [1,2,3], Jun Xing [1,4], Xi Li [1,5], Chao Zhang [1,2,3,*], Wieslaw K. Binienda [6] and Yulong Li [1,2,3]

[1] Department of Aeronautical Structure Engineering, Northwestern Polytechnical University, Youyi West Road 127#, Xi'an 710072, Shaanxi, China; zhaozhenqiang@mail.nwpu.edu.cn (Z.Z.); haoyuandang@mail.nwpu.edu.cn (H.D.); xingjun_acc@caac.gov.cn (J.X.); xili@mail.nwpu.edu.cn (X.L.); liyulong@nwpu.edu.cn (Y.L.)

[2] Shaanxi Key Laboratory of Impact Dynamics and Its Engineering Applications, Youyi West Road 127#, Xi'an 710072, Shaanxi, China

[3] Joint International Research Laboratory of Impact Dynamics and Its Engineering Applications, Youyi West Road 127#, Xi'an 710072, Shaanxi, China

[4] Airworthiness Certification Center, Civil Aviation Administration of China, Beijing 100102, China

[5] Faculty of Aerospace Engineering, Delft University of Technology, 2629 Delft, The Netherlands

[6] Department of Civil Engineering, The University of Akron, Akron, OH 44325, USA; wbinienda@uakron.edu

* Correspondence: chaozhang@nwpu.edu.cn; Tel.: +86-29-88460786

Received: 15 February 2019; Accepted: 6 March 2019; Published: 12 March 2019

Abstract: The mechanical characterization of textile composites is a challenging task, due to their nonuniform deformation and complicated failure phenomena. This article introduces a three-dimensional mesoscale finite element model to investigate the progressive damage behavior of a notched single-layer triaxially-braided composite subjected to axial tension. The damage initiation and propagation in fiber bundles are simulated using three-dimensional failure criteria and damage evolution law. A traction–separation law has been applied to predict the interfacial damage of fiber bundles. The proposed model is correlated and validated by the experimentally measured full field strain distributions and effective strength of the notched specimen. The progressive damage behavior of the fiber bundles is studied by examining the damage and stress contours at different loading stages. Parametric numerical studies are conducted to explore the role of modeling parameters and geometric characteristics on the internal damage behavior and global measured properties of the notched specimen. Moreover, the correlations of damage behavior, global stress–strain response, and the efficiency of the notched specimen are discussed in detail. The results of this paper deliver a throughout understanding of the damage behavior of braided composites and can help the specimen design of textile composites.

Keywords: braided composites; mesoscale model; notched specimen; damage evolution

---

1. Introduction

Carbon fiber reinforced composite materials has some distinctive features in physical, mechanical, and thermal properties, such as high stiffness and strength to weight ratio, excellent resistance to fatigue, and corrosion. Traditional carbon fiber composite structures have layers of unidirectional fiber lamina and each layer can have a different direction of fiber lay-up, which is able to produce desired specific mechanical properties. However, a weak interlaminar plane where damage can initiate and cause delamination as in case of foreign object impact has limited the use of fiber composites in a variety of structures [1,2].

Textile composites such as braided or woven composites are known to have excellent damage tolerance and impact resistance and are increasingly used in aircraft structures [3]. For example, the two-dimensional triaxially-braided composite is introduced to fabricate the engine fan case structure, which is mainly designed to contain the fan blade and its fragments during a blade failure event. Apart from its superior impact resistance property [4], the two-dimensional triaxially-braided composite also shows excellent specific energy absorption property and is considered as an alternative material system for front rail structures of vehicles [5,6]. Two-dimensional triaxially-braided fabrics are made by three distinct sets of yarns, which are intertwined to form a single layer of fabrics. Figure 1c shows the architecture of a typical $0°/\pm 60°$ braided fabric, bias fiber bundles undulate over and under each alternatively, while $0°$ yarns are straight and define the axial direction of the composite. The rectangle in Figure 1c indicates the size of a unit cell, which is considered as the smallest repeating element of a composite that can represent the composite's geometric features in particular and its mechanical response as a whole. The length of a unit cell is the axial distance between center lines of two neighboring bias yarns, and the width is twice the transverse distance between the center lines of two neighboring axial yarns.

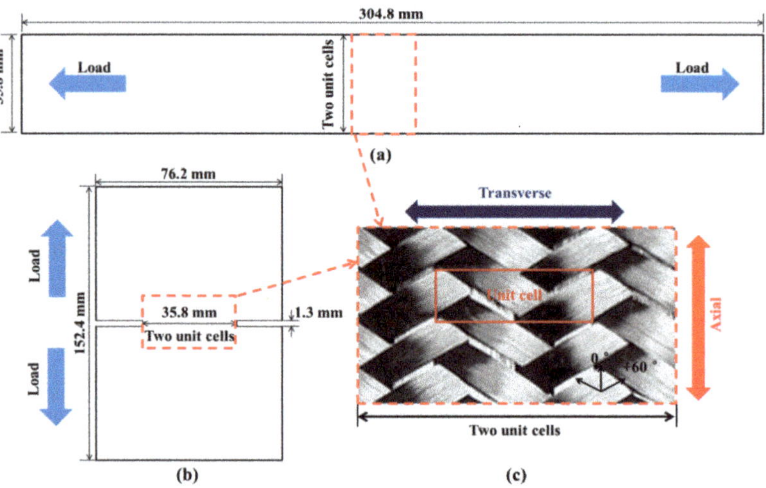

Figure 1. (**a**) Dimensions of the straight-side coupon. (**b**) Dimensions of double edge notch specimen. (**c**) Representative architecture of triaxially-braided composite.

Due to the more complicated mesoscopic structure, the complexity of deformation and damage process for textile composites is greatly increased compared to that of laminates. Thus, the determination of mechanical properties for textile composites has drawn a lot of attention and raised significant challenges on the experiment techniques [7–9]. This paper focuses mainly on the tension failure behavior of triaxially-braided composite and investigates specifically the progressive failure process of a notched tensile specimen.

Waas and coworkers [7,10,11] studied extensively the compressive properties of a $0°/\pm 45°$ triaxially-braided composite using experimental, analytical, and numerical approaches. Goldberg et al. [12] identified that a $0°/\pm 60°$ braided composite offers improved impact resistance because of its quasi-isotropic nature (properties are balanced in all directions). Littell [13] and Kohlman et al. [14] studied experimentally the mechanical performance of a $0°/\pm 60°$ triaxially-braided composite using different kinds of experimental methods. Littell [13] conducted comprehensive tests to measure the quasi-static responses of triaxially-braided composites, including tension, compression, and shear. Littell's results led to the conclusion that there were different damage mechanisms affecting the material response, including inherent damage accumulations (fiber bundle cracking and interface delamination) and geometry-induced premature failure behaviors (free-edge effect induced edge delamination). The presence of premature edge damage behavior in the standard straight-sided coupon specimen results in lower measured mechanical properties of the material.

For composite materials, it is difficult to avoid the possible premature failure caused by interlaminar stress concentration at the free edges of the specimen, which is more thought-provoking to accurately test the textile or braided composites. One major limitation is the local variation of properties for the fabrics since the methods for calculating lamina properties rely on the assumption of homogeneous strain and stress distribution in a uniaxial specimen [14]. For the triaxially-braided composite, the internal damage and its propagation depend significantly on the mesoscopic architecture of the material; the initiation of new damage will cause redistribution of internal loads, resulting in an inhomogeneous stress state. Through a combined experimental and numerical approach, Zhang et al. [15] investigated the mechanism of free-edge effect and the size-dependent mechanical properties of triaxially-braided composites. It was identified that the free-edge effect is an elastic behavior resulting from the termination of bias fiber bundles and affecting continuously the material response. Kueh et al. [16] identified the relationship of effective elastic properties of triaxially-braided composite against specimen size using an analytical approach.

To examine the realistic effective strength properties of the triaxially-braided composite, Kohlman et al. [14] designed several kinds of improved specimens to measure the mechanical properties of $0°/\pm 60°$ triaxially-braided composite, including both tube and notch geometries. The results further prove the sensitivity of measured properties to specimen shape and the significance of free-edge effect in triaxially-braided composites. It was also concluded that the notched coupon specimen produces higher measured strength values because of the enforced tensile failure of fiber bundles at the notched gauge section. Compared with the straight-sided coupon specimen, the damage behavior of notched specimens is more complicated, due to the presence of stress concentration in the notched zone. Thus, it is necessary to develop representative numerical models to analyze and elucidate the progressive failure behavior of notched tensile specimen. Using a numerical model as a virtual testing tool of composites can also provide insights in revealing the localized mechanical response and exploring damage mechanism at meso and microscopic scale, which can then facilitate the development of experimental techniques.

Mesoscale finite element (FE) is known for its capability in predicting the local response and damage events of textile composite [17–19]. Lomov et al. [20] conducted a comprehensive study on the mesoscale finite element modeling approach of textile composites. Especially for triaxially-braided composites, Zhang et al. [15,18] established a mesoscale finite element framework, with emphasize on imposing representative loading/boundary conditions against an experimental set-up; Zhao [21] utilizes the mesoscale FE model to study intensively the failure behavior under transverse tension and compression, and its damage behavior under high-speed impact has been exactly captured by proposing a multiscale modeling framework based on a fully validated mesoscale FE model [22]. Apart from these, the fracture process of triaxially-braided composite for straight-sided coupon specimens also can be simulated by means of the mesoscale FE model [23,24].

However, there is no reported work applying the mesoscale FE model to the analysis of specimens with more complicated shapes, e.g., notched specimen, tube specimen, and specimen with hole. This limits the confidence of the community on the feasibility of meso-FE model for virtual testing. On the other hand, the presence of challenges in characterizing the mechanical properties of 2DTBC requires further efforts in investigating the failure mechanism and optimizing the test specimens. Thus, in this work, the mesoscale finite element method with three-dimensional damage model is introduced to investigate the progressive failure behavior of notched specimen of the triaxially-braided composite under axial tension. The presented model intends to simulate the damage initiation, damage propagation, and ultimate fracture of the notched specimen, as well as to predict the effective strength of the triaxially-braided composite. The results demonstrate the accessibility of using mesoscale finite element model as virtual testing for textile composites, which can significantly enhance the design efficiency of composite structures. This research paper firstly describes the material system and experimental details followed by the progressive damage model of the composite (which consists of damage initiation criteria and its subsequent evolution). Then, the mesoscale finite element model is introduced. The Section 5 of this paper examines the capability of the mesoscale model through correlation with experiments conducted by Kohlman et al. [14] and presents the predicted results of local initiation and progression of damage. Additionally, the parameters study and geometric characteristic analysis of notched specimen are also discussed in this section. The conclusions are listed in the last section of this paper.

2. Materials and Experiment

The $0°/\pm 60°$ triaxially-braided composite studied in this paper was fabricated with Toray 24 K T700 s axial tows and Toray 12 K T700 s bias tows. Epon's 862 epoxy resin was chosen as matrix material, which is a thermoset resin with low viscosity. The composite panels were processed through resin transfer molding (RTM). Table 1 presents the properties of each component, which are obtained from Littell [13].

Table 1. Properties of composite components.

Property	Fiber	Matrix
Material type	T700 s	E862 epoxy
Density (g/cm^3)	1.8	1.2
Axial modulus (MPa)	230,000	2700
Transverse modulus (MPa)	15,000	2700
Shear modulus (MPa)	27,000	1000
Tensile strength (MPa)	4900	61
Poisson's ratio	0.2	0.363

The sample considered in the present study is a single-layer panel (a composite containing only one braided ply through thickness) with double edge notches (shown in Figure 1b), which was designed and tested by Kohlman [14] to address deficiencies of straight-side tension coupons. The thickness of the single-layer specimen is 0.65 mm. Other dimensions of the notched and straight-sided coupon specimens are shown in Figure 1. A diamond saw was used to cut the notches and kept the same width of gauge region to compare with the test results of straight-sided coupon specimens. Tensile tests were performed using a servohydraulic tension/torsion test frame capable of loading to 220KN (MTS Systems Corporation, Eden Prairie, MN, USA). The specimens were stretched under displacement-controlled load until fracture of the specimen. 3D digital image correlation (GOM, Braunschweig, Germany) technique was used to obtain the full field displacement and strain data on the surface of the notched specimens.

3. Progressive Damage Model of Braided Composite

The fiber bundle of textile composites is generally considered as a transversely isotropic unidirectional lamina in numerical simulation [25–27]. In this part, a progressive damage model of the unidirectional lamina is formulated in terms of damage initiation and damage evolution. This research aims to investigate the internal damage initiation and propagation of a notched specimen, therefore material failure (element deletion) is not introduced in this damage model.

3.1. Damage Initiation

A three-dimensional failure criterion for the fiber bundle was adopted based on Hashin's [28] and Hou's [29] criteria and was incorporated with continuum damage laws. Four distinct failure modes are considered: fiber tensile failure, fiber compression failure, matrix tension failure, and matrix compression failure. The damage initiation criteria are formulated below.

For fiber tension failure ($\sigma_{11} > 0$),

$$f_{1t} = \left(\frac{\sigma_{11}}{S_{1t}}\right)^2 + \alpha\left(\frac{\sigma_{12} + \sigma_{31}}{S_{12}}\right) \geq 1 \tag{1}$$

For fiber compression failure ($\sigma_{11} < 0$),

$$f_{1c} = \left(\frac{\sigma_{11}}{S_{1c}}\right)^2 \geq 1 \tag{2}$$

For matrix tension failure ($\sigma_{22} > 0$),

$$f_{2t} = \left(\frac{\sigma_{22}}{S_{2t}}\right)^2 + \left(\frac{\sigma_{12}}{S_{12}}\right)^2 + \left(\frac{\sigma_{23}}{S_{23}}\right)^2 \geq 1 \tag{3}$$

For matrix compression failure ($\sigma_{22} < 0$),

$$f_{2c} = \frac{1}{4}\left(\frac{-\sigma_{22}}{S_{12}}\right)^2 + \frac{\sigma_{22}}{S_{2c}}\left(\left(\frac{S_{2c}}{2S_{12}}\right)^2 - 1\right) + \left(\frac{\sigma_{12}}{S_{23}}\right)^2 \geq 1 \tag{4}$$

where f_{1t}, f_{1c}, f_{2t}, and f_{2c} are failure indices corresponding to each damage mode, respectively. The first subscripts, 1, 2, and 3, indicate the fiber axial direction, in-plane transverse direction, and out-of-plane direction, respectively. When the stress state of an element make one of the four failure indices larger than 1, the corresponding damage mode will be initiated in this element and constitutive law entering the stage of damage evolution. S_{1t}, S_{1c}, S_{2t}, S_{2c}, S_{12}, and S_{23} are axial tensile strength, axial compressive strength, transverse tensile strength, transverse compressive strength, longitudinal shear strength, and transverse shear strength of the fiber bundle. α is the shear failure coefficient which plays an important role in failure prediction of textile composite. The previous research conducted by Zhang et al. [18] indicated that the coefficient α has an obvious impact on the global stress–strain response and mainly on the failure prediction. The value of α was determined to be 0.06 for fiber bundles of triaxially-braided composite through correlation with an experimental ultimate strength of straight-sided coupon specimens [18,22]. The same value of parameter α is adopted in the present study in consideration that the studied materials are totally the same.

3.2. Damage Evolution

For the damage evolution behavior, the Murakami–Ohno [30] damage theory is adopted to predict the post-peak softening, and the crack band model developed by Bazant and Oh [30] is also employed to mitigate the mesh size dependency of the proposed mesoscale model in this study. A characteristic element length is introduced into damage evolution expression, aiming to dissipate the constant energy release rate per unit area in the solid element [31,32], and the element dissipated energy can be expressed as

$$G_{f,I} = \frac{1}{2}\sigma_{eq}^f \varepsilon_{eq}^f l_c \tag{5}$$

where l_c is the characteristic length of element, which calculates by extracting the cubic root of the volume of each element; $G_{f,I}$ is fracture energy of fiber bundles corresponding to the specific damage mode I. The values for the fracture energies of axial and bias fiber bundles used in this study (listed in Table 2) were cited from Li et al. [17]. σ_{eq}^f and ε_{eq}^f are the equivalent peak stress and equivalent failure strain, respectively. The evolution of each damage variable is governed by an equivalent displacement expressed by the following equation.

$$d_I = \frac{\delta^f_{I,eq}\left(\delta_{I,eq} - \delta^0_{I,eq}\right)}{\delta_{I,eq}\left(\delta^f_{I,eq} - \delta^0_{I,eq}\right)}, \quad I = ft,\ fc,\ mt,\ mc \tag{6}$$

where $\delta^f_{I,eq}$ is the fully damaged equivalent displacement of the corresponding failure mode and $\delta^0_{I,eq}$ is the equivalent displacement at which the failure criterion is satisfied. For a certain failure mode, the equivalent displacement used in the initiation criteria is expressed in terms of the components corresponding to the effective stress components. Detailed algorithm equations of equivalent displacement and stress for each failure mode can be found in Zhang and coworkers' work [18]. As material parameters of fiber bundles, $\delta^0_{I,eq}$ and $\delta^f_{I,eq}$ be computed by the following equations.

$$\delta^f_{I,eq} = \frac{2G_f}{\sigma^0_{I,eq}} \tag{7}$$

$$\delta^0_{I,eq} = \frac{\delta_{I,eq}}{\sqrt{f_I}} \tag{8}$$

Table 2. Fracture energies of the fiber bundle.

G_{ft} (mJ/mm²)	G_{fc} (mJ/mm²)	G_{mt} (mJ/mm²)	G_{mc} (mJ/mm²)
12.5	12.5	1	1

Here, $\sigma^0_{I,eq}$ denotes the equivalence stress when each kind of damage criteria is satisfied. Meanwhile, the value of f_I can be obtained from Equations (1)–(4).

To simulate the softening process of damage element, a second-order symmetric tensor is used to describe the damage state. The corresponding damaged compliance matrix $S(d)$ is obtained as Equation (9), and the damaged stiffness matrix $C(d)$ is the inverse of $S(d)$.

$$S(d) = \begin{bmatrix} \frac{1}{d_f E_{11}} & -\frac{\nu_{21}}{E_{22}} & -\frac{\nu_{31}}{E_{33}} & & & zero \\ -\frac{\nu_{12}}{E_{11}} & \frac{1}{d_m E_{22}} & -\frac{\nu_{32}}{E_{33}} & & & \\ -\frac{\nu_{13}}{E_{11}} & -\frac{\nu_{23}}{E_{22}} & \frac{1}{E_{33}} & & & \\ & & & \frac{1}{d_f d_m G_{12}} & & \\ & & & & \frac{1}{d_m G_{23}} & \\ & zero & & & & \frac{1}{d_f G_{13}} \end{bmatrix} \tag{9}$$

where d_f and d_m are global damage variables associated with fiber and matrix failure, which are introduced to control the degree of stiffness degeneration of damaged elements, and also satisfy the following equations, respectively.

$$d_f = 1 - \min\left\{\max\left(d_{ft}, d_{fc}\right), \gamma_f\right\} \tag{10}$$

$$d_m = 1 - \min\{\max(d_{mt}, d_{mc}), \gamma_m\} \tag{11}$$

Here γ_f and γ_m are defined as the damage thresholds of global fiber and matrix damage, respectively. Numerically, the nonzero constants γ_f and γ_m address the singularity issue, and physically represent the effective resultant resistance of the homogeneous damaged elements. The effect of γ_f and γ_m on the global stress–strain responses will be discussed in a later section. Also, the Duvaut and Lions regularization model [33] is applied to promote the numerical computation and smooth the stiffness degradation process.

3.3. Cohesive Element Model for Interface

Interface is the bridge between the fiber bundle and matrix, which determines how stresses are transferred. The damage status of interface influences significantly the damage initiation and propagation of the composite material [34]. Littell [13] indicates that failure in the transverse tests was a result of edge delamination which occurred quickly and propagated along the bias fibers; and in the axial tensile direction, the subsurface delamination caused the nonlinearities in the global stress–strain response curves. In this study, the tow-to-tow interface is simulated by using the cohesive zone modeling approach, which has been embedded into ABAQUS as an optional element type. The responses of cohesive elements are governed by a typical bilinear traction–separation law, and a quadratic nominal stress criterion is used to describe interfacial damage initiation [32,34,35]. Besides, a power law criterion is adopted, which claims that failure under mixed-mode conditions is governed by a second-order power law interacting of the energies required to cause failure in the individual (normal and two shear) modes. The quadratic nominal stress criterion for damage initiation and a power law criterion for failure are represented in Equations (12) and (13).

$$\left(\frac{\langle t_n \rangle}{t_n^0} \right)^2 + \left(\frac{\langle t_s \rangle}{t_s^0} \right)^2 + \left(\frac{\langle t_t \rangle}{t_t^0} \right)^2 = 1 \tag{12}$$

$$\left(\frac{G_n}{G_n^c} \right)^2 + \left(\frac{G_s}{G_s^c} \right)^2 + \left(\frac{G_t}{G_t^c} \right)^2 = 1 \tag{13}$$

where t_n denotes the traction normal stress, and t_s and t_t denote shear stresses. The Macaulay brackets are used to signify that a pure compressive deformation or stress state does not initiate damage. t_n^0, t_s^0, and t_t^0 represent the interface strength in normal and two shear directions. Similarly, G_n, G_s, and G_t refer to the work done by the traction and its conjugate relative displacement in the normal, first, and second shear directions, respectively; and G_n^c, G_s^c, and G_t^c are critical fracture energies required to cause failure in each of the three directions. Table 3 presents the interface properties, and the values of interface strengths and fracture toughness, which are cited from Zhang et al. [18]. Detailed formulations of the mixed-mode cohesive zone model can be found in the ABAQUS user's manual.

Table 3. Strengths and fracture toughness of cohesive elements.

t_n^0 (MPa)	t_s^0 (MPa)	t_t^0 (MPa)	G_n^0 (mJ/mm^2)	G_n^0 (mJ/mm^2)	G_n^0 (mJ/mm^2)
122	136	136	0.268	1.45	1.45

4. Finite Element Model Development

The mesoscale finite element (FE) model simulates explicitly the fiber bundles of a braided composite structure and defines locally the realistic local fiber volume ratios and bundle orientations of the impregnated bundles. The advantage of the mesoscale model is its ability to analyze the local damage and failure of each component implemented individually through specific failure models for the various constituents and to predict the response of each constituent and their contribution to the global behavior. In this work, a mesoscale finite element model is introduced to study the internal damage and failure mechanism of the notched tensile specimen for $0°/\pm 60°$ triaxially-braided composite.

Based on the geometric parameters identified by Zhang et al. [18], a FE model of a single unit cell, which can represent the composite's geometric features, was generated using 8-node solid element. As shown in Figure 2a, the mesh of a unit cell was constructed through TexGEN software by keying the dimensions of the unit cell and fiber bundles. In Figure 2, the axial fiber bundle, $+60°$ and $-60°$

bias fiber bundles and matrix elements, which fill the space between fiber bundles to form plate, are represented by light blue, dark blue, yellow, and green colors, respectively. Cohesive element layers (colored red), which have the same in-plane size as brick elements but zero thickness, are manually inserted between each two fiber bundles, and fiber bundle and matrix, see Figure 2a. It should be pointed out that the pure matrix is modeled as an elastic perfectly-plastic material. It is assumed that the pure matrix elements will not fail before the fracture of the specimen, due to the much larger failure strain of matrix than that of the fiber bundle. The resultant mesh may have different fiber bundle volume values than the real material, so the fiber volume ratio in each fiber bundle is adjusted to match the real fiber volume content. As identified by Zhang et al. [18], the realistic fiber volume ratios for axial and bias fiber bundles are 77% and 74.5%, respectively. The resultant fiber volume ratio in the present FE model is 86% for the axial fiber bundles and 69% for the bias fiber bundles. Mechanical properties of fiber bundles are listed in Table 4 referring to Zhang's work [18] for the same materials.

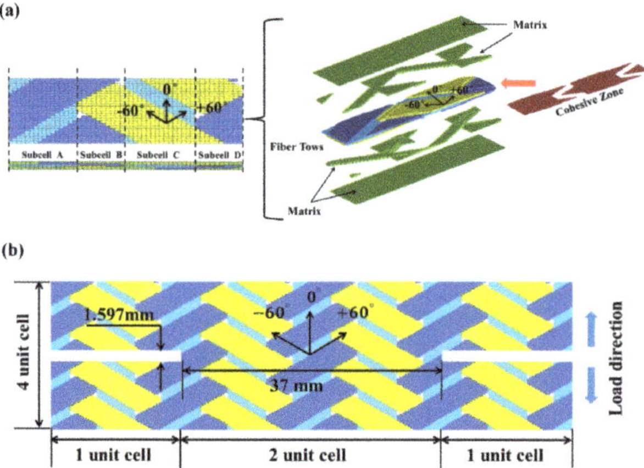

Figure 2. (a) Composition of a unit cell finite element model. (b) Finite element mesh of axial tension notched model.

Table 4. Mechanical properties of axial and bias fiber tows.

	Axial Fiber Tows	Bias Fiber Tows
Fiber volume fraction V_f	86%	69%
E_{11} (GPa)	198.18	159.54
$E_{22} = E_{33}$ (GPa)	11.22	8.30
$G_{12} = G_{13}$ (GPa)	8.58	4.48
G_{23} (GPa)	3.71	2.71
$v_{12} = v_{13}$	0.29	0.30
v_{23}	0.51	0.53
S_{1t} (MPa)	4222	3398.8
S_{1c} (MPa)	1478.48	1386.21
$S_{2t} = S_{3t}$ (MPa)	49.87	49.70
$S_{2c} = S_{3c}$ (MPa)	122.80	124.64
$S_{12} = S_{13} = S_{23}$ (MPa)	80.60	78.53

Figure 2b shows the detailed dimensions of the FE model for the notched specimen. The FE model of the notched tension specimen consists of 208,000 linear brick elements (C3D8R) and 29,412 eight-node three-dimensional cohesive elements (COH3D8), with four unit cells through the width and length direction, respectively. Two whole unit cells (37 mm) are assembled through the width direction of the notched gauge for the axial tension model, which is consistent with the experimental specimen. The length of the finite element model is 20.68 mm with four unit cells for axial tension, which are shorter than the gauge length of experimental specimens fabricated by Kohlman [14] (Figure 1b). These reasonable simplifications of model size are intended to reduce the computational time, in consideration of the minor effect of the remote sections. The numerical models are solved by ABAQUS Explicit using a 24-core workstation and each job costs ~20 h. By evaluating the various energies generated during the computation process, the accumulated kinetic energy was always less than 1% of the internal energy of the model, which ensures a quasi-static loading status of the problem.

5. Results and Discussion

In this section, the damage initiation and propagation behavior of the notched tension specimen modeled using the proposed mesoscale FE scheme is correlated with experimental results. The effect of specimen geometry on the internal damage evolution behavior and the effective strength of this notched specimen are further discussed through numerical parametric studies. These results show the applicability and reliability of the mesoscale FE model for failure study of braided composites.

5.1. Model Correlation

In our previous works [15,18,21,22], the failure behavior of straight-side coupon specimens of 2DTBC has been intensively studied. Figure 3 shows the comparison of the experimental and numerical predicted results for the single-layer straight-side coupon specimen under axial tension. The results indicate that the mesoscale FE model can not only predict well the global stress–strain curve, but also the strain distributions which are highly sensitive to the local braided architecture.

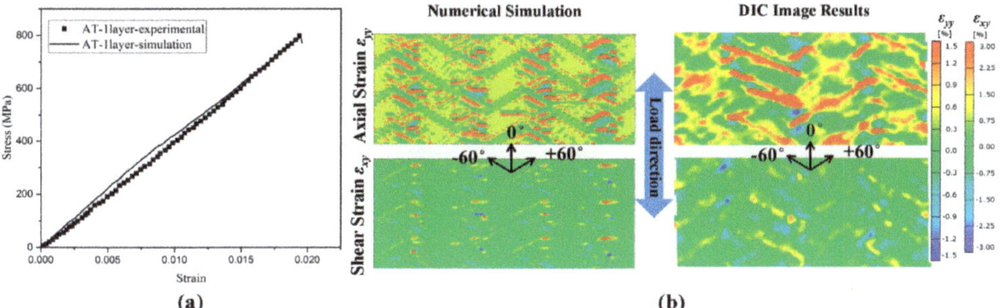

Figure 3. Comparison of the experimental and numerical predicted results for straight-side coupon specimen. (**a**) Global stress–strain curves. (**b**) Distribution of the axial and in-plane shear strain at global strain level of 2.0%.

The applicability of the mesoscale FE model is further demonstrated through the model validation for the notched specimen. For both the experimental characterization and numerical simulations, the effective strength of the tensile specimen is determined as the ratio of total reaction force at the loading section against the cross-section area at the gauge section (two unit cells wide for both notched and straight-sided coupon specimen). Table 5 compares the numerical predicted and experimental measured effective strength of triaxially-braided composite under axial tension, for both straight-sided coupon and notched specimens. The measured strength values of the straight-sided coupon and notched specimen are referred to as Kohlman [14]. It is evident from Table 5 that the mesoscale FE model predicted effective strength values are in good agreement with the experimental values, suggesting the accuracy of this modeling scheme. The effective strength of the notched specimen tends to be lower than the straight-sided coupon specimen, which is due to the presence of stress concentration at the notched section.

Table 5. Comparison of simulation predicted and experimental measured the effective tensile strength of the triaxially-braided composite.

Type of Specimen	Ultimate Strength	
	Simulation (MPa)	Experiment (MPa)
Straight-side coupon	800	814 ± 30 *
Double-notched	751	765 *

* Refer to Kohlman et al. [14].

To further demonstrate the reliability of the meso-FE model, the numerically predicted strain distributions are compared with DIC results. As seen in Figure 4, the simulation results match well with the local strain distribution of the notched specimen and additionally capturing the local strain concentration around the notch area and the diagonal distribution along the bias fiber bundles. The axial strain (ε_{yy}) and shear strain (ε_{xy}) distributions show higher strain at the pure matrix zone between bias fiber bundles, which coincides with the damage propagation paths of fiber bundle damage. The damage of fiber bundles will induce the decrease of fiber's capability in carrying the load and as a result, more load will be transferred to the matrix material, thereby causing the strain concentration in this area (red and blue in Figure 4). On the other hand, the shear strain shows more obvious strain concentration effect at the notched section. Unlike the traditional isotropic material where strain concentration effect propagates in a circular shape, the shear strain concentration effect of this braided composite propagates along the fiber bundle directions. The mesoscale FE model simulations capture the strain distribution and its magnitude accurately for the axial tension of the notched specimen. The result also reflects the advantages of mesoscale finite element model in studying the internal damage of textile composite that is difficult to achieve from experiments.

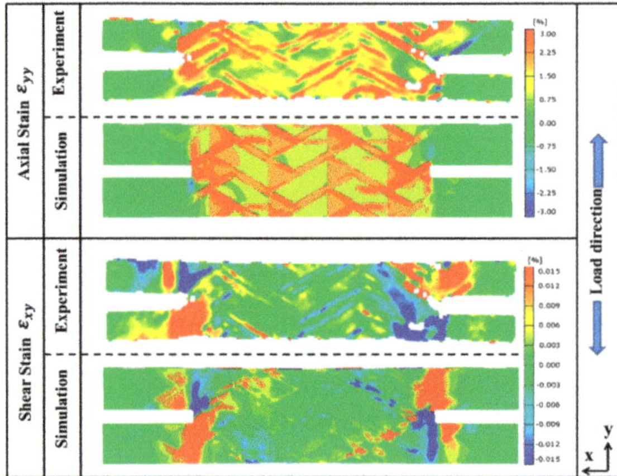

Figure 4. Comparison of numerical predicted and experimentally measured strain contours of the notched specimen under axial tension.

Furthermore, another reason for the localized strain concentration effect is because of the lower local fiber volume ratio at the corresponding regions. The inconspicuous discrepancy between numerically predicted and experimental axial strain contours is very likely results from the idealization of the geometry in the FE model. However, the manufacturing defects like axial fiber bundle undulation are inevitable in actual specimens. The cutting process may also introduce unexpected fiber damage, which could lead to more significant strain concentration area near the notch. Regardless of these factors, the capability and accuracy of this 3D mesoscale model are highly acceptable.

As mentioned before in the introduction section, both Littell [13] and Zhang et al. [15] observed out-of-plane warping behavior along the free edges in their tensile tests using the straight-side coupon specimen, which would lead to the premature damage initiation from the free edges. Kolhman et al. [14] also discovered the same phenomena in the notched specimen. Figure 5 compares the out-of-plane displacement (U3) contours from simulation and experiment for the notched tension test. The out-of-plane displacement is formed due to the tension–torsion coupling resulting from the termination of fiber bundles at the free edges. In the simulation results, the specimen shows the antisymmetric distribution of out-of-plane warping at the four corners of the notched specimen, which is related to the antisymmetric mesoscopic structure of this material. In addition, no obvious out-of-plane deformation was observed in the gauge section of the specimen indicates that the failure of the specimen is not affected by the free-edge effect. However, the out-of-plane deformation still consumes partially the energy from external loading, leading to a relatively lower measured modulus of the specimen. Thus, the notched specimen may not be suitable for modulus measurement. This is one of the contents which need to be studied in the future.

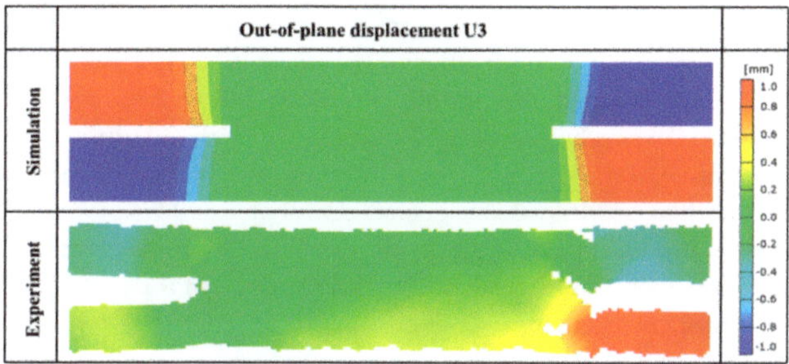

Figure 5. Comparison of numerical predicted and experimental measured out-of-plane displacement contours of notched tension specimen.

5.2. Progressive Damage Analysis

The progressive damage process is then studied using the correlated mesoscale FE model, as shown in Figure 6. The status of damage corresponds to the value of the particular damage variable ranging from 0 to 1, where a value of 0 indicates that damage has not occurred yet while a value of 1 indicates that the element is totally damaged.

Figure 6. Comparison of numerical predicted damage development of axial tension.

The matrix damage initiated at the global strain level of ~0.5% and spread to the whole central region at the global strain level of 0.80%. The distribution of internal damage inside the bundles reveals that there is no fiber tension damage occurring at these strain levels while matrix tension damage (matrix cracking in fiber bundles) is observed to start from the notched area. The matrix tension damage propagates along the notches and the bias fiber bundles; while in the meantime, in the central region matrix tension damage occurs mainly where bias fiber bundles intersect. The interfaces are found to be intact at the first two status of Figure 6 as there is no damage in cohesive elements. This behavior is similar to the observations during the test of straight-sided coupon specimen [13].

At the global strain level of 1.8%, unloading is identified followed by fiber tensile damage of axial fiber bundles and almost all elements of bias fiber bundles are dominated by matrix tension damage. The fiber tension damage initiates at the notched area same as the matrix tension damage. It is also found that interface failure appears only near the notches due to localized shear stress concentration. The inert of interface for all other areas indicates the excellent interface properties of this triaxially-braided composite, which is consistent with the previous experimental examinations where delamination is rarely observed during axial tension tests of this material [14].

Figure 7 shows the numerically predicted stress distributions of axial and biaxial tows before the specimen failure. The stresses on both sides of the notches are not symmetrical because of the asymmetry geometry feature. As seen from the stress contours, most of the applied loads are afforded by the axial bundles across the notched sections. Similar to the strain distribution contours in Figure 4, the shear stress concentration at the notched area is more significant than the normal stress. The shear stress concentration will then result in shear tension-dominated failure of axial fiber bundles at the notched area and tension dominated failure of axial fiber bundles at the central area, corresponding to the inclined (along bias fiber bundle direction) failure pattern of the two axial fiber bundles near the notches and horizontal failure pattern of fiber bundles at other region (fiber tension damage at global strain level 1.8% as shown in Figure 6). Similarly, the stress distribution of bias fiber bundles also explains the initiation and propagation behavior of matrix tension damage behavior at global strain levels 0.5% and 0.8% (Figure 6).

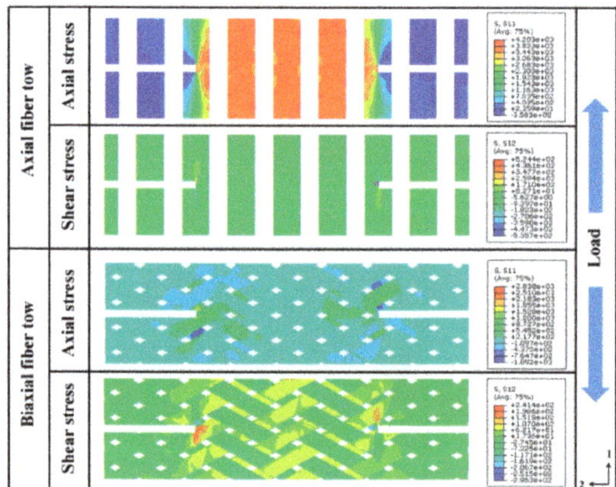

Figure 7. Numerical predicted stress contours before the instant of failure under axial tension.

Overall, matrix tension damage appears firstly around the notched areas and propagates along the bias fiber bundles, resulting in a slight decrease of effective stiffness. With the increase of external loads, the fiber tension damage and delamination will initiate around the notches due to shear stress and strain concentration in this region. The notch-induced stress/strain concentration effect disturbs only a limited region of the specimen and the failure of the specimen is mainly due to normal tension stress-induced fiber tension damage in the axial fiber bundles. Thus, the measured tensile strength of a notched specimen can be representative and can be a lower bound of the realistic strength.

The damage areas at strain level 1.8% of Figure 6 are in good correlation with the strain concentration region of the experimental results reported by Kolhman et al. [14]. As observed by Kolhman, the highest strain appears near the notch tips, which mainly attribute to the fiber tension failure. The relatively high strain concentrated along bias fiber is a result of matrix damage within the fiber bundles. The results of this section demonstrate the capability of a mesoscale model in predicting the internal damage initiation and propagation behavior of textile composites using various specimen shapes.

5.3. Numerical Parametric Study

Numerical studies were carried out to further investigate the features of the proposed mesoscale FE model and how the specific material parameters contribute to the global response of the model. This is a difficult task mainly because of the tedious modeling process and the considerable computational quantity. Due to nonuniformity of the strain distribution for the entire specimen, smooth stress–strain curves can hardly be produced in the notched tension tests, which were used mainly for strength measurement in Kohlman's work [14]. For numerical comparison, in the meso-FE simulations, stress–strain curves are generated based on the method of "digital strain gauge" proposed by Littell [13]. The macroscopic effective stress for the analyzed section is computed by determining the summation of the reaction forces on the face of the loaded cross-section divided by the area of the cross-section between two notches. The effective strain is calculated by dividing the relative displacement along the loading direction between two nodes by the initial distance between these two nodes before loading, where the relative displacement is calculated as the midpoints on each end (through load direction) section of the gauge region.

Figure 8a shows the variation of global stress–strain responses with different damage thresholds (γ_f and γ_m). The dashed line in the figure corresponds to the experimental measured effective strength of this notched specimen by Kohlman [14]. As mentioned previously, γ_f and γ_m are the damage thresholds of d_f and d_m, which are the global damage variables associated with fiber-dominated and matrix-dominated failure, respectively. d_f and d_m control the descent of element stiffness along with damage accumulation, while γ_f and γ_m determine the extent of stiffness degradation. As we can see in Figure 8a, there is a sudden force drop with damage accumulation when γ_f is lower than 0.3 and γ_m is lower than 0.5, which is not physically consistent with the experimental stress–strain response of this material where the slope of the stress–strain curve tends to degrade gradually. Numerically, the stress–strain curves become smooth when γ_f and γ_m are larger than 0.3 and 0.5, respectively. Combined with the progressive damage behavior of this specimen discussed in the previous sections, the effect of γ_f and γ_m is concluded as follows; a smaller value of γ_f leads to premature unloading of the stress–strain curve and γ_m impacts the extent of nonlinearity of the curve. This is because the external load is mainly carried by axial fiber bundles under axial tension and the fiber tension damage mainly presences in the axial bundles. Then a rapid degradation of stiffness caused by fiber damage will result in a loss of load-bearing capacity instantaneously in the local area and correspond to a sharp drop of the stress–strain curve. On the other hand, γ_m may not affect much the global responses at the initial stage due to the anisotropic feature of fiber bundles. However, as the matrix damage accumulates and propagates, γ_m becomes more significant and could result in unexpected slope change followed by the initiation of fiber damage (see Figure 8a $\gamma_f = 0.3$ and $\gamma_m = 0.5$). Also, the predicted effective strength is found to be sensitive to the parameters γ_f and γ_m. In this work, the numerical predicted effective strength correlates with experimental results when $\gamma_f = 0.3$ and $\gamma_m = 0.5$.

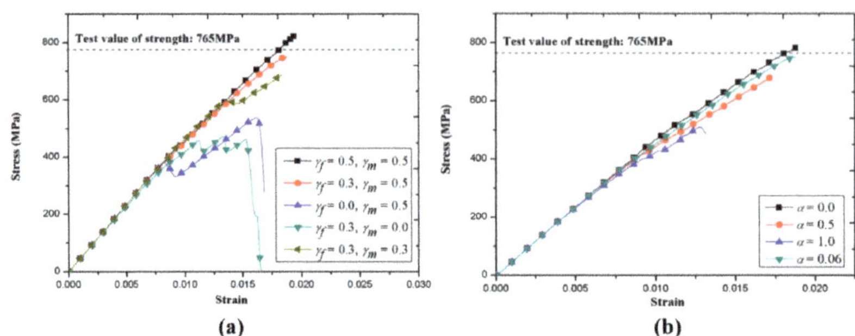

Figure 8. Numerical predicted stress-stain responses of notched tension specimen with: (**a**) different damage thresholds and (**b**) different shear failure coefficients.

The shear failure coefficient α controls the contribution of shear stress on fiber tension damage. The results exhibited in Figure 8b are similar to the conclusion for the same material reported in Zhang [18]. It is found that the effect of α on effective strength is not apparent when the value of α is lower than 0.5. As expected, the notch induced shear stress/strain concentration produces slightly higher sensitivity of the global stress–strain responses to the shear coefficient. The results of parameter analysis also declare that the present constitutive model of fiber bundles needs further improvement to enhance accuracy and applicability.

5.4. Effect of Specimen Shape

The mesoscale FE model proposed in this study can also be used to assess the rationality of the designed specimen. The effect of geometrical characteristics of the notched specimen on the test results was not considered by Kohlmen [14], due to the enormous consumption of time and availability of specimens. In this work, the correlated mesoscale FE model is employed to address this concern.

Referring to the $0°/\pm 60°$ braided architecture shown in Figure 2a, a unit cell of this braided composite can be discretized into four adjacent subcell regions depending on the presence of axial and braided tows or lack thereof. A subcell-based modeling approach has been used by many researchers [12,36] to investigate the static and impact behavior of this triaxially-braided composite. Subcells A and C contain both axial and bias bundles while subcells B and D contain only bias bundles, so the distinction of fiber volume in each subcell leads to the different local mechanical properties in a unit cell. As from the top view and front-side view of the unit cell (Figure 2a), subcells C and D are antisymmetric against subcells A and B in the thickness direction. The section studies the relationship of different subcell distributions in the gauge region with damage evolution behavior and the ultimate strength property of the notched specimen. Figure 9 compares the geometry and damage contours of different notched specimens, where the subcells adjacent to the notches of the specimens are inequitable. The label above each specimen represents the two subcells which are closest to the two notches. For instance, "A-D" indicates that subcell A and D are located in the ends of gauge region while "A/2-A/2" expresses that half of subcell A connects with the notches.

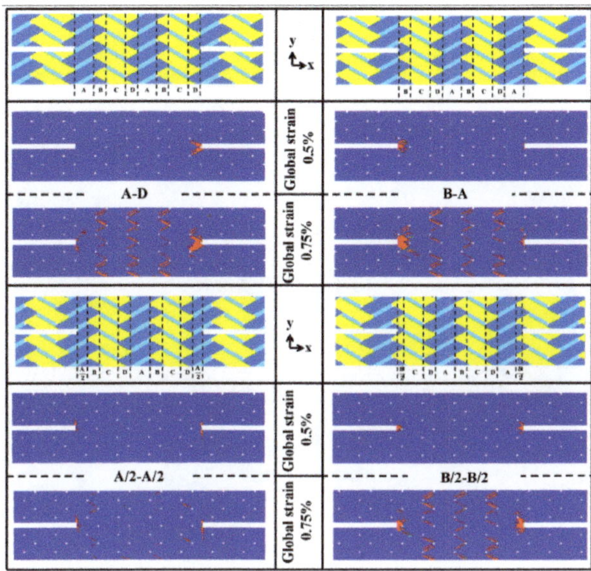

Figure 9. Numerical predictions of matrix damage in fiber bundles for notched tension specimen with different architectures across the gauge section.

As from Figure 9, the damages of all specimens initiate and concentrate at the notched area at a global strain level of 0.5%. For specimens "A-D" and "B-A", the damage initiates and concentrates at one of the two notched zones, and the damage initiation locations of the two specimens are asymmetrically suggesting the asymmetric fabric architecture of the two specimens. This damage behavior is related to the fabrics subcells layout (Figure 9), where subcells A and C consist of both axial and bias bundles and are supposed to be stronger in the axial direction than subcells B and D. Thus, when subcells A/C and B/D are subjected to the same extent of load, subcells B and D are likely to damage earlier. Similarly, specimens "B/2-B/2" and "A/2-A/2" show symmetric damage progression behavior, as they both have the same fabric subcells layout at the notched area. Besides, comparing the specimens "B/2-B/2" and "A/2-A/2", it is found that the presence of axial fiber bundles along the notches restrict the amount of matrix damage and its propagation in fiber bundles, especially obvious at the global strain level of 0.75% where the matrix damage area is much smaller than that in the other three specimens. However, matrix damage in fiber bundles has little influence on the axial stiffness of fiber bundles and shows the negligible impact to the ultimate strength of the notched specimen. Table 6 lists the predicted effective strength of the four different notched specimens shown in Figure 9. While all specimens have more or less the same effective strength, specimen "A/2-A/2" shows relatively lower strength attributing to the incomplete axial fiber bundles in the gauge region. In the numerical simulations, the crossing of the fiber damage through the width (*x*-direction) of a single fiber bundle corresponds to the fracture of the specimen. Thus, the presence of incomplete axial fiber bundles at both ends of the gauge region in specimen "A/2-A/2" results in an earlier failure of the specimen. It should be noted that the fiber volume ratio in the gauge region of these four kinds of specimens is the same to ensure the comparability.

Table 6. The effective strength of notched specimen with different architecture across the gauge section.

Specimen	A-D	B-A	A/2-A/2	B/2-B/2
Strength	764 MPa	772 MPa	751 MPa	768 MPa

To investigate the effect of notch geometry on the effective strength of the specimen numerically, specimens with three different notch sizes (dimension in *y*-direction as shown in Figure 10a) and five different widths of the gauge section (dimension in *x*-direction as shown in Figure 10b) were studied.

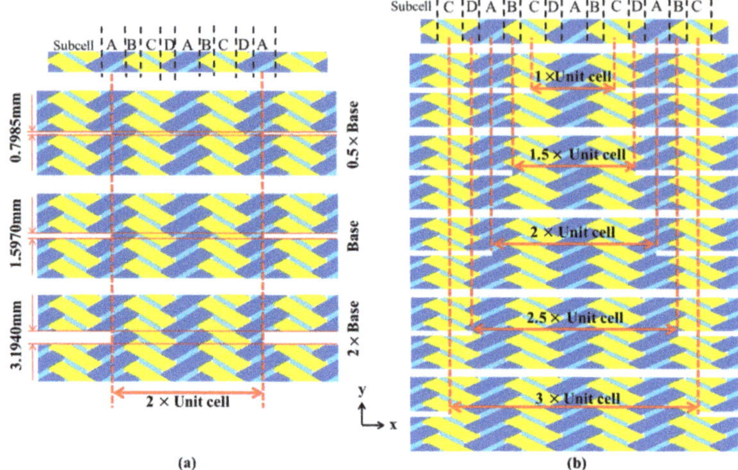

Figure 10. Mesoscale models of notched specimens with different geometrical characteristics. (**a**) Different notch sizes. (**b**) Different widths of the gauge section (with complete axial tows between notches).

The numerically predicted effective strength results are also listed in Table 7, where the effective strength values are found to be insensitive to the height (*y*-direction value) of the notch size for the cases studied in this work. However, the axial ultimate strength for the specimen changes with the changing of widths for the notch, as shown in Table 7. The predicted values of specimens denoted by "1.5UC" and "2.5UC" are slightly higher than the other three models, which may attribute to the integrity of axial tows in the gauge region. On the other hand, axial fibers among gauge regions of the other three specimens are all incomplete resulting in earlier failure of the specimens than that of specimens with complete fiber bundles.

Table 7. The predicted strength of notched specimens with different geometrical characteristics.

Notch Size	Predicted Strength (MPa)	Section Size	Predicted Strength (MPa)
0.5 × Base	762	1 UC	756
Base	751	1.5 UC *	780 *
2 × Base	755	2 UC	751
		2.5 UC *	772 *
		3 UC	760

* With complete axial tows between notches.

The fiber damage behavior of two different specimens ("1.5UC" and "2UC") is shown in Figure 11. As we can see, fiber damage starts in the specimen "2UC" at the global strain level of 1.4%. The damage will spread quickly across the axial fiber bundles of half-width and propagate into the central area. By contrast, fiber damage is not observed in the "1.5UC" specimens at the global strain level of 1.4%. Fiber damage at a global strain level of 1.8% was found to be less significant compared to that in the specimen "2UC". The results further confirm the necessity of including complete axial fiber bundles in the gauge section when preparing the notched specimens, which could avoid possible variation from cutting and provides more stable and reliable measured properties of the material.

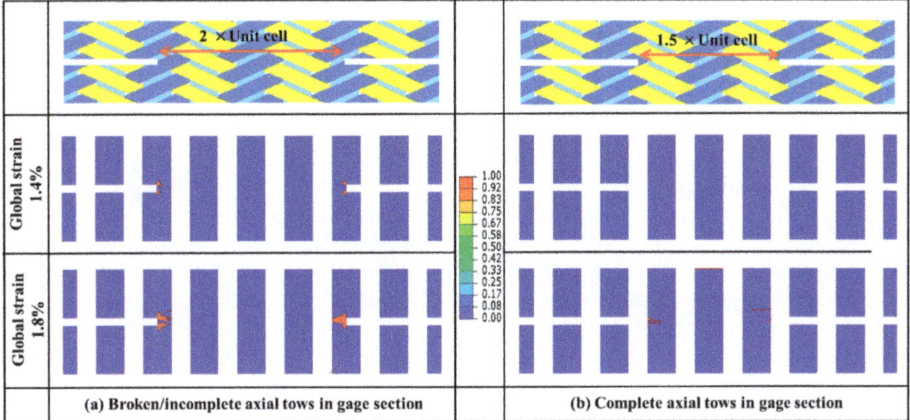

Figure 11. Comparison of predicted damage in in-complete (**a**) and complete (**b**) axial fiber bundles in gauge region.

The appropriate design of the specimen is the prerequisite for determining accurately the mechanical properties of braided composites, which is a big challenge due to the lack of advanced test standards. It is an onerous and costly process to optimize the specimen through an experiment only, and the presented numerical method is an alternative and efficient tool for specimen design.

Overall, the fabrics architecture across the gauge section has little influence on the effective strength of the notched sample. However, the simulation capability of the mesomechanical model in predicting the impacts of mesoscale geometric features on the global effective properties of the samples is valuable and advantageous.

6. Conclusions

The progressive damage behavior of a notched single-layer triaxially-braided composite under axial tension is analyzed using a three-dimensional mesoscale FE model with anisotropic damage model and an interlaminar tow-to-tow cohesive zone. The proposed model is correlated and validated against full field strain distributions and strength value acquired in the open literature. This mesoscale model is successfully applied to predict the damage propagation of each constituent, including an axial fiber bundle, bias fiber bundle, and interface.

The nonlinearity of global effective stress–strain curve of this notched specimen under axial tension is caused by matrix damage among bias fiber bundles, and the final unloading is identified followed by the fiber tensile damage of axial fiber bundle. The numerical parametric studies identify the sensitivity of stress–strain response to damage parameters. Through geometric characteristic analysis, the different subcell arrangements in the gauge region and dimensions of the notched region are further investigated. The integrity of axial fiber bundles in the test region is considered as the key factor which affects obviously the effective strength of notched specimen. The geometric layout of the subcells in the gauge region shows the negligible impact on the effective strength.

The develop mesoscale FE model could be extremely useful in understanding the failure behavior of this braided composite material. For further studies, this mesoscale model could be used to investigate notched specimens under different loading conditions, like transverse tension. The results of this work demonstrate the feasibility of using a mesoscale FE model as a virtual testing tool framework for braided composites.

Author Contributions: Conceptualization, C.Z.; Formal Analysis, Y.L.; Funding Acquisition, J.X.; Methodology, Z.Z.; Software, Z.Z., H.D., and X.L.; Supervision, C.Z.; Writing—Original Draft, Z.Z.; Writing—Review & Editing, C.Z., W.K.B., and Y.L.

Funding: This research was funded by the National Nature Science Foundation of China (NSFC), grant number 11772267.

Conflicts of Interest: The authors declare no conflict of interest.

References

1. Marrey, R.V.; Sankar, B.V. A Micromechanical Model for Textile Composite Plates. *J. Compos. Mater.* **1997**, *31*, 1187–1213. [CrossRef]
2. Chen, Z.; Yunfei, R.; Zhe, L.; Wei, L. Low-Velocity Impact Behavior of Interlayer/Intralayer Hybrid Composites Based on Carbon and Glass Non-Crimp Fabric. *Materials* **2018**, *11*, 2472.
3. Khan, A.I.; Borowski, E.C.; Soliman, E.M.; Reda Taha, M.M. Examining Energy Dissipation of Deployable Aerospace Composites Using Matrix Viscoelasticity. *J. Aerosp. Eng.* **2017**, *30*, 04017040. [CrossRef]
4. Roberts, G.; Pereira, J.M.; Revilock, D.; Binienda, W.; Xie, M.; Braley, M. Ballistic Impact of Composite Plates and Half-Rings with Soft Projectiles. In Proceedings of the AIAA/ASME/ASCE/AHS/ASC Structures, Structural Dynamics, & Materials Conference, Norfolk, Virginia, 7–10 April 2003.
5. Mcgregor, C.J.; Vaziri, R.; Poursartip, A.; Xiao, X. Simulation of progressive damage development in braided composite tubes under axial compression. *Compos. Part A* **2007**, *38*, 2259. [CrossRef]
6. Mcgregor, C.; Vaziri, R.; Poursartip, A.; Xiao, X. Axial Crushing of Triaxially Braided Composite Tubes at Quasi-static and Dynamic Rates. *Compos. Struct.* **2016**, *157*, 197–206. [CrossRef]
7. Shu, C.Q.; Waas, A.; Shahwan, K.W.; Agaram, V. Compressive response and failure of braided textile composites: Part 1—Experiments. *Int. J. Non-Linear Mech.* **2004**, *39*, 635–648.
8. Liebig, W.V.; Leopold, C.; Hobbiebrunken, T.; Fiedler, B. New test approach to determine the transverse tensile strength of CFRP with regard to the size effect. *Compos. Commun.* **2016**, *1*, 54–59. [CrossRef]
9. Wehrkamp-Richter, T.; Hinterhölzl, R.; Pinho, S.T. Damage and failure of triaxial braided composites under multi-axial stress states. *Compos. Sci. Technol.* **2017**, *150*, 32–44. [CrossRef]
10. Quek, S.C.; Waas, A.; Shahwan, K.W.; Agaram, V. Compressive response and failure of braided textile composites: Part 2—Computations. *Int. J. Non-Linear Mech.* **2004**, *39*, 649–663. [CrossRef]
11. Song, S.; Waas, A.M.; Shahwan, K.W.; Xiao, X.; Faruque, O. Braided textile composites under compressive loads: Modeling the response, strength and degradation. *Compos. Sci. Technol.* **2007**, *67*, 3059–3070. [CrossRef]
12. Goldberg, R.K.; Blinzler, B.J.; Binienda, W.K. Investigation of a Macromechanical Approach to Analyzing Triaxially-Braided Polymer Composites. *AIAA J.* **2011**, *49*, 205–215. [CrossRef]
13. Littell, J.D. The Experimental and Analytical Characterization of the Macromechanical Response for Triaxial Braided Composite. Ph.D. Thesis, The University of Akron, Akron, OH, USA, 2008.
14. Kohlman, L.W. Evaluation of Test Methods for Triaxial Braid Composites and the Development of A Large Multiaxial Test Frame for Validation Using Braided Tube Specimens. Ph.D. Thesis, The University of Akron, Akron, OH, USA, 2012.
15. Zhang, C.; Binienda, W.K.; Goldberg, R.K. Free-edge effect on the effective stiffness of single-layer triaxially braided composite. *Compos. Sci. Technol.* **2015**, *107*, 145–153. [CrossRef]
16. Kueh, A.B.H. Size-influenced mechanical isotropy of singly-plied triaxially woven fabric composites. *Compos. Part A* **2014**, *57*, 76–87. [CrossRef]
17. Li, X.; Binienda, W.K.; Goldberg, R.K. Finite Element Model for Failure Study of Two Dimensional Triaxially Braided Composite. *J. Aerosp. Eng.* **2010**, *24*, 170–180. [CrossRef]
18. Zhang, C.; Li, N.; Wang, W.; Binienda, W.K.; Fang, H. Progressive damage simulation of triaxially braided composite using a 3D meso-scale finite element model. *Compos. Struct.* **2015**, *125*, 104–116. [CrossRef]
19. Wang, C.; Roy, A.; Silberschmidt, V.V.; Chen, Z. Modelling of Damage Evolution in Braided Composites: Recent Developments. *Mech. Adv. Mater. Mod. Process.* **2017**, *3*, 15. [CrossRef]
20. Lomov, S.V.; Ivanov, D.S.; Verpoest, I.; Zako, M.; Kurashi, T.; Nakai, H.; Hirosawa, S. Meso-FE modelling of textile composites: Road map, data flow and algorithms. *Compos. Sci. Technol.* **2007**, *67*, 1870–1891. [CrossRef]

21. Zhao, Z.; Liu, P.; Chen, C.; Zhang, C.; Li, Y. Modeling the transverse tensile and compressive failure behavior of triaxially braided composites. *Compos. Sci. Technol.* **2019**, *172*, 96–107. [CrossRef]
22. Zhao, Z.; Dang, H.; Zhang, C.; Yun, G.J.; Li, Y. A multi-scale modeling framework for impact damage simulation of triaxially braided composites. *Compos. Part A* **2018**, *110*, 113–125. [CrossRef]
23. Smilauer, V.; Hoover, C.G.; Bazant, Z.P.; Caner, F.C.; Waas, A.M.; Shahwan, K.W. Multiscale simulation of fracture of braided composites via repetitive unit cells. *Eng. Fract. Mech.* **2011**, *78*, 901–918. [CrossRef]
24. García-Carpintero, A.; Van den Beuken, B.N.G.; Haldar, S.; Herráez, M.; González, C.; Lopes, C.S. Fracture behaviour of triaxial braided composites and its simulation using a multi-material shell modelling approach. *Eng. Fract. Mech.* **2018**, *188*, 268–286. [CrossRef]
25. Ren, Y.; Zhang, S.; Jiang, H.; Xiang, J. Meso-Scale Progressive Damage Behavior Characterization of Triaxial Braided Composites under Quasi-Static Tensile Load. *Appl. Compos. Mater.* **2018**, *25*, 335–352. [CrossRef]
26. Wehrkamp-Richter, T.; Carvalho, N.V.D.; Pinho, S.T. A meso-scale simulation framework for predicting the mechanical response of triaxial braided composites. *Compos. Part A* **2018**, *107*, 489–506. [CrossRef]
27. Ge, J.; He, C.; Liang, J.; Chen, Y.; Fang, D. A coupled elastic-plastic damage model for the mechanical behavior of three-dimensional (3D) braided composites. *Compos. Sci. Technol.* **2018**, *157*, 86–98. [CrossRef]
28. Hashin, Z. Failure Criteria for Unidirectional Fiber Composites. *J. Appl. Mech.* **1980**, *47*, 329–334. [CrossRef]
29. Hou, J.P.; Petrinic, N.; Ruiz, C.; Hallett, S.R. Prediction of impact damage in composite plates. *Compos. Sci. Technol.* **2000**, *60*, 273–281. [CrossRef]
30. Zdeněk, P.; Oh Bažant, B.H. Crack band theory for fracture of concrete. *Matériaux Constr.* **1983**, *16*, 155–177.
31. Hongkarnjanakul, N.; Bouvet, C.; Rivallant, S. Validation of low velocity impact modelling on different stacking sequences of CFRP laminates and influence of fibre failure. *Compos. Struct.* **2013**, *106*, 549–559. [CrossRef]
32. Li, X.; Ma, D.; Liu, H.; Tan, W.; Gong, X.; Zhang, C.; Li, Y. Assessment of failure criteria and damage evolution methods for composite laminates under low-velocity impact. *Compos. Struct.* **2019**, *207*, 727–739. [CrossRef]
33. Duvaut, G.; Lions, J.L.; John, C.W. Inequalities in Mechanics and Physics. *J. Appl. Mech.* **1977**, *44*, 364. [CrossRef]
34. Zhang, C.; Curiel-Sosa, J.L.; Bui, T.Q. A novel interface constitutive model for prediction of stiffness and strength in 3D braided composites. *Compos. Struct.* **2017**, *163*, 32–43. [CrossRef]
35. Barile, C.; Casavola, C. Mode-I Fracture Behavior of CFRPs: Numerical Model of the Experimental Results. *Materials* **2019**, *12*, 513. [CrossRef] [PubMed]
36. Cheng, J.; Binienda, W.K. Simplified Braiding through Integration Points Model for Triaxially Braided Composites. *J. Aerosp. Eng.* **2008**, *21*, 152–161. [CrossRef]

MDPI
St. Alban-Anlage 66
4052 Basel
Switzerland
Tel. +41 61 683 77 34
Fax +41 61 302 89 18
www.mdpi.com

Materials Editorial Office
E-mail: materials@mdpi.com
www.mdpi.com/journal/materials